W9-CLQ-522

CONTROL AND DYNAMIC SYSTEMS

Advances in Theory and Applications

Volume 53

CONTRIBUTORS TO THIS VOLUME

C. ABDALLAH

VENKATARAMANAN BALAKRISHNAN

MICHELE BONOLLO

STEPHEN BOYD

ANTONIO CANTONI

LIANG-WEY CHANG

MUNTHER A. DAHLEH

HEPING DAI

A. DATTA

D. DAWSON

P. DORATO

MENG H. ER

PETROS A. IOANNOU

M. JAMSHIDI

WOLFGANG J. RUNGGALDIER

MEHRDAD SAIF

NARESH K. SINHA

A. R. STUBBERUD

GANG TAO

H. M. WABGAONKAR

CONTROL AND
DYNAMIC SYSTEMS

ADVANCES IN THEORY
AND APPLICATIONS

Edited by

C. T. LEONDES

School of Engineering and Applied Science
University of California, Los Angeles
Los Angeles, California
        and
Department of Electrical Engineering
and Computer Science
University of California, San Diego
La Jolla, California

VOLUME 53: HIGH PERFORMANCE SYSTEMS
TECHNIQUES AND APPLICATIONS

ACADEMIC PRESS, INC.
Harcourt Brace Jovanovich, Publishers
San Diego   New York   Boston
London   Sydney   Tokyo   Toronto

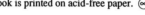
Academic Press, Inc.
1250 Sixth Avenue, San Diego, California 92101-4311

United Kingdom Edition published by
Academic Press Limited
24–28 Oval Road, London NW1 7DX

Library of Congress Catalog Number: 64-8027

International Standard Book Number: 0-12-012753-9

PRINTED IN THE UNITED STATES OF AMERICA
92 93 94 95 96 97   QW   9 8 7 6 5 4 3 2 1

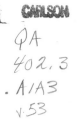
CONTENTS

CONTRIBUTORS vii
PREFACE ix

Global Optimization in Control System Analysis and Design 1

    Venkataramanan Balakrishnan and Stephen Boyd

Robust Techniques for Combined Filtering and Parameter Estimation ... 57

    Wolfgang J. Runggaldier and Michele Bonollo

Stability and Robustness of Multivariable Model Reference Adaptive
Control Schemes 99

    Gang Tao and Petros A. Ioannou

Decentralized Adaptive Control 125

    A. Datta and P. Ioannou

Robust Recursive Estimation of States and Parameters of Bilinear
Systems 173

    Heping Dai and Naresh K. Sinha

Robust Controller Design: A Bounded-Input, Bounded-Output
Worst-Case Approach 219

    Munther A. Dahleh

Techniques in Robust State Estimation Theory with Applications 261

    Mehrdad Saif

Sliding Control Design in Robust Nonlinear Control Systems 299

    Liang-Wey Chang

Techniques in Robust Broadband Beamforming 321

    Meng H. Er and Antonio Cantoni

Techniques for the Robust Control of Rigid Robots 387

    C. Abdallah, D. Dawson, P. Dorato, and M. Jamshidi

Introduction to Non-linear Control Using Artificial Neural Networks 427

    H. M. Wabgaonkar and A. R. Stubberud

INDEX 523

CONTRIBUTORS

Numbers in parentheses indicate the pages on which the authors' contributions begin.

C. Abdallah (387), *CAD Laboratory for Systems and Robotics, Department of Electrical and Computer Engineering, University of New Mexico, Albuquerque, New Mexico 87131*

Venkataramanan Balakrishnan (1), *Information Systems Laboratory, Department of Electrical Engineering, Stanford University, Stanford, California 94305*

Michele Bonollo (57), *Dipartimento di Matematica, Pura ed Applicata, Universitá di Padova, 35131-Padova, Italy*

Stephen Boyd (1), *Information Systems Laboratory, Department of Electrical Engineering, Stanford University, Stanford, California 94305*

Antonio Cantoni (321), *Department of Electrical & Electronic Engineering, University of Western Australia, Nedlands 6009, W.A., Australia*

Liang-Wey Chang (299), *Department of Mechanical Engineering, Naval Postgraduate School, Monterey, California 93943*

Munther A. Dahleh (219), *Department of Electrical Engineering and Computer Science, Massachusetts Institute of Technology, Cambridge, Massachusetts 02139*

Heping Dai (173), *Department of Electrical and Computer Engineering, McMaster University, Hamilton, Ontario L8S 4L7, Canada*

A. Datta (125), *Department of Electrical Engineering Systems, University of Southern California, Los Angeles, California 90089*

D. Dawson (387), *Department of Electrical and Computer Engineering, Clemson University, Clemson, South Carolina 29634*

P. Dorato (387), *CAD Laboratory for Systems and Robotics, Department of Electrical and Computer Engineering, University of New Mexico, Albuquerque, New Mexico 87131*

Meng H. Er (321), *School of Electrical & Electronic Engineering, Nanyang Technological University, Singapore 2263*

Petros A. Ioannou (99, 125), *Department of Electrical Engineering Systems, University of Southern California, Los Angeles, California 90089*

M. Jamshidi (387), *CAD Laboratory for Systems and Robotics, Department of Electrical and Computer Engineering, University of New Mexico, Albuquerque, New Mexico 87131*

Wolfgang J. Runggaldier (57), *Dipartimento di Matematica, Pura ed Applicata, Universitá di Padova, 35131-Padova, Italy*

Mehrdad Saif (261), *School of Engineering Science, Simon Fraser University, Burnaby, British Columbia V5A 1S6, Canada*

Naresh K. Sinha (173), *Department of Electrical and Computer Engineering, McMaster University, Hamilton, Ontario L8S 4L7, Canada*

A. R. Stubberud (427), *Department of Electrical Engineering, University of California, Irvine, Irvine, California 92717*

Gang Tao (99), *Department of Electrical and Computer Engineering, Washington State University, Pullman, Washington 99164*

H. M. Wabgaonkar (427), *Department of Electrical Engineering, University of California, Irvine, Irvine, California 92717*

PREFACE

Effective control concepts and applications go back over millennia. One very familiar example of this is the windmill. It was designed so as to derive maximum benefit from windflow, a simple but highly effective optimization technique. Harold Hazen's paper of 1932 in the *Journal of the Franklin Institute* was one of the earlier reference points wherein an analytical framework for modern control theory would start to be established. There were many other notable items along the way, including the MIT Radiation Laboratory Series volume on servomechanisms, the Brown and Campbell book, and Bode's book, all published shortly after mid-1945. However, it remained for Kalman's papers of the late 1950s (wherein a foundation for modern state space techniques was established) and the tremendous evolution of digital computer technology (which was underpinned by the continuous giant advances in integrated electronics) for truly powerful control systems techniques for increasingly complex systems to be developed. Today we can look forward to a future that is rich in possibilities in a wide variety of areas of major significance, including manufacturing systems, electric power systems, robotics, aerospace systems, and many others with significant economic, safety, cost-effectiveness, reliability, and many other implications. Thus, this volume is devoted to the most timely theme of "Techniques in High Performance Control Systems Design."

The first contribution in this volume is "Global Optimization in Control System Analysis and Design," by V. Balakrishnan and Stephen Boyd. The main goal of this contribution is the presentation of techniques that combine recent (and continuing) gains in computing power with advances in theory to answer questions that were previously unanswerable. As a result, this is a most appropriate contribution with which to begin this theme volume.

The next contribution is "Robust Techniques for Combined Filtering and Parameter Estimation," by Wolfgang J. Runggaldier and Michele Bonollo. In control practice, design techniques rest on a satisfactory knowledge of the system model, something that is not always readily available. In this contribution, robust solution techniques are presented for combined filtering and parameter estimation in discrete time, thus presenting means for effectively dealing with this fundamental problem.

The next contribution is "Stability and Robustness of Multivariable Model Reference Adaptive Control Schemes," by Gang Tao and Petros A.

Ioannou. This contribution discusses a unified framework for the design and analysis of model reference adaptive control techniques for both continuous and discrete time multivariable plants with additive and multiplicative unmodeled dynamics. Issues such as plant and controller parametrizations, design of adaptive laws, stability, robustness, and performance are clarified.

The next contribution is "Decentralized Adaptive Control," by A. Datta and P. Ioannou. It presents techniques for the design of continuous time model reference adaptive controllers for large scale systems composed of N interconnected linear subsystems with unknown parameters. In today's world, with such complex systems, these techniques can often be important.

The next contribution is "Robust Recursive Estimation of States and Parameters of Bilinear Systems," by Heping Dai and Naresh K. Sinha. It illustrates effective techniques for robust estimation in dynamic systems in the event of failures in system elements, analog to digital conversion errors, system disturbances, errors in system data transmission, and other events called outliers that can produce large errors. The companion chapter to this contribution by Dai and Sinha will be found in Volume 55, whose theme is "Digital and Numerical Techniques in Control Systems."

The next contribution is "Robust Controller Design: A Bounded-Input, Bounded-Output Worst-Case Approach," by Munther A. Dahleh. It provides a general framework for designing robust control systems in the presence of uncertainty when the specifications are posed in the time domain.   —

The next contribution is "Techniques in Robust State Estimation Theory with Applications," by Mehrad Saif. It presents techniques for state estimation for linear stationary dynamic systems that are subject to unknown time varying plant and output disturbances. Two highly effective approaches are presented, one of which is more general, and they are illustrated with various applications.

The next contribution is "Sliding Control Design in Robust Nonlinear Control System," by Liang-Wey Chang. A sliding control algorithm is presented that is effective in dealing with filtering uncertainty for time-varying nonlinear dynamic systems. One of the important by-products of this algorithm is that the trade-off between tracking accuracy and the robustness to uncertainties no longer exists.

The next contribution is "Techniques in Robust Broadband Beamforming," by Meng H. Er and Antonio Cantoni. Degradations in such systems occur because of the great multiplicity of the elements required by them and the continual failure of some of these elements. Because of the paucity of the work in the literature on robustness of broadband array processors, this contribution constitutes a valuable reference source.

The next contribution is "Techniques for the Robust Control of Rigid Robots," by C. Abdallah, D. Dawson, P. Dorato, and M. Jamshidi. Five categories of robust robotic controllers are presented and their comparative performance is analyzed. All categories can achieve boundedness of position-tracking error for a range of uncertain parameters, and some are asymptotically convergent.

The final contribution to this volume is "Introduction to Non-linear Control Using Artificial Neural Networks," by H. M. Wabgaonker and A. R. Stubberud. The increasing demands for performance and versatility that are being placed on engineering systems clearly suggest, in many cases, the crucial requirement that systems be able to perform autonomously under uncertain environments. In this contribution, the powerful avenue for dealing with this problem through the use of artificial neural networks is presented. It is an in-depth treatment with illustrations of the potentially great effectiveness of artificial neural networks.

This volume rather clearly manifests the significance and power of the techniques that are available and under continuing development for the design of high performance control systems. The coauthors are all to be commended for their splendid contributions to this volume, which will provide a significant reference source for workers on the international scene for years to come.

Global Optimization in Control System Analysis and Design

Venkataramanan Balakrishnan
Stephen Boyd

Information Systems Laboratory
Department of Electrical Engineering
Stanford University
Stanford, CA 94305
U. S. A.

I. INTRODUCTION

Many problems in control system analysis and design can be posed in a setting where a system with a fixed model structure and nominal parameter values is affected by parameter variations. An example is parametric robustness analysis, where the parameters might represent physical quantities that are known only to within a certain accuracy, or vary depending on operating conditions etc. Frequently asked questions here deal with performance issues: "How bad can a certain performance measure of the system be over all possible values of the parameters?" Another example is parametric controller design, where the parameters represent degrees of freedom available to the control system designer. A typical question here would be: "What is the best choice of parameters, one that optimizes a certain design objective?"

Many of the questions above may be directly restated as optimization problems: If q denotes the vector of parameters, $\mathcal{Q}_{init}$ the set of values that q might assume and $f(q)$ an objective function, then the questions above translate into solving one of the following optimization problems:

$$\text{P1:} \quad \min_{q \in \mathcal{Q}_{init}} f(q),$$

or

$$\text{P2:} \quad \max_{q \in \mathcal{Q}_{init}} f(q).$$

In general, (P1) and (P2) are *non-convex* optimization problems, and are much harder[1] to solve than say, *convex* optimization problems, for which there exist a number of effective algorithms. Solving (P1) or (P2) where $\mathcal{Q}_{init}$ is a set with finite or countably infinite elements is the well-studied combinatorial optimization problem [2].

Since solving (P1) or (P2) is hard in general, it is worth discussing the costs and benefits associated with approximate or suboptimal solutions such as local optimization methods, Monte Carlo methods, gridding etc. The attractiveness of such methods stems from the ease with which they may be performed — they typically require much less computation than global optimization methods. The cost associated with a suboptimal solution depends upon the underlying physical problem. For example, suppose that problem (P1) arises from robustness analysis, where one seeks the worst (smallest) possible value of a certain performance measure $f(q)$. Then confidence misplaced on the local minimum returned by a local optimization procedure might be potentially disastrous, that is, local optimization might not find the worst-case parameter which might, for example, render the system unstable. In this case, the cost associated with not finding the global minimum would be high. (In such situations, methods that yield lower bounds are of greater value, evidenced by the vast research into conservative analytical techniques for robustness analysis.) On the other hand, if (P1) arises from a design problem, where one seeks the parameters that yield the smallest value of a design objective, a local minimization method would yield a possibly conservative upper bound for the global minimum. In this case, the cost associated with not finding the global minimum would usually be acceptable.

---

[1] For a quantitative description of the term "hard", see for example, [1].

There exist several popular methods for solving global optimization problems (P1) or (P2) (see [3], for example). *Simulated Annealing* (see [4] and the references therein) describes a family of iterative methods where every iteration consists of taking a step in parameter space with a probability that decays exponentially with an "energy" function associated with the new parameter value. This technique reportedly performs well and has been applied widely to computer-aided design of electronic integrated circuits, design of error-correcting codes etc. It must be mentioned, however, that simulated annealing does not maintain both upper and lower bounds of the global optimum that it seeks. As a result, the algorithm has no stopping criterion; and its termination, at any time, does not yield any bounds for the optimum. This is not a serious drawback in the many applications where it has been successfully used, since it has led to designs that are significantly better than any found before; the cost associated with not finding the globally optimal design in these cases is acceptable. However, since annealing does not yield any guarantees about the solutions it yields — though some probabilistic statements can be made about the convergence to the global optimum — it cannot be used for problems such as robustness analysis where the cost associated with not finding the global optimum is usually high. Another approach to global optimization is to apply interior point algorithms [5], which have been reported to perform well on some integer programming problems. Here again, as with annealing, there are no guaranteed bounds on the optimum.

In contrast with the above techniques for global optimization, *branch and bound* algorithms, as they progress, do maintain upper and lower bounds for the global optimum; thus termination at any time yields guaranteed bounds for the optimum. These algorithms derive their name from the way they proceed: They break up the parameter region into subregions ("branching") to derive bounds for the global optimum over the original region ("bounding"). The branching is done based on some heuristic rules. Though these heuristics often work well, it must be emphasized that these algorithms are worst-case combinatoric. Thus they may require unacceptably long computation times on some simple problems.

Traditionally, branch and bound algorithms have been used in discrete programming problems (see [6, 7] for early articles, [8, 9] for surveys and [10, 11, 3] for texts). A recent application of a branch and bound algorithm on a parametric robustness problem arising in control systems analysis is in [12], where De Gaston and Safonov use a branch and bound algorithm for computing the robust stability margin for systems with uncorrelated uncertain parameters. Sideris and Peña [13] extend this algorithm to the case when the parameters are real and may be correlated. In [14], Chang *et al.* describe a similar branch and bound algorithm for computing the real structured singular value and the real multivariable stability margin. Vicino *et al.* [15] use a branch and bound algorithm with geometric programming ideas to compute the robust stability margin. Demarco *et al.* [16] use a branch and bound algorithm to study stability problems arising in power systems.

In this chapter, we restrict our attention to following setup: *We consider linear systems with a number of constant, unknown parameters that lie in intervals.* Thus, the parameter region $\mathcal{Q}_{init}$ is a rectangle. For such systems, we consider problems of parameter robustness analysis and parameter selection (that is, design). We show how a branch and bound technique may be used to solve these problems.

The organization of the chapter is as follows. Section II describes the basic branch and bound algorithm, its convergence properties and a simple extension. Section III discusses some problems that arise in parameter dependent linear systems, Section IV discusses the computation of bounds for these problems and Section V presents some simple examples that illustrate the performance of the branch and bound algorithm on these problems. Section VI makes some closing remarks.

Some Notation

R (**C**) denotes the set of real (complex) numbers. For $c \in \mathbf{C}$, Re c is the real part of c. The set of $m \times n$ matrices with real (complex) entries is denoted $\mathbf{R}^{m \times n}$ ($\mathbf{C}^{m \times n}$). P^T stands for the transpose of P, and P^*, the complex conjugate transpose. I denotes the identity matrix, with size determined from context.

For a matrix $P \in \mathbf{R}^{n \times n}$ (or $\mathbf{C}^{n \times n}$), $\lambda_i(P)$, $1 \le i \le n$ denotes the ith eigenvalue of P (with no particular ordering). Tr(P) stands for the trace (sum of the diagonal entries) of P. $\overline{\sigma}(P)$ denotes the maximum singular value of P, defined as

$$\overline{\sigma}(P) = \max_{1 \leq i \leq n} \sqrt{\lambda_i(P^*P)},$$

and $\underline{\sigma}(P)$ the minimum singular value of P, defined as

$$\underline{\sigma}(P) = \min_{1 \leq i \leq n} \sqrt{\lambda_i(P^*P)}.$$

The *condition number* of a matrix P with positive minimum singular value is the ratio $\overline{\sigma}(P)/\underline{\sigma}(P)$ (it is defined to be ∞ if $\underline{\sigma}(P) = 0$). $\|P\|_F$ is the Frobenius norm of P, given by $\sqrt{\mathrm{Tr}(P^*P)}$. The definitions for $\overline{\sigma}(P)$ and $\|P\|_F$ hold also for $P \in \mathbf{R}^{m \times n}$ (or $\mathbf{C}^{m \times n}$).

II. A BRANCH AND BOUND ALGORITHM

Most of the material (Subsections A and B) in this section is from [17]. We reproduce it here for completeness.

The branch and bound algorithm we present here finds the global minimum of a function $f : \mathbf{R}^m \rightarrow \mathbf{R}$ over an m-dimensional rectangle $\mathcal{Q}_{\mathrm{init}}$. (Of course, by replacing f by $-f$, the algorithm can also be used to find the global maximum.)

For a rectangle $\mathcal{Q} \subseteq \mathcal{Q}_{\mathrm{init}}$ we define

$$\Phi_{\min}(\mathcal{Q}) = \min_{q \in \mathcal{Q}} f(q).$$

Then, the algorithm computes $\Phi_{\min}(\mathcal{Q}_{\mathrm{init}})$ to within an absolute accuracy of $\epsilon > 0$, using two functions $\Phi_{\mathrm{lb}}(\mathcal{Q})$ and $\Phi_{\mathrm{ub}}(\mathcal{Q})$ defined over $\{\mathcal{Q} | \mathcal{Q} \subseteq \mathcal{Q}_{\mathrm{init}}\}$ (which, presumably, are easier to compute than $\Phi_{\min}(\mathcal{Q})$). These two functions satisfy the following conditions.

(R1) $$\Phi_{\mathrm{lb}}(\mathcal{Q}) \leq \Phi_{\min}(\mathcal{Q}) \leq \Phi_{\mathrm{ub}}(\mathcal{Q}).$$

Thus, the functions Φ_{lb} and Φ_{ub} compute a lower and upper bound on $\Phi_{\min}(\mathcal{Q})$, respectively.

(R2) As the maximum half-length of the sides of $\mathcal{Q}$, denoted by $\mathrm{size}(\mathcal{Q})$, goes to zero, the difference between upper and lower bounds *uniformly* converges to zero, *i.e.*,

$$\forall \epsilon > 0 \; \exists \delta > 0 \text{ such that}$$
$$\forall \mathcal{Q} \subseteq \mathcal{Q}_{\mathrm{init}}, \; \mathrm{size}(\mathcal{Q}) \leq \delta \Longrightarrow \Phi_{\mathrm{ub}}(\mathcal{Q}) - \Phi_{\mathrm{lb}}(\mathcal{Q}) \leq \epsilon.$$

Roughly speaking, then, the bounds Φ_{lb} and Φ_{ub} become sharper as the rectangle shrinks to a point.

We now describe the algorithm. We start by computing $\Phi_{lb}(\mathcal{Q}_{init})$ and $\Phi_{ub}(\mathcal{Q}_{init})$. If $\Phi_{ub}(\mathcal{Q}_{init}) - \Phi_{lb}(\mathcal{Q}_{init}) \leq \epsilon$, the algorithm terminates. Otherwise we partition $\mathcal{Q}_{init}$ as a union of subrectangles as $\mathcal{Q}_{init} = \mathcal{Q}_1 \cup \mathcal{Q}_2 \cup \ldots \cup \mathcal{Q}_N$, and compute $\Phi_{lb}(\mathcal{Q}_i)$ and $\Phi_{ub}(\mathcal{Q}_i)$, $i = 1, 2, ..., N$. Then

$$\min_{1 \leq i \leq N} \Phi_{lb}(\mathcal{Q}_i) \leq \Phi_{min}(\mathcal{Q}_{init}) \leq \min_{1 \leq i \leq N} \Phi_{ub}(\mathcal{Q}_i),$$

so we have new bounds on $\Phi_{min}(\mathcal{Q}_{init})$. If the difference between the new bounds is less than or equal to ϵ, the algorithm terminates. Otherwise, the partition of $\mathcal{Q}_{init}$ is further refined and the bounds updated.

If a partition $\mathcal{Q}_{init} = \cup_{i=1}^{N} \mathcal{Q}_i$ satisfies size$(\mathcal{Q}_i) \leq \delta, i = 1, 2, ..., N$, then by condition (R2) above,

$$\min_{1 \leq i \leq N} \Phi_{ub}(\mathcal{Q}_i) - \min_{1 \leq i \leq N} \Phi_{lb}(\mathcal{Q}_i) \leq \epsilon;$$

thus a "δ-grid" ensures that $\Phi_{min}(\mathcal{Q}_{init})$ is determined to within an absolute accuracy of ϵ. However, for the "δ-grid", the number of rectangles forming the partition (and therefore the number of upper and lower bound calculations) grows exponentially with $1/\delta$. The branch and bound algorithm applies a heuristic rule for partitioning $\mathcal{Q}_{init}$, which in most cases leads to a reduction of the number of calculations required to solve the problem compared to the δ-grid. The heuristic is this: Given any partition $\mathcal{Q}_{init} = \cup_{i=1}^{N} \mathcal{Q}_i$ that is to be refined, pick a rectangle $\mathcal{Q}$ from the partition such that $\Phi_{lb}(\mathcal{Q}) = \min_{1 \leq i \leq N} \Phi_{lb}(\mathcal{Q}_i)$, and split it into two halves. The rationale behind this rule is that since we are trying to find the minimum of a function, we should concentrate on the "most promising" rectangle. We must emphasize that this is a heuristic, and in the worst case will result in a δ-grid.

A. The general branch and bound algorithm

In the following description, k stands for the iteration index. $\mathcal{L}_k$ denotes the list of rectangles, L_k the lower bound and U_k the upper bound for $\Phi_{min}(\mathcal{Q}_{init})$, at the end of k iterations.

Algorithm I

$k = 0;$
$\mathcal{L}_0 = \{\mathcal{Q}_{\text{init}}\};$
$L_0 = \Phi_{\text{lb}}(\mathcal{Q}_{\text{init}});$
$U_0 = \Phi_{\text{ub}}(\mathcal{Q}_{\text{init}});$
while $U_k - L_k > \epsilon,$ {
       pick $\mathcal{Q} \in \mathcal{L}_k$ *such that* $\Phi_{\text{lb}}(\mathcal{Q}) = L_k;$
       split $\mathcal{Q}$ *along one of its longest edges into* $\mathcal{Q}_I$ *and* $\mathcal{Q}_{II};$
       $\mathcal{L}_{k+1} := (\mathcal{L}_k - \{\mathcal{Q}\}) \cup \{\mathcal{Q}_I, \mathcal{Q}_{II}\};$
       $L_{k+1} := \min_{\mathcal{Q} \in \mathcal{L}_{k+1}} \Phi_{\text{lb}}(\mathcal{Q});$
       $U_{k+1} := \min_{\mathcal{Q} \in \mathcal{L}_{k+1}} \Phi_{\text{ub}}(\mathcal{Q});$
       $k := k + 1;$

}

The requirement that we split the chosen rectangle along a longest edge may seem mysterious at this point. This splitting rule controls the condition number of the rectangles in the partition; see the proof of convergence in Subsection B.

At the end of k iterations, U_k and L_k are upper and lower bounds respectively for $\Phi_{\text{min}}(\mathcal{Q}_{\text{init}})$. We prove in Subsection B that if the bounds $\Phi_{\text{lb}}(\mathcal{Q})$ and $\Phi_{\text{ub}}(\mathcal{Q})$ satisfy condition (R2), $U_k - L_k$ is guaranteed to converge to zero, and therefore the branch and bound algorithm will terminate in a finite number of steps.

It is clear that in the branching process described above, the number of rectangles is equal to the number of iterations N. However, we can often eliminate some rectangles from consideration; they may be *pruned* since $\Phi_{\text{min}}(\mathcal{Q}_{\text{init}})$ cannot be achieved in them. This is done as follows. At each iteration:

Eliminate from list $\mathcal{L}_k$ *the rectangles* $\mathcal{Q} \in \mathcal{L}_k$ *that satisfy*

$$\Phi_{\text{lb}}(\mathcal{Q}) > U_k.$$

If a rectangle $\mathcal{Q} \in \mathcal{L}_k$ satisfies this condition, then $q \in \mathcal{Q} \Rightarrow f(q) > U_k;$ however the minimum of $f(q)$ over $\mathcal{Q}_{\text{init}}$ is *guaranteed* to be less then U_k, and therefore cannot be found in $\mathcal{Q}$.

Though pruning is not necessary for the algorithm to work, it does reduce storage requirements. The algorithm often quickly prunes a large portion of Q_{init}, and works with only a small remaining subset. The set $\mathcal{L}_k$, the union of the rectangles in the pruned list, acts as an approximation of the set of minimizers of f. In fact, every minimizer of f is guaranteed to be in $\mathcal{L}_k$.

The term *pruning* comes from the following. The algorithm can be viewed as growing a binary tree of rectangles representing the current partition $\mathcal{L}_k$, with the nodes corresponding to rectangles and the children of a given node representing the two halves obtained by splitting it. By removing a rectangle from consideration, we prune the tree.

As we noted at the beginning of this section, the above algorithm can also be used for global maximization, merely by minimizing $-f$. However, we will find it convenient to have a version of the algorithm for directly finding $\Psi_{max}(Q_{init}) = \max_{q \in Q_{init}} f(q)$. Here, for $Q \subseteq Q_{init}$, $\Psi_{lb}(Q)$ and $\Psi_{ub}(Q)$ denote lower and upper bounds for $\Psi_{max}(Q)$, and are required to satisfy

(R1′)
$$\Psi_{lb}(Q) \leq \Psi_{max}(Q) \leq \Psi_{ub}(Q).$$

(R2′)

$$\forall \epsilon > 0 \; \exists \delta > 0 \text{ such that}$$
$$\forall Q \subseteq Q_{init}, \; size(Q) \leq \delta \implies \Psi_{ub}(Q) - \Psi_{lb}(Q) \leq \epsilon.$$

In the following, L_k and U_k give lower and upper bounds for $\Psi_{max}(Q_{init})$ at the end of k iterations.

Algorithm II

> $k = 0;$
> $\mathcal{L}_0 = \{Q_{init}\};$
> $L_0 = \Psi_{lb}(Q_{init});$
> $U_0 = \Psi_{ub}(Q_{init});$
> *while* $U_k - L_k > \epsilon,$ {
> > *pick* $Q \in \mathcal{L}_k$ *such that* $\Psi_{ub}(Q) = U_k;$
> > *split* Q *into* Q_I *and* Q_{II} *along the longest edge;*
> > $\mathcal{L}_{k+1} := (\mathcal{L}_k - \{Q\}) \cup \{Q_I, Q_{II}\};$
> > $L_{k+1} := \max_{Q \in \mathcal{L}_{k+1}} \Psi_{lb}(Q);$
> > $U_{k+1} := \max_{Q \in \mathcal{L}_{k+1}} \Psi_{ub}(Q);$
> > $k := k + 1;$
>
> }

The corresponding pruning step is

Eliminate from list $\mathcal{L}_k$, the rectangles $Q \in \mathcal{L}_k$ that satisfy

$$\Psi_{ub}(Q) < L_k.$$

B. Analysis of Convergence of the Branch and Bound Algorithm

We now show that the branch and bound algorithm converges in a finite number of steps, provided the bound functions $\Phi_{lb}(\cdot)$ and $\Phi_{ub}(\cdot)$ satisfy conditions (R1) and (R2) listed at the beginning of this section. (We will only consider Algorithm I, since the proof for Algorithm II then follows analogously.)

An upper bound on the number of branch and bound iterations

The derivation of an upper bound on the number of iterations of the branch and bound algorithm involves the following steps. We first show that after a large number of iterations k, the partition $\mathcal{L}_k$ must contain a rectangle of small volume. (The volume of a rectangle is defined as the product of the lengths of its sides.) We then show that this rectangle has a small size, and this in turn implies that $U_k - L_k$ is small.

First, we observe that the number of rectangles in the partition $\mathcal{L}_k$ is just k (without pruning, which in any case does not affect the number of iterations). The total volume of these rectangles is $\text{vol}(Q_{init})$, and therefore

$$\min_{Q \in \mathcal{L}_k} \text{vol}(Q) \leq \frac{\text{vol}(Q_{init})}{k}. \tag{1}$$

Thus, after a large number of iterations, at least one rectangle in the partition has small volume.

Next, we show that small volume implies small size for a rectangle in any partition. We define the *condition number* of a rectangle $Q = \prod_i [l_i, u_i]$ as

$$\text{cond}(Q) = \frac{\max_i (u_i - l_i)}{\min_i (u_i - l_i)}.$$

We then observe that our splitting rule, which requires that we split rectangles along a longest edge, results in an upper bound on the condition number of rectangles in our partition.

Lemma 1 *For any k and any rectangle $Q \in \mathcal{L}_k$,*

$$\text{cond}(Q) \le \max\{\text{cond}(Q_{\text{init}}), 2\}. \tag{2}$$

Proof

It is enough to show that when a rectangle Q is split into rectangles Q_1 and Q_2,

$$\text{cond}(Q_1) \le \max\{\text{cond}(Q), 2\}, \quad \text{cond}(Q_2) \le \max\{\text{cond}(Q), 2\}.$$

Let ν_{max} be the maximum edge length of Q, and ν_{min}, the minimum. Then $\text{cond}(Q) = \nu_{\text{max}}/\nu_{\text{min}}$. When Q is split into Q_1 and Q_2, our splitting rule requires that Q be split along an edge of length ν_{max}. Thus, the maximum edge length of Q_1 or Q_2 can be no larger than ν_{max}. Their minimum edge length could be no smaller than the minimum of $\nu_{\text{max}}/2$ and ν_{min}, and the result follows. ∎

We note that there are other splitting rules that also result in a uniform bound on the condition number of the rectangles in any partition generated. One such rule is to cycle through the index on which we split the rectangle. If Q was formed by splitting its parent along the ith coordinate, then when we split Q, we split it along the $(i + 1)$ modulo m coordinate.

We can bound the size of a rectangle Q in terms of its volume and condition number, since

$$
\begin{aligned}
\text{vol}(Q) &= \prod_i (u_i - l_i) \\
&\ge \max_i(u_i - l_i)\left(\min_i(u_i - l_i)\right)^{m-1} \\
&= \frac{(2\,\text{size}(Q))^m}{\text{cond}(Q)^{m-1}} \\
&\ge \left(\frac{2\,\text{size}(Q)}{\text{cond}(Q)}\right)^m
\end{aligned}
$$

Thus,

$$\text{size}(Q) \le \frac{1}{2}\,\text{cond}(Q)\text{vol}(Q)^{1/m}. \tag{3}$$

Combining equations (1), (2) and (3) we get

$$\min_{Q \in \mathcal{L}_k} \text{size}(Q) \le \frac{1}{2}\max\{\text{cond}(Q_{\text{init}}), 2\}\left(\frac{\text{vol}(Q_{\text{init}})}{k}\right)^{1/m}. \tag{4}$$

Thus, for large k, the partition $\mathcal{L}_k$ must contain a rectangle of small size.

Finally, we show that if a partition has a rectangle of small size, the upper and lower bounds cannot be too far apart. More precisely, we show that given some $\epsilon > 0$, there is some N such that $U_N - L_N \leq \epsilon$ for some $N \leq k$.

First, let δ be small enough such that if size$(\mathcal{Q}) \leq 2\delta$ then $\Phi_{\mathrm{ub}}(\mathcal{Q}) - \Phi_{\mathrm{lb}}(\mathcal{Q}) \leq \epsilon$ (recall requirement (R2) at the beginning of this section). Let k be large enough such that

$$\max\{\mathrm{cond}(\mathcal{Q}_{\mathrm{init}}), 2\} \left(\frac{\mathrm{vol}(\mathcal{Q}_{\mathrm{init}})}{k} \right)^{1/m} \leq 2\,\delta. \tag{5}$$

Then from equation (4), some $\mathcal{Q} \in \mathcal{L}_k$ satisfies size$(\mathcal{Q}) \leq \delta$. Then the rectangle $\tilde{\mathcal{Q}}$, one of whose halves is $\mathcal{Q}$, must satisfy size$(\tilde{\mathcal{Q}}) \leq 2\delta$, and therefore

$$\Phi_{\mathrm{ub}}(\tilde{\mathcal{Q}}) - \Phi_{\mathrm{lb}}(\tilde{\mathcal{Q}}) \leq \epsilon.$$

However, since $\tilde{\mathcal{Q}}$ was split at some previous iteration, it must have satisfied $\Phi_{\mathrm{lb}}(\tilde{\mathcal{Q}}) = L_N$ for some $N \leq k$. Thus

$$U_N - L_N \leq \Phi_{\mathrm{ub}}(\tilde{\mathcal{Q}}) - L_N \leq \epsilon,$$

or we have an upper bound on the number of branch and bound iterations.

C.   Simultaneous maximization of multiple objectives

In many cases, the maximization (or minimization) of several objectives at the same time and over the same parameter rectangle $\mathcal{Q}_{\mathrm{init}}$ may be of interest. In this setting, we have M objective functions $f^{(1)}, f^{(2)}, \ldots, f^{(M)}$, defined over the same parameter region $\mathcal{Q}_{\mathrm{init}}$, and we seek to maximize each of these objectives over $\mathcal{Q}_{\mathrm{init}}$. One way to do this is to simply apply the Algorithm II M times, maximizing one objective function each time. However, if the functions $f^{(i)}$, $i = 1, \ldots, M$ are "correlated" — this is often the case in robustness analysis where the same set of parameters leads to the worst possible performance with several different objectives, or in controller design where the same set of parameters is optimal for more than one performance measure — this sequential approach would prove wasteful. We now present a heuristic method that exploits any correlation between the objectives; in the worst-case, this method behaves like the branch and bound algorithm applied M times, without pruning.

We use $\Psi_{lb}^{(j)}(Q)$ and $\Psi_{ub}^{(j)}(Q)$ respectively, to denote the lower and upper bounds of the maximum $\Psi_{max}^{(j)}(Q)$ of the jth objective function $f^{(j)}$ over the parameter rectangle Q. Then our heuristic algorithm for simultaneous maximization runs as follows: We start by computing the M upper and lower bounds over the initial parameter rectangle, that is, we compute $\Psi_{lb}^{(j)}(Q_{init})$ and $\Psi_{ub}^{(j)}(Q_{init})$ for $i = 1, \ldots, M$. If $\Psi_{ub}^{(j)}(Q_{init}) - \Psi_{lb}^{(j)}(Q_{init}) \le \epsilon$, for every j, the algorithm terminates. Otherwise, we partition Q_{init} and proceed to refine the bounds.

The major difference between the multiple objective maximization and the single objective maximization is this: In the maximization of a single objective, given any partition $Q_{init} = \cup_{i=1}^{N} Q_i$ that is to be refined, we pick a rectangle Q from the partition such that $\Psi_{ub}(Q) = \max_{1 \le i \le N} \Psi_{ub}(Q_i)$, and split it into two halves. However, in the multiple objective case, there might be no such candidate rectangle to be split, since in general there is no rectangle Q all of whose upper bounds satisfy $\Psi_{ub}^{(j)}(Q) = \max_{1 \le i \le N} \Psi_{ub}^{(j)}(Q_i)$, $j = 1, \ldots, M$. To address this issue, we propose the following heuristic rule: We *cycle* through the objective functions to determine which rectangle to split. More precisely, if at some iteration, we split a rectangle Q that satisfies $\Psi_{ub}^{(j)}(Q) = \max_{1 \le i \le N} \Psi_{ub}^{(j)}(Q_i)$, then at the next iteration, we split a rectangle Q that satisfies $\Psi_{ub}^{(j_{new})}(Q) = \max_{1 \le i \le N} \Psi_{ub}^{(j_{new})}(Q_i)$, where $j_{new} = (j + 1) \bmod M$. The iterations continue till the difference between the upper and lower bounds for every objective is less than or equal to ϵ.

Since the original branch and bound algorithm converges, so does this new algorithm. In the following description of the algorithm, L_k and U_k now denote *vectors* of length m. The jth component of L_k, denoted $L_k^{(j)}$, is the lower bound of $\Psi_{max}^{(j)}(Q_{init})$ at the end of k iterations, and similarly for U_k.

Algorithm III

$k = 0;\ \mathcal{L}_0 = \{\mathcal{Q}_{\text{init}}\};$
for $j=1,...,M$ {
$\qquad L_0^{(j)} = \Psi_{\text{lb}}^{(j)}(\mathcal{Q}_{\text{init}});$
$\qquad U_0^{(j)} = \Psi_{\text{ub}}^{(j)}(\mathcal{Q}_{\text{init}});$
}
until $\max_j \{U_k^{(j)} - L_k^{(j)}\} \le \epsilon,$ {
$\qquad j_{\text{new}} = k \bmod M;$
$\qquad pick\ \mathcal{Q} \in \mathcal{L}_k\ such\ that\ \Psi_{\text{ub}}^{(j_{\text{new}})}(\mathcal{Q}) = U_k^{(j_{\text{new}})};$
$\qquad split\ \mathcal{Q}\ along\ one\ of\ its\ longest\ edges\ into\ \mathcal{Q}_I\ and\ \mathcal{Q}_{II};$
$\qquad \mathcal{L}_{k+1} := (\mathcal{L}_k - \{\mathcal{Q}\}) \cup \{\mathcal{Q}_I, \mathcal{Q}_{II}\};$
$\qquad for\ j = 1, ..., M,$ {
$\qquad\qquad L_{k+1}^{(j)} := \max_{\mathcal{Q} \in \mathcal{L}_{k+1}} \Psi_{\text{lb}}^{(j)}(\mathcal{Q});$
$\qquad\qquad U_{k+1}^{(j)} := \max_{\mathcal{Q} \in \mathcal{L}_{k+1}} \Psi_{\text{ub}}^{(j)}(\mathcal{Q});$
$\qquad$ }
$\qquad k := k + 1;$
}

The pruning step needs to modified, so that the rectangles that are eliminated from the rectangle list are those where *none* of the functions $f^{(j)}$ can achieve their maximum:

Eliminate from list $\mathcal{L}_k$ *the rectangles* $\mathcal{Q} \in \mathcal{L}_k$ *that satisfy*

$$\Psi_{\text{ub}}^{(j)}(\mathcal{Q}) < L_k^{(j)}, \text{ for every } j = 1, 2, ..., M.$$

III. SOME PARAMETER PROBLEMS IN LTI CONTROLLER ANALYSIS AND DESIGN

A. Our Framework

We now apply the branch and bound algorithms of Section II to the computation of some quantities that arise in the analysis and design of parameter-dependent linear systems. Our framework consists of a family of linear time-invariant systems described by

$$\begin{aligned}
\dot{x} &= Ax + B_u u + B_w w, & x(0) = x_0, \\
y &= C_y x + D_{yu} u + D_{yw} w, \\
z &= C_z x + D_{zu} u + D_{zw} w, \\
u &= \Delta y,
\end{aligned} \tag{6}$$

where $x(t) \in \mathbf{R}^n$, $w(t) \in \mathbf{R}^{n_i}$, $z(t) \in \mathbf{R}^{n_o}$, $u(t), y(t) \in \mathbf{R}^p$, and A, B_u, B_w, C_y, C_z, D_{yu}, D_{yw}, D_{zu} and D_{zw} are real matrices of appropriate sizes. Δ is a diagonal matrix, parametrized by a vector of parameters $q = [q_1, q_2, \ldots, q_m]$, and is given by

$$\Delta = \text{diag}(q_1 I_1, q_2 I_2, \ldots, q_m I_m), \tag{7}$$

where I_i is an identity matrix of size p_i. Of course, $\sum_i^m p_i = p$. The rectangle in which q lies is given by $Q_{\text{init}} = [l_1, u_1] \times [l_2, u_2] \times \cdots \times [l_m, u_m]$. Figure 1 shows a block diagram of our setup.

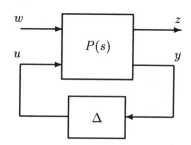

Fig. 1. System in standard form.

We also define

$$\begin{aligned}
P_{yu} &= C_y (sI - A)^{-1} B_u + D_{yu}, \\
P_{yw} &= C_y (sI - A)^{-1} B_w + D_{yw}, \\
P_{zu} &= C_z (sI - A)^{-1} B_u + D_{zu}, \\
P_{zw} &= C_z (sI - A)^{-1} B_w + D_{zw}.
\end{aligned} \tag{8}$$

P_{yu} is the (open-loop, *i.e.* $\Delta = 0$) transfer matrix from u to y and so on. Eliminating u and y from equations (6) yields the closed-loop system equations:

$$\dot{x} = \left(A + B_u\Delta(I - D_{yu}\Delta)^{-1}C_y\right)x + \left(B_u\Delta(I - D_{yu}\Delta)^{-1}D_{yw} + B_w\right)w,$$

$$z = \left(C_z + D_{zu}\Delta(I - D_{yu}\Delta)^{-1}C_y\right)x + \left(D_{zu}\Delta(I - D_{yu}\Delta)^{-1}D_{yw} + D_{zw}\right)w. \qquad (9)$$

For convenience, we let

$$\begin{aligned}
\mathcal{A}(q) &= A + B_u\Delta(I - D_{yu}\Delta)^{-1}C_y, \\
\mathcal{B}(q) &= B_u\Delta(I - D_{yu}\Delta)^{-1}D_{yw} + B_w, \\
\mathcal{C}(q) &= C_z + D_{zu}\Delta(I - D_{yu}\Delta)^{-1}C_y, \\
\mathcal{D}(q) &= D_{zu}\Delta(I - D_{yu}\Delta)^{-1}D_{yw} + D_{zw}.
\end{aligned}$$

We note that the entries of $\mathcal{A}(q)$, $\mathcal{B}(q)$, $\mathcal{C}(q)$ and $\mathcal{D}(q)$ are *rational functions* of the parameter vector q.

The closed-loop transfer matrix from w to z is denoted $P_{cl}(q)$ and is given by

$$P_{cl}(q) = P_{zw} + P_{zu}\Delta(I - P_{yu}\Delta)^{-1}P_{yw}. \qquad (10)$$

Loosely speaking, the above framework describes linear systems with fixed, unknown gains that lie in intervals. $P(s)$ is often called the *open-loop system*, and corresponds to the case when all the gains are set to zero. Δ, on one hand, might represent unknown parameters, in which case it has the interpretation of a perturbation to a linear system; on the other hand, the entries of Δ might represent gains which a designer may choose at will, in which case Δ has the interpretation of a design variable. w consists of the inputs to the system and z the outputs, and the closed loop transfer matrix $P_{cl}(q)$ consists of all transfer functions of interest.

A number of parameter problems in control can be addressed in this setting: study of Kharitonov polynomials and interval matrices, parametric controller design etc. (We refer the reader to [17] for how the first two problems can be cast into our setup.) Roughly speaking, any system with state-space matrices whose entries are rational functions of the uncertain parameters can be considered in our framework[2]. However, we wish to emphasize the restriction that the uncertain parameters lie in a rectangle;

---

[2]The precise condition is that none of the rational functions that make up the state space entries have a singularity at $q = 0$.

we cannot, for example, directly consider a situation where the uncertain parameters lie in a general polytope or an ellipsoid.

There are a number of important quantities associated with systems described by (6). We describe some of them below.

B.   Well-posedness

One of the most fundamental properties of the feedback system in Figure 1 is well-posedness: We say that the system is *well-posed* if it is well-defined for all $q \in \mathcal{Q}_{\text{init}}$, that is, if

$$\det(I - D_{yu}\Delta) \neq 0 \quad \text{for all} \quad q \in \mathcal{Q}_{\text{init}}. \tag{11}$$

Condition (11) is necessary and sufficient for equations (9) to be well-defined for all $q \in \mathcal{Q}_{\text{init}}$. Obviously, this condition is equivalent to none of the rational matrices $\mathcal{A}(q)$, $\mathcal{B}(q)$, $\mathcal{C}(q)$ and $\mathcal{D}(q)$ having singularities in $\mathcal{Q}_{\text{init}}$.

If Δ has the interpretation of an uncertainty, then the question of well-posedness is one of robustness analysis, where one asks if there is any choice of parameters that makes $I - D_{yu}\Delta$ singular. If Δ, on the other hand, contains design parameters, then the question is whether there is any choice of parameters that makes $I - D_{yu}\Delta$ nonsingular. Of course, this question has a particularly simple answer: $I - D_{yu}\Delta$ is either nonsingular for almost all values of q or it is singular for *every* q — this follows from the fact that $\det(I - D_{yu}\Delta)$ is a rational function of q — and therefore very simple algorithms can be devised to answer the design question.

The answer to the question of well-posedness is a Boolean "yes" or "no". One may also define a quantitative measure of well-posedness as, say, the condition number of $I - D_{yu}\Delta$. In that case, the robustness analysis question is the worst (*i.e.* largest) possible condition number of $I - D_{yu}\Delta$ over all possible Δ. The corresponding design problem would seek the choice of parameters that minimizes the condition number of $I - D_{yu}\Delta$.

For simplicity, we will henceforth assume that the system in Figure 1 is well-posed over $\mathcal{Q}_{\text{init}}$.

C.   Stability degree

The *stability degree* of an LTI system $\dot{x} = Fx$ where $F \in \mathbf{R}^{n \times n}$ is denoted $\lambda_{sd}(F)$ and defined as

$$\lambda_{sd}(F) = - \max_{1 \leq i \leq n} \operatorname{Re} \lambda_i(F).$$

The stability degree gives the slowest possible decay rate of the solutions of the system and thus may be regarded as a stability measure for the system:

$$\lambda_{sd}(F) = \sup\Big\{ \alpha \ \Big| \ \lim_{t \to \infty} e^{\alpha t} x(t) = 0 \text{ whenever } \dot{x} = Fx \Big\}.$$

We note that the stability degree equals the negative of the largest Lyapunov exponent of the system [18].

For a linear system, a "large" stability degree means that the solutions decay "fast". Based on this observation, the stability degree may be used to define a robustness measure for an uncertain system, or a design goal for parametric design. We describe these below.

1.   Minimum Stability Degree ($\mathcal{D}_{min}$)

A robustness measure for the parameter-dependent system in Figure 1 is the *minimum stability degree* ($\mathcal{D}_{min}$), the smallest or worst-case stability degree of the system over all possible value of the parameters:

$$\mathcal{D}_{min}(\mathcal{Q}_{init}) = \min_{q \in \mathcal{Q}_{init}} \lambda_{sd} \mathcal{A}(q).$$

$\mathcal{D}_{min}$ and other quantities that we describe below are functions of the system given by equations (6); we will not show this dependence explicitly for convenience.

The system in Figure 1 is robustly stable, *i.e.* the eigenvalues of the system have negative real parts for all Δ if and only if its $\mathcal{D}_{min}$ is positive. Moreover, $\mathcal{D}_{min}$ gives the worst-case decay rate of the solutions $x(t)$ of the state equations:

$$\mathcal{D}_{min}(\mathcal{Q}_{init}) = \min_{q \in \mathcal{Q}_{init}} \sup\Big\{ \alpha \ \Big| \ \lim_{t \to \infty} x(t) e^{\alpha t} = 0 \text{ whenever } \dot{x} = \mathcal{A}(q)x \Big\}.$$

From the point of view of robustness analysis, it is desirable to have a large, positive $\mathcal{D}_{min}$. This ensures that the states decay fast, irrespective of the uncertain parameter vector q and the initial condition.

2.   Maximum Stability Degree ($\mathcal{D}_{\max}$)

Treating the stability degree of the system as a design objective, one may define the *maximum stability degree* ($\mathcal{D}_{\max}$) as

$$\mathcal{D}_{\max}(\mathcal{Q}_{\text{init}}) = \max_{q \in \mathcal{Q}_{\text{init}}} \lambda_{\text{sd}}\mathcal{A}(q).$$

Thus, with the components of q regarded as design parameters, the problem is one of finding the set of parameters that maximizes the slowest possible decay rate of the solutions to the system, that is,

$$\mathcal{D}_{\max}(\mathcal{Q}_{\text{init}}) = \max_{q \in \mathcal{Q}_{\text{init}}} \sup\left\{ \alpha \; \middle| \; \lim_{t \to \infty} x(t)e^{\alpha t} = 0 \text{ whenever } \dot{x} = \mathcal{A}(q)x \right\}.$$

Computing $\mathcal{D}_{\max}$ checks *stabilizability*: There exists a choice of parameters that stabilizes the system if and only if $\mathcal{D}_{\max}$ is positive.

D. $\mathbf{H}_\infty$ norm

For the system in Figure 1, another quantity of interest is $\|P_{\text{cl}}\|_\infty$, the closed-loop $\mathbf{H}_\infty$ norm from w to z (see equation (10)), where $\|\cdot\|_\infty$ refers to the $\mathbf{H}_\infty$ norm:

$$\|G\|_\infty = \sup_{\text{Re } s > 0} \overline{\sigma}(G(s)).$$

$\|P_{\text{cl}}\|_\infty$ is just the *root mean square gain* (RMS-gain) of the system between the input w and the output z, *i.e.*,

$$\|P_{\text{cl}}\|_\infty = \max_{w(t) \neq 0} \frac{\|z\|_{\text{rms}}}{\|w\|_{\text{rms}}},$$

where the RMS value of a signal $w(t)$ is defined as

$$\|w\|_{\text{rms}} = \left(\lim_{T \to \infty} \frac{1}{T} \int_0^T w(t)^2 \, dt \right)^{1/2},$$

provided the limit exists.

Often w has the interpretation of a disturbance and z, that of some error signal; it is then desirable to have $\|P_{\text{cl}}\|_\infty$ small.

1.   Maximum $\mathbf{H}_\infty$ norm ($\mathcal{H}_{\infty,\max}$)

With q regarded as an uncertain parameter vector, we define the *maximum* $\mathbf{H}_\infty$ *norm* from w to z (denoted $\mathcal{H}_{\infty,\max}$) for the system in Figure 1 as

$$\mathcal{H}_{\infty,\max}(\mathcal{Q}_{\text{init}}) = \max_{q \in \mathcal{Q}_{\text{init}}} \|P_{\text{cl}}(q)\|_\infty.$$

Thus $\mathcal{H}_{\infty,\max}$ is the worst-case RMS gain of the system from w to z over all possible parameters.

We note that system in Figure 1 is robustly stable if and only if $\mathcal{H}_{\infty,\max}$ is finite; if the system is unstable for some choice of parameters q, then $\|P_{\text{cl}}(q)\|_\infty = \infty$. Moreover, $\mathcal{H}_{\infty,\max}$ then serves as a measure of robust stability: A smaller $\mathcal{H}_{\infty,\max}$ means a "more robustly stable" system.

2.   Minimum $\mathbf{H}_\infty$ norm ($\mathcal{H}_{\infty,\min}$)

In contrast with $\mathcal{H}_{\infty,\max}$, we may regard q as a design parameter and ask what choice of parameters yields the smallest possible RMS-gain between w and z. This is the the *minimum* $\mathbf{H}_\infty$ *norm* from w to z (denoted $\mathcal{H}_{\infty,\min}$) in Figure 1:

$$\mathcal{H}_{\infty,\min}(\mathcal{Q}_{\text{init}}) = \min_{q \in \mathcal{Q}_{\text{init}}} \|P_{\text{cl}}(q)\|_\infty.$$

Clearly, the system is stabilizable if and only if $\mathcal{H}_{\infty,\min}$ is finite.

E. $\mathbf{H}_2$ norm

We consider finally the $\mathbf{H}_2$ norm of the closed-loop transfer matrix from w to z, which, for a stable, strictly proper[3] transfer matrix P_{cl} is defined as

$$\|P_{\text{cl}}\|_2 = \left(\text{Tr}\frac{1}{2\pi}\int_{-\infty}^{\infty} P_{\text{cl}}(j\omega)P_{\text{cl}}(j\omega)^* \, d\omega\right)^{1/2}.$$

$\|P_{\text{cl}}\|_2 = \infty$ if P_{cl} is either unstable or not strictly proper. $\|P_{\text{cl}}\|_2$ has the interpretation of the RMS value of the output z when the components of the input w are independent white noises with unit power spectral density.

---

[3] P_{cl} is said to be strictly proper if $P_{\text{cl}}(\infty) = 0$.

1. Maximum $\mathbf{H}_2$ norm ($\mathcal{H}_{2,\max}$)

When q represents uncertainties, we define the *maximum* $\mathbf{H}_2$ *norm* from w to z (denoted $\mathcal{H}_{2,\max}$) in Figure 1 as

$$\mathcal{H}_{2,\max}(\mathcal{Q}_{\text{init}}) = \max_{q \in \mathcal{Q}_{\text{init}}} \|P_{\text{cl}}(q)\|_2.$$

Thus $\mathcal{H}_{2,\max}$ is the worst-case RMS value of z when w is driven by white noise whose power spectral density is the identity.

We note that $\mathcal{H}_{2,\max}$ is finite if and only if the system in Figure 1 is robustly stable and strictly proper for all $q \in \mathcal{Q}_{\text{init}}$. Moreover, a smaller $\mathcal{H}_{2,\max}$ means that the output is less susceptible to noises at the input; thus $\mathcal{H}_{2,\max}$ serves a measure of robust performance.

We also note that the computation of the maximum total state co-variance of a system driven by white noise can be cast as a problem of computing $\mathcal{H}_{2,\min}$ [19].

2. Minimum $\mathbf{H}_2$ norm ($\mathcal{H}_{2,\min}$)

The design problem corresponding to the $\mathbf{H}_2$ norm measuring the size of the closed-loop transfer matrix is that of finding the minimum $\mathbf{H}_2$ norm ($\mathcal{H}_{2,\min}$). This is the choice of parameters q that minimizes the closed-loop $\mathbf{H}_2$ norm:

$$\mathcal{H}_{2,\min}(\mathcal{Q}_{\text{init}}) = \min_{q \in \mathcal{Q}_{\text{init}}} \|P_{\text{cl}}(q)\|_2.$$

F. Remarks on Complexity

We now make some general observations regarding the computation of the quantities defined so far. First, we note that the fundamental problem of well-posedness (*cf.* equation (11)) is NP-hard in general [20, 21, 22, 23]; roughly speaking, in the worst case, the number of computations required to establish well-posedness increases more than polynomially with the problem size m (which is the size of D_{yu} and Δ). This makes it likely that *any* algorithm that computes any of the six quantities above to within some fixed accuracy also requires, in the worst case, computations that increase more than polynomially with the problem size. This conjecture is especially interesting in light of the fact that maximum number of branch and bound algorithm iterations increases exponentially with m for a given accuracy (see Subsection II B); therefore, if the conjecture were true, no algorithm would perform substantially better than a branch and bound algorithm on these problems.

We know of no existing algorithms that compute any of the quantities $\mathcal{D}_{min}$, $\mathcal{D}_{max}$, $\mathcal{H}_{\infty,max}$, $\mathcal{H}_{\infty,min}$, $\mathcal{H}_{2,max}$ or $\mathcal{H}_{2,min}$ for the general framework in equations (6). However, much work has been done in special cases. Kharitonov's theorem [24, 25] gives a very efficient method for determining robust stability for the special case when the coefficients of the characteristic polynomial of $\mathcal{A}(q)$ are just the uncertain parameters q_i. Kharitonov's theorem has been extended to cover the case in which the characteristic polynomial is an affine function of q [26, 27]. Another problem that may be considered in our setup is the study of interval matrices [28, 29, 30].

In [31], Anderson *et al.* observe that the robust stability question is *decidable*, which means that by evaluating a finite number of polynomial functions of the input data (the entries of the state-space matrices, and the l_i, u_i), we can determine whether the system is robustly stable. It turns out, however, that these decision procedures involve an extraordinarily large number of polynomials, even for small systems with few parameters. Moreover the number of polynomials that need to be checked grows very rapidly (more than exponentially) with system size and number of parameters (see also [32]).

Though most of the methods described above do not directly consider the computation of the six quantities described in our framework, they do provide useful *bounds* for these quantities. Local optimization procedures provide lower (upper) bounds for the maximization (minimization) problems; there exist many analytical techniques (small gain theorem, Lyapunov theory based methods etc.) that yield bounds in the other direction. We will use some of these techniques to derive bounds for $\mathcal{D}_{min}$, $\mathcal{D}_{max}$, $\mathcal{H}_{\infty,max}$, $\mathcal{H}_{\infty,min}$, $\mathcal{H}_{2,max}$ and $\mathcal{H}_{2,min}$ in the next section.

IV.   COMPUTATION OF BOUNDS

A.   A Loop Transformation

Before we go on to describing the computation of bounds, we describe a loop transformation that converts the problem of finding bounds over an arbitrary rectangle to that of finding bounds over the cube $\mathcal{U} = [-1, 1]^m$. We refer the reader to [33] for a complete discussion of loop transformations.

The loop transformation is best explained through Figure 2, where the symbols $\tilde{H}(s)$ and $\tilde{\Delta}$ refer to the "new" system and the "normalized" perturbation.

The loop transformation can be interpreted as translating Q to the origin, and then scaling it to the cube $[-1, 1]^m$.

$$K = \mathrm{diag}(\frac{u_1 + l_1}{2}I_1, \frac{u_2 + l_2}{2}I_2, \ldots, \frac{u_m + l_m}{2}I_m),$$
$$F = \mathrm{diag}(\frac{u_1 - l_1}{2}I_1, \frac{u_2 - l_2}{2}I_2, \ldots, \frac{u_m - l_m}{2}I_m)$$

are the matrices that accomplish this.

A state-space representation of the loop-transformed system $\tilde{P}(s)$ is given by $\{\tilde{A}, \tilde{B}, \tilde{C}, \tilde{D}\}$, where

$$\tilde{A} = A + B_u TKC_y; \quad \tilde{B} = \left[\begin{array}{cc} \underbrace{B_w + B_u TKD_{yw}}_{\tilde{B}_z} & \underbrace{B_u TF^{\frac{1}{2}}}_{\tilde{B}_u} \end{array} \right];$$

$$\tilde{C} = \left[\begin{array}{c} \overbrace{\underbrace{\begin{array}{c} C_z + D_{zu} TKC_y \end{array}}_{}}^{\tilde{C}_z} \\ \underbrace{F^{\frac{1}{2}}(I + D_{yu} TK)C_y}_{\tilde{C}_y} \end{array} \right]; \qquad (12)$$

$$\tilde{D} = \left[\begin{array}{cc} \overbrace{\underbrace{D_{zw} + D_{zu} TKD_{yw}}_{\tilde{D}_{zw}}}^{\tilde{D}_{zw}} & \overbrace{D_{zu} TKF^{\frac{1}{2}}}^{\tilde{D}_{zu}} \\ \underbrace{F^{\frac{1}{2}}(I + D_{yu} TK)D_{yw}}_{\tilde{D}_{yw}} & \underbrace{F^{\frac{1}{2}} D_{yu} TF^{\frac{1}{2}}}_{\tilde{D}_{yu}} \end{array} \right].$$

$T = (I - KD_{yu})^{-1}$, and I is the identity matrix of appropriate size. We remind the reader of the assumption that the system is well-posed, which guarantees that $(I - KD_{yu})$ is invertible.

We make the obvious but important remark that all the six quantities we have defined in the previous subsection remain invariant under the loop transformation. Thus any bounds for these quantities computed with the loop transformed system in Figure 2 are valid for the system in Figure 1.

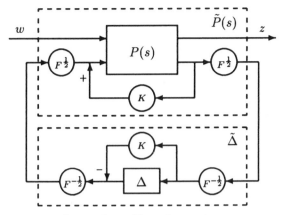

Fig. 2. Loop Transformation.

B. Bounds for $\mathcal{D}_{\min}$

Computation of $\mathcal{D}_{\min}$ is a global minimization problem and we will therefore use Algorithm I of Section II. In the notation used to describe Algorithm I, we have $f(q) = \lambda_{sd}(\mathcal{A}(q))$ and $\Phi_{\min}(\mathcal{U}) = \mathcal{D}_{\min}(\mathcal{U})$. We now need to compute a lower bound $\Phi_{lb}(\mathcal{U})$ and an upper bound $\Phi_{ub}(\mathcal{U})$ for $\mathcal{D}_{\min}(\mathcal{U})$.

Upper bounds

One simple upper bound on $\mathcal{D}_{\min}$ over the cube $\mathcal{U}$ is just the stability degree of the system evaluated at the *midpoint* of the cube. Thus:

$$\Phi_{ub}(\mathcal{U}) = \lambda_{sd}(\mathcal{A}(0)) = \lambda_{sd}(A). \tag{13}$$

This upper bound can be improved quite easily by local optimization methods such as a gradient search (see any text on optimization, [34] for example). Another heuristic is to set the upper bound over $\mathcal{U}$ to be equal to the minimum of the stability degrees over the vertices and the center:

$$\Phi_{ub}(\mathcal{U}) = \min_{q \in \{\text{vertices of } \mathcal{U}\} \cup \{0\}} \lambda_{sd}(\mathcal{A}(q)). \tag{14}$$

There are two justifications for this heuristic: First, in many special cases, $\mathcal{D}_{\min}$ is always achieved at a vertex; secondly, since the number of local minima for $\mathcal{D}_{\min}$ is finite, as the size of $\mathcal{Q}$ (before loop-transformation) becomes small, $\mathcal{D}_{\min}$ is more likely to be achieved on the boundary of $\mathcal{Q}$ than in the interior; therefore the vertices are more likely to yield better bounds on $\mathcal{D}_{\min}$ than the center.

Lower bounds

Computation of lower bounds is a little more involved. It is based on the application of a generalized *small gain theorem* (SGT) which states that for every $q \in \mathcal{U}$, the (closed-loop) system in Figure 1 has the same number of stable poles as $P(s)$, the open-loop system, provided $\|P_{yu}\|_{\mathbf{L}_\infty} < 1$, where

$$\|H\|_{\mathbf{L}_\infty} = \sup_{\omega \in \mathbf{R}} \overline{\sigma}(H(j\omega))$$

is the $\mathbf{L}_\infty$ norm of the transfer matrix H. This theorem is a simple extension of the conventional small gain theorem which can be found, for example, in [33]. Thus, if $\|P_{yu}\|_{\mathbf{L}_\infty} < 1$,

$$A \text{ stable} \implies \mathcal{D}_{\min}(\mathcal{U}) > 0, \tag{15}$$

and

$$A \text{ unstable} \implies \mathcal{D}_{\max}(\mathcal{U}) < 0. \tag{16}$$

Thus, if A is stable and $\|P_{yu}\|_{\mathbf{L}_\infty} < 1$, (15) gives a lower bound of zero for $\mathcal{D}_{\min}(\mathcal{U})$. Otherwise, we may conclude nothing.

In order to derive a better lower bound on $\mathcal{D}_{\min}(\mathcal{U})$, we consider the exponentially time-weighted system

$$
\begin{aligned}
\dot{x}_e &= (A + \alpha I)x_e &+\quad B_u u &+\quad B_w w, &\qquad x_e(0) = x_0, \\
y &= C_y x_e &+\quad D_{yu} u &+\quad D_{yw} w, \\
z &= C_z x_e &+\quad D_{zu} u &+\quad D_{zw} w, \\
u &= \Delta\, y,
\end{aligned}
$$

$$\tag{17}$$

where $\alpha < \lambda_{\mathrm{sd}}(A)$. Note that $(A + \alpha I)$ is guaranteed to be *stable*. The solutions of equations (17) and (6) are simply related by $x_e(t) = e^{\alpha t} x(t)$, and therefore we conclude that

$$\mathcal{D}_{\min}(\mathcal{U}) > \alpha, \quad \text{whenever } \|P_{yu}\|_{\infty,\alpha} < 1,$$

where

$$\|H\|_{\infty,\alpha} = \sup_{\{s = -\alpha + j\omega \,|\, \omega \in \mathbf{R}\}} \overline{\sigma}(H(s))$$

is the α-shifted $\mathbf{L}_\infty$ norm of H [35]. Therefore, we define $\Phi_{\text{lb}}(\mathcal{U})$ as

$$\Phi_{\text{lb}}(\mathcal{U}) = \sup\{\alpha \mid \alpha < \lambda_{\text{sd}}(A),\ \|P_{yu}\|_{\infty,\alpha} < 1\}. \qquad (18)$$

Note that if $\|P_{yu}\|_{\infty,\alpha} \geq 1$ for all α, then $\Phi_{\text{lb}}(\mathcal{U}) = -\infty$. (This occurs only if $\overline{\sigma}(D_{yu}) \geq 1$.)

The condition in (18) is readily checked by forming an appropriate Hamiltonian matrix and checking its eigenvalues (see [36, 37]); a simple bisection can be used to compute Φ_{lb}.

We note that our procedure for computing $\Phi_{\text{lb}}(\mathcal{U})$ is just an application of the "shifted circle criterion" (Anderson and Moore [38]).

Very often, the lower bound above turns out to be too conservative; the reason for this lies in the application of the SGT in (15) (and (16)). The very special structure of Δ (real, diagonal, and with possibly many repeated entries) has been ignored, and this means that the SGT may be extremely conservative in guaranteeing the robust stability of the exponentially weighted closed-loop system. Eliminating or reducing this conservatism of the SGT that arises due to "structured perturbations" is a major area of research in itself (structured singular value [39], scaling or the scaled singular value [40]). The scaled singular value (which we will abbreviate as SSV) is directly relevant to our problem, and we will give a brief and informal discussion here.

The motivation for the SSV arises from the following simple observation: The system shown in Figure 3 is equivalent to the system in Figure 1 (in the sense that the solution to the closed-loop state equations as well as the closed-loop transfer matrices from w to z are equal) for all nonzero $\alpha \in \mathbf{C}$ and invertible matrices[4] $S_\Delta \in \mathbf{C}^{p \times p}$ such that $S_\Delta \Delta = \Delta S_\Delta$. In other words, the structure of Δ makes the closed-loop system invariant under the scaling of the open loop transfer matrix described by

$$\begin{bmatrix} P_{zw} & P_{zu} \\ P_{yw} & P_{yu} \end{bmatrix} \longrightarrow \begin{bmatrix} \alpha I_1 & 0 \\ 0 & S_\Delta \end{bmatrix} \begin{bmatrix} P_{zw} & P_{zu} \\ P_{yw} & P_{yu} \end{bmatrix} \begin{bmatrix} \alpha I_2 & 0 \\ 0 & S_\Delta \end{bmatrix}^{-1},$$

where $\alpha \in \mathbf{C}$, I_1 and I_2 are identity matrices of sizes n_o and n_i respectively, and $S_\Delta \in \mathbf{C}^{p \times p}$ is an invertible matrix that commutes with Δ.

---

[4]Both α and S_Δ can be functions of s rather than simply constants, but we will not consider this more general setting here.

We let

$$S_{\text{left}} = \begin{bmatrix} \alpha I_1 & 0 \\ 0 & S_\Delta \end{bmatrix} \quad \text{and} \quad S_{\text{right}} = \begin{bmatrix} \alpha I_2 & 0 \\ 0 & S_\Delta \end{bmatrix}.$$

S_{left} and S_{right} are referred to as the *left* and *right scalings*. The set of matrices that commute with Δ is denoted $\mathbf{S}_\Delta$:

$$\mathbf{S}_\Delta = \left\{ S_\Delta \mid S_\Delta \Delta = \Delta S_\Delta, \ S_\Delta \in \mathbf{C}^{p \times p} \right\}.$$

$\mathbf{S}_\Delta$ determines the set of left and right scalings, denoted $\mathbf{S}_{\text{left}}$ and $\mathbf{S}_{\text{right}}$ respectively, that leave the closed-loop system invariant:

$$\mathbf{S}_{\text{left}} = \left\{ S_{\text{left}} \mid S_{\text{left}} = \begin{bmatrix} \alpha I_1 & \\ & S_\Delta \end{bmatrix}, \ \alpha \in \mathbf{C}, \ S_\Delta \in \mathbf{S}_\Delta \right\}, \tag{19}$$

$$\mathbf{S}_{\text{right}} = \left\{ S_{\text{right}} \mid S_{\text{right}} = \begin{bmatrix} \alpha I_2 & \\ & S_\Delta \end{bmatrix}, \ \alpha \in \mathbf{C}, \ S_\Delta \in \mathbf{S}_\Delta \right\}. \tag{20}$$

It is not hard to see that for our case, every $S_\Delta \in \mathbf{S}_\Delta$ is of the form

$$S_\Delta = \begin{bmatrix} D_1 & & & \\ & D_2 & & \\ & & \ddots & \\ & & & D_p \end{bmatrix},$$

where $D_i \in \mathbf{C}^{p_i \times p_i}$ and invertible, $i = 1, 2, ..., p$.

In the computation of a lower bound for $\mathcal{D}_{\min}$ through equation (18), the transfer matrix from u to y is the only one of interest. The scaling of $P(s)$ to $S_{\text{left}} P(s) S_{\text{right}}^{-1}$ results in the scaling of $P_{yu}(s)$ to $S_\Delta P_{yu}(s) S_\Delta^{-1}$. Then we may define a new, possibly improved lower bound as

$$\Phi_{\text{lb}}(\mathcal{U}) = \sup \left\{ \alpha \ \middle| \ \alpha < \lambda_{\text{sd}}(A), \ \inf_{S_\Delta \in \mathbf{S}_\Delta} \|S_\Delta P_{yu} S_\Delta^{-1}\|_{\infty, \alpha} < 1 \right\}. \tag{21}$$

There are many heuristics for performing the optimization in equation (21). We will not describe any of them here. However, we note that for a fixed α, computation of the scaling that minimizes $\|S_\Delta P_{yu} S_\Delta^{-1}\|_{\infty, \alpha}$ can be formulated as a quasi-convex optimization problem (see [35], chapters 13 and 14), and therefore can be performed by effective methods. Scaling a *constant* matrix (as opposed to a transfer matrix) optimally is a

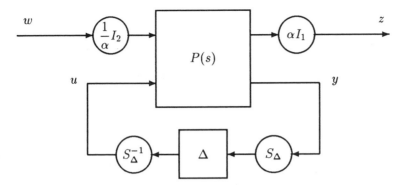

Fig. 3.   The standard form with scaling. $\alpha \in \mathbf{C}$, I_1 and I_2 are identity
matrices of sizes n_i and n_o respectively, $S_\Delta \in \mathbf{C}^{p \times p}$ is invertible
and commutes with Δ.

well studied problem. A (by no means exhaustive) list of references is
[41, 42, 43, 44, 45, 46, 47].

Note that the SSV has ignored the fact that Δ is real. This aspect has
been addressed in a recent work that accounts for real perturbations [48].
The results therein may be incorporated into the bounds in (18) to improve
the lower bound further.

For more details, including a proof that the bounds we have described for
$\mathcal{D}_{min}$ satisfy conditions (R1) and (R2) stated at the beginning of Section II,
see [17].

C.   Bounds for $\mathcal{D}_{max}$

Computation of $\mathcal{D}_{max}$ is a global maximization problem, so we will use
Algorithm II of Section II. Our notation is then $f(q) = \lambda_{sd}(\mathcal{A}(q))$ and
$\Psi_{max}(\mathcal{U}) = \mathcal{D}_{max}(\mathcal{U})$; we now describe the computation of a lower bound
$\Psi_{lb}(\mathcal{U})$ and an upper bound $\Psi_{ub}(\mathcal{U})$ for $\mathcal{D}_{max}(\mathcal{U})$.

Lower bounds

A simple lower bound on $\mathcal{D}_{max}$ over the cube $\mathcal{U}$ is just the stability degree
of the system evaluated at the *midpoint* of the cube. Thus:

$$\Psi_{lb}(\mathcal{U}) = \lambda_{sd}(\mathcal{A}(0)) = \lambda_{sd}(A). \tag{22}$$

The above lower bound can be improved by the same heuristic as in equation (14):

$$\Psi_{lb}(\mathcal{U}) = \max_{q \in \{\text{vertices of } \mathcal{U}\} \cup \{0\}} \lambda_{sd}(\mathcal{A}(q)). \qquad (23)$$

Upper bounds

Our upper bound for $\mathcal{D}_{max}(\mathcal{U})$ is

$$\Psi_{ub}(\mathcal{U}) = \inf \left\{ \alpha \mid \alpha > \lambda_{sd}(A), \|P_{yu}\|_{\infty,\alpha} < 1 \right\}. \qquad (24)$$

Note that if $\|P_{yu}\|_{\infty,\alpha} \geq 1$ for all α, then $\Phi_{lb}(\mathcal{U}) = \infty$.

This bound can be established using a derivation identical to that leading to the bound in equation (18). Instead, let us give a brief intuitive explanation of this bound.

The bound in equation (18) is just the largest amount of "anti-damping" α we may add to system in Figure (1) with the SGT proving robust stability. In contrast, the bound in equation (24) is just the smallest anti-damping α we must add to system in Figure (1) for the SGT to prove robust *instability*, *i.e.* to guarantee that the closed-loop system in Figure 1 is unstable for all $q \in \mathcal{U}$; the negative of the maximum real part of the eigenvalues of the closed-loop system can be no larger than this anti-damping.

This upper bound can be improved using the same scaling techniques as in equation (21):

$$\Psi_{ub}(\mathcal{U}) = \inf \left\{ \alpha \mid \alpha > \lambda_{sd}(A), \inf_{S_\Delta \in \mathbf{S}_\Delta} \|S_\Delta P_{yu} S_\Delta^{-1}\|_{\infty,\alpha} < 1 \right\}. \qquad (25)$$

It can be shown that the bounds for $\mathcal{D}_{max}$ satisfy conditions (R1') and (R2') stated in Section II.

D.   Bounds for $\mathcal{H}_{\infty,max}$

We next describe the computation of a lower bound $\Psi_{lb}(\mathcal{U})$ and an upper bound $\Psi_{ub}(\mathcal{U})$ for $\mathcal{H}_{\infty,max}$.

Lower Bounds

A simple lower bound for $\mathcal{H}_{\infty,max}(\mathcal{U})$ is just the $\mathbf{H}_\infty$ norm of the closed-loop system with the parameter vector set to the midpoint of the parameter region $\mathcal{U}$:

$$\Psi_{\text{lb}}(\mathcal{U}) = \|P_{\text{cl}}(0)\|_\infty = \|P_{zw}\|_\infty. \tag{26}$$

This bound may be improved using the same heuristic as in in equations (14) and (23), by setting the lower bound of $\mathcal{H}_{\infty,\text{max}}$ over $\mathcal{U}$ to be equal to the maximum of the $\mathbf{H}_\infty$ norm from w to z computed with the parameters assuming the values at the vertices and the center:

$$\Psi_{\text{lb}}(\mathcal{U}) = \max_{q \in \{\text{vertices of } \mathcal{U}\} \cup \{0\}} \|P_{\text{cl}}(q)\|_\infty. \tag{27}$$

Lower bounds

We now describe a simple scheme for computing an upper bound that is based on a small gain based robust stability condition due to Doyle [39] and Safonov [44] (see [35, p239-241]).

We define

$$P_\beta = \begin{bmatrix} \dfrac{P_{zw}}{\beta} & \dfrac{P_{zu}}{\sqrt{\beta}} \\[2mm] \dfrac{P_{yw}}{\sqrt{\beta}} & P_{yu} \end{bmatrix}, \tag{28}$$

where $\beta > 0$. Then

$$\|P_\beta\|_\infty < 1 \implies \sup_{\|\Delta\|_\infty \leq 1} \left\| \left[P_{zw} + P_{zu}\Delta(I - P_{yu}\Delta)^{-1}P_{yw} \right] \right\|_\infty < \beta.$$

Our upper bound is:

$$\Psi_{\text{ub}}(\mathcal{U}) = \inf \left\{ \beta \mid \|P_\beta\|_\infty < 1 \right\}, \tag{29}$$

with the convention that the infimum of a function over the empty set is infinity.

As with the lower bounds for $\mathcal{D}_{\text{min}}$, Ψ_{ub} may be rapidly computed using a bisection on Hamiltonian matrices [36, 37]. We may also use scaling the improve the upper bound for $\mathcal{H}_{\infty,\text{max}}$ in equation (29). The possibly improved upper bound is then given by

$$\Psi_{\text{ub}}(\mathcal{U}) = \inf \left\{ \beta \mid \inf_{S_{\text{left}} \in \mathbf{S}_{\text{left}}, S_{\text{right}} \in \mathbf{S}_{\text{right}}} \|S_{\text{left}} P_\beta S_{\text{right}}^{-1}\|_\infty < 1 \right\}, \tag{30}$$

where $\mathbf{S}_{\text{left}}$ and $\mathbf{S}_{\text{right}}$ are given by equations (20) and (19).

We refer the reader to [49] for a proof that the bounds for $\mathcal{H}_{\infty,\text{max}}$ satisfy conditions (R1$'$) and (R2$'$) stated in Section II.

E.   Bounds for $\mathcal{H}_{\infty,\min}$

We describe next the computation of an upper bound $\Phi_{\mathrm{ub}}(\mathcal{U})$ and a lower bound $\Phi_{\mathrm{lb}}(\mathcal{U})$ for $\Phi_{\min}(\mathcal{U}) = \mathcal{H}_{\infty,\min}(\mathcal{U})$. The bounds we present here make the possibly unrealistic assumption that system in Figure 1 is robustly stable. In terms of design, this requires the designer to apply the algorithm only over parameter ranges where the system is guaranteed to be stable.

Upper bounds

A simple upper bound for $\mathcal{H}_{\infty,\min}(\mathcal{U})$ is just the $\mathbf{H}_\infty$ norm of the closed-loop system with the parameter vector set to the midpoint of the parameter region $\mathcal{U}$:

$$\Phi_{\mathrm{lb}}(\mathcal{U}) = \|P_{\mathrm{cl}}(0)\|_\infty = \|P_{zw}\|_\infty. \tag{31}$$

This upper bound may be improved by employing the same heuristic as in equations (14), (23) and (27):

$$\Phi_{\mathrm{lb}}(\mathcal{U}) = \min_{q \in \{\text{vertices of } \mathcal{U}\} \cup \{0\}} \|P_{\mathrm{cl}}(q)\|_\infty. \tag{32}$$

Lower bound

If $\|P_{yu}\|_\infty < 1$, we may compute a lower bound based on simple norm inequalities:

$$
\begin{aligned}
\|P_{\mathrm{cl}}(q)\|_\infty &= \|P_{zw} + P_{zu}\Delta(I - P_{yu}\Delta)^{-1}P_{yw}\|_\infty \\
&\geq \|P_{zw}\|_\infty - \frac{\|P_{zu}\|_\infty \|P_{yw}\|_\infty}{1 - \|P_{yu}\|_\infty}.
\end{aligned}
\tag{33}
$$

If $\|P_{yu}\|_\infty \geq 1$, our lower bound is merely 0.

We refer the reader to [49] for more details, including a proof that the bounds for $\mathcal{H}_{\infty,\min}$ satisfy conditions (R1) and (R2) stated in Section II. .

F.   Bounds for $\mathcal{H}_{2,\max}$

We describe next the computation of an upper bound $\Psi_{\mathrm{ub}}(\mathcal{U})$ and a lower bound $\Psi_{\mathrm{lb}}(\mathcal{U})$ for $\Psi_{\max}(\mathcal{U}) = \mathcal{H}_{2,\max}(\mathcal{U})$. The bounds are based on the observation that $\mathbf{H}_2$ norms can be computed by solving Lyapunov equations (see for example, [35]). More precisely, if the stable, strictly proper transfer function $G(s)$ has a state-space realization $\{A, B, C\}$, then $\|G\|_2 =$

$\sqrt{\text{Tr}(CW_cC^T)}$, where $W_c = W_c^T > 0$ is the unique solution of the Lyapunov equation

$$AW_c + W_cA^T + BB^T = 0. \tag{34}$$

($\|G\|_2 = \infty$ if A is unstable or if G is not strictly proper.)

Lower bounds

A simple lower bound $\Psi_{\text{lb}}(\mathcal{U})$ for $\mathcal{H}_{2,\text{max}}$ is given by the $\mathbf{H}_2$ norm of the system evaluated with parameters assuming values at the center of $\mathcal{U}$:

$$\Psi_{\text{lb}}(\mathcal{U}) = \|P_{\text{cl}}(0)\|_2. \tag{35}$$

This lower bound may be improved by simple heuristics as in equation (14) or by local optimization methods; in fact, there is a whole body of research on this problem. We refer the reader to the survey by Toivonen and Mäkilä [50].

Upper bound

Noting that (34) is just a system of linear equations, we may compute an upper bound $\Psi_{\text{ub}}(\mathcal{U})$ based on a simple perturbation analysis. We present the bound below, omitting tedious details.

In what follows, L_A is the Lyapunov operator associated with A, given by $A \otimes I + I \otimes A$ ("$\otimes$" is the Kronecker product, see for example, [51]), $N_1 = \min\{n_i, n\}$, $N_2 = \min\{n_o, n\}$. For convenience, we let

$$
\begin{aligned}
a &= \frac{\overline{\sigma}(B_u)\overline{\sigma}(C_y)}{1 - \overline{\sigma}(D_{yu})}, \\
b &= \frac{\overline{\sigma}(B_u)\overline{\sigma}(D_{yw})}{1 - \overline{\sigma}(D_{yu})}, \\
c &= \frac{\overline{\sigma}(C_y)\overline{\sigma}(D_{zu})}{1 - \overline{\sigma}(D_{yu})}, \\
d &= \frac{1}{\underline{\sigma}(L_A) - 2a}\left(2a\|W_c\|_F + b^2\sqrt{N_1} + 2b\|B_w\|_F\right), \\
e &= \left(2c\sqrt{N_2}\|C_zW_c\|_F + c^2N_2\|W_c\|_F + dN_2\left(\overline{\sigma}(C_z) + c\right)^2\right)^{1/2}.
\end{aligned}
\tag{36}
$$

W_c is the controllability Gramian of the system with $\Delta = 0$ and satisfies $AW_c^T + W_cA + B_wB_w^T = 0$.

Then, if $\overline{\sigma}(D_{yu}) < 1$ and $\underline{\sigma}(L_A) > 2a$,

$$\Psi_{ub}(\mathcal{U}) = \left((\Psi_{lb}(\mathcal{U}))^2 + e^2\right)^{1/2}. \tag{37}$$

If $\overline{\sigma}(D_{yu}) \geq 1$ or $\underline{\sigma}(L_A) \leq 2a$, the upper bound is only ∞.

It can be shown that the bounds for $\mathcal{H}_{2,\max}$ satisfy conditions (R1′) and (R2′) of Section II.

G.   Bounds for $\mathcal{H}_{2,\min}$

An upper bound $\Phi_{ub}(\mathcal{U})$ and a lower bound $\Phi_{lb}(\mathcal{U})$ for $\Phi_{\min}(\mathcal{U}) = \mathcal{H}_{2,\min}(\mathcal{U})$ can be derived analogously to those for $\mathcal{H}_{2,\max}$.

Upper bounds

An upper bound $\Phi_{ub}(\mathcal{U})$ for $\mathcal{H}_{2,\min}$ is given by the $\mathbf{H}_2$ norm of the system evaluated with parameters assuming values at the center of $\mathcal{U}$:

$$\Phi_{ub}(\mathcal{U}) = \|P_{cl}(0)\|_2. \tag{38}$$

This upper bound may be improved by local optimization methods.

Lower bound

If $\overline{\sigma}(D_{yu}) < 1$, $\underline{\sigma}(L_A) > 2a$ and $\Phi_{ub}(\mathcal{U}) > e$,

$$\Phi_{lb}(\mathcal{U}) = \max\left\{0, \left((\Phi_{ub}(\mathcal{U}))^2 - e^2\right)^{1/2}\right\}, \tag{39}$$

where a, b, c, d and e are given by equations (36). If $\overline{\sigma}(D_{yu}) \geq 1$ or $\underline{\sigma}(L_A) \leq 2a$ or $\Phi_{ub}(\mathcal{U}) \leq e$, the lower bound is merely 0. As with the lower bound for $\mathcal{H}_{\infty,\min}$, this bound requires that the system (6) be robustly stable.

It can be shown that the bounds for $\mathcal{H}_{2,\max}$ satisfy conditions (R1) and (R2) of Section II.

V.   EXAMPLES

We consider a mechanical plant consisting of two masses connected by a spring with the left-hand mass driven by a force, as shown in Figure 4.

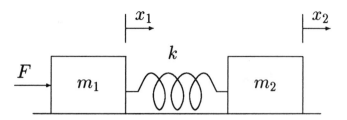

Fig. 4. The plant consisting of two masses connected by a spring.

We note that this example is so simple that many of the quantities associated with robustness analysis and design could be computed by hand; of course, the value of the methods that we present is in solving more complex problems that cannot be solved by hand or simple *ad hoc* procedures.

A.   Examples for Analysis

We will first study the robustness properties of this system under variations of m_2 and k. More specifically, we will examine the common perception that an LQR-optimal state-feedback is "robust". To this end, we let $[x_1 \ \dot{x}_1 \ x_2 \ \dot{x}_2]^T$ be the state vector, and employ a state-feedback law $F = -k_{\text{LQR}}x$ which is LQR optimal for the parameter values $m_1 = m_2 = k = 1$ (with weights $Q = I$, $\rho = 1$). In other words, $F(t) = -k_{\text{LQR}}x(t)$ is the input that minimizes the cost function

$$\int_0^\infty \left(x(t)^T x(t) + (F(t))^2 \right) dt,$$

(irrespective of the initial condition) where the state equations are those of the system in Figure 4.

We will now study the robustness of this LQR-optimal closed-loop system with respect to variations in the parameters m_2 and k in a range between 2/3 and 3/2:

$$2/3 \le m_2 \le 3/2, \qquad 2/3 \le k \le 3/2. \tag{40}$$

Thus, these physical parameters can vary over a range exceeding $2 : 1$. The closed-loop transfer function P_{cl} that we will examine is the closed-loop

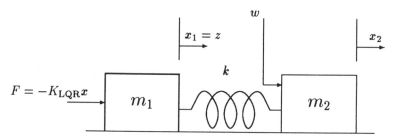

Fig. 5. The closed-loop system with LQR-optimal state feedback.
$x = [x_1 \ \dot{x}_1 \ x_2 \ \dot{x}_2]^T$ is the state vector. $m_1 = 1$ is fixed, and m_2
and k are uncertain parameters known to lie in $[2/3, 3/2]$. w is
the force on the second mass and z is the position of the first
mass.

transfer function from the force on the second mass to the position of the
first mass (see Figure 5).

1. $\mathcal{D}_{\min}$ of the system with the LQR-optimal controller

For this nominally LQR-optimal system, we compute the first measure of
robustness, *i.e.* $\mathcal{D}_{\min}$ over the parameter ranges in (40).

Figure 6 shows the results obtained from applying the branch and bound
algorithm to this problem. The solid lines correspond to the bounds in (18)
and (13), and the dashed lines to the "improved" bounds[5] in (21) and (14).
We note the following:

- *The system is robustly stable.* The system is robustly stable if and
  only if $\mathcal{D}_{\min}$ is positive. The lower bound for $\mathcal{D}_{\min}$ is positive at
  the end of 8 iterations using the lower bound (18) and at the end
  of 6 iterations using the lower bound (21), which establishes robust
  stability in each case.

- $0.1853 \le \mathcal{D}_{\min} \le 0.1862$. Thus the algorithm can prove that the
  decay rate of the state $x(t)$ is at least 0.1853, irrespective of the
  initial state x_0 and q. (For reference, the stability degree (λ_{sd}) of the

---

[5]The optimization problem to perform the scaling in equation (21) was not solved
exactly; instead, a lower bound to the optimum was used.

nominal system is 0.3738. Thus the parameter variations can degrade the stability degree by a factor of about 2.)

- With the bounds in (18) and (13), the algorithm takes 307 iterations to determine $\mathcal{D}_{\min}$ to within an absolute accuracy of 0.001, whereas with the improved bounds in (21) and (14), it takes only 176 iterations; this is offset by the increased computation accompanying the improved bounds.

- The algorithm returns the worst-case parameters $m_2 = 2/3$ and $k = 3/2$, which happen to lie on a vertex of the parameter box. This is not the case in general.

- Figure 6 also shows the number of rectangles in the partition as well as the pruned volume percentage as a function of iterations, for the two different sets of bounds. It is clear that the "improved" bounds do show improved performance.

- Figure 7 shows the parameter region under consideration at various stages of the algorithm. The algorithm can *prove* that $\mathcal{D}_{\min}$ cannot be achieved outside the shaded region.

2. $\mathcal{H}_{\infty,\max}$ of the system with the LQR-optimal controller

We now turn to our second measure of robustness: We will compute the largest possible $\mathbf{H}_{\infty}$ norm of the closed-loop transfer function P_{cl} due to the parameter variations in (40).

Figure 8 shows the results from the branch and bound algorithm applied to the computation of $\mathcal{H}_{\infty,\max}$. The solid lines correspond to the bounds in (26) and (29), and the dashed lines once again to the "improved" bounds[6] (in (27) and in (30)). We note the following:

- *The system is robustly stable.* Either upper bound (from (29) or (30)) is finite only if the system is robustly stable. The upper bound on $\mathcal{H}_{\infty,\max}$ from (29) is finite at the end of 8 iterations, (and the upper

---

[6] The optimal scaling problem in (30) was again not solved exactly; instead, an upper bound to the optimum was used.

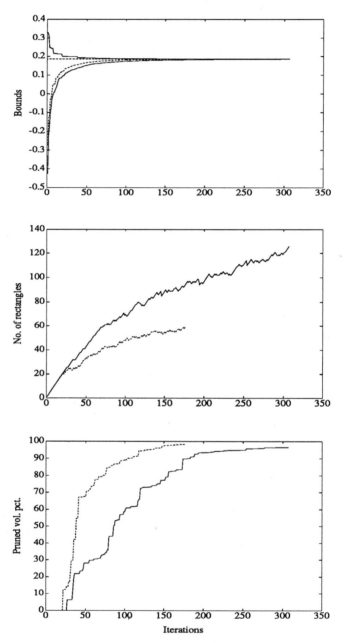

Fig. 6. Results from the branch and bound algorithm for $\mathcal{D}_{\min}$. Solid
lines correspond to bounds from equations (18) and (13). Dotted
lines correspond to bounds from equations (21) and (14).

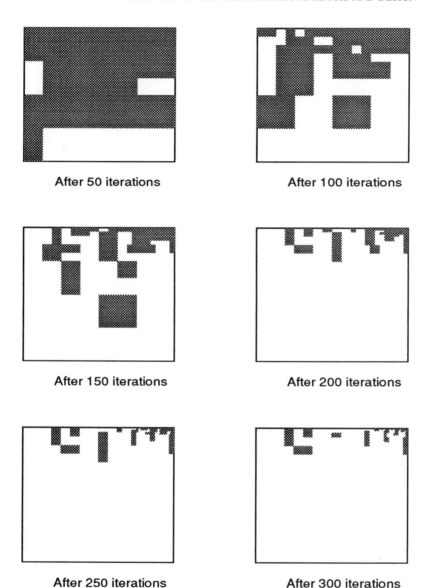

Fig. 7. The unpruned parameter region at various stages of the algorithm during the computation of $\mathcal{D}_{\min}$. The x- and y-coordinates are k and $1/m_2$ respectively. The algorithm can guarantee that $\mathcal{D}_{\min}$ cannot be achieved outside the shaded region.

bound from (30) by 6 iterations), indicating that the system is robustly stable. Note that this is consistent with our observations from $\mathcal{D}_{\min}$ computation.

- With bounds from (26) and (29), at the end of 122 iterations, the algorithm guarantees that $2.499 \leq \mathcal{H}_{\infty,\max} \leq 2.500$.

  The performance with the bounds from (27) and (30) is considerably better; at the the the end of only 40 iterations, the algorithm guarantees that $2.499 \leq \mathcal{H}_{\infty,\max} \leq 2.500$. For reference, the $\mathbf{H}_{\infty}$ norm from w to z for the nominal system is 1.008.

- The algorithm returns worst-case parameters $m_2 = 3/2$ and $k = 2/3$, which are different than the worst-case parameters for $\mathcal{D}_{\min}$; thus, $\mathcal{D}_{\min}$ and $\mathcal{H}_{\infty,\max}$ are inequivalent measures of robustness for our problem.

- Figure 8 also shows the number of rectangles in the partition as well as the pruned volume percentage as a function of iterations. It is again clear that the "improved" bounds do show improved performance.

3. $\mathcal{H}_{2,\max}$ of the system with the LQR-optimal controller

We finally consider the worst-case $\mathbf{H}_2$ norm of the closed-loop transfer function, our third measure of robustness.

The results from the branch and bound algorithm are shown in Figure 9 corresponding to the bounds in (35) and (37). We note that:

- *The system is robustly stable.* The upper bound from (37) is finite only if the system is robustly stable. The upper bound on $\mathcal{H}_{2,\max}$ from (37) is finite at the end of 467 iterations, indicating that the system is robustly stable.

  The branch and bound algorithm applied towards computing $\mathcal{D}_{\min}$ and $\mathcal{H}_{\infty,\max}$ establishes robust stability at the end of just 8 iterations, while it takes 467 iterations with $\mathcal{H}_{2,\max}$. The reason is that the bounds for $\mathcal{D}_{\min}$ and $\mathcal{H}_{\infty,\max}$ use the same test for robust stability while the bound for $\mathcal{H}_{2,\max}$ uses a different, more conservative test.

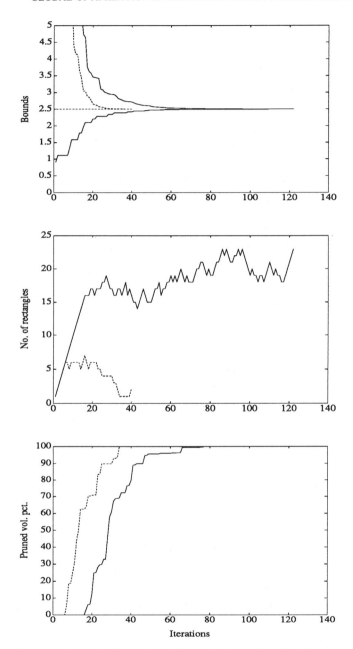

Fig. 8.   Results from the branch and bound algorithm for $\mathcal{H}_{\infty,\mathrm{max}}$. Solid lines correspond to bounds from equations (26) and (29). Dotted lines correspond to bounds from equations (27) and (30).

- The algorithm takes 15,000 iterations to return $1.1304 \le \mathcal{H}_{2,\max} \le 1.1404$. (For reference, the nominal $\mathbf{H}_2$ norm from w to z is 0.6922.) Clearly, the progress here is much slower than with $\mathcal{D}_{\min}$ or $\mathcal{H}_{\infty,\max}$. This is because the upper bound for $\mathcal{H}_{2,\max}$ is rather poor.

  This illustrates an important point about the branch and bound algorithm: If the bounds are bad, the algorithm may take a very long time to converge.

- The algorithm returns the same set of worst-case parameters ($m_2 = 3/2$ and $k = 2/3$) as with $\mathcal{H}_{\infty,\max}$.

B. Examples for Design

We consider the system shown in Figure 4 with the parameters assuming the nominal values $m_1 = 1$, $m_2 = 1$ and $k = 1$; for this system, we consider the problem of designing a state feedback that does not use the position or velocity of the second mass m_2, that is, we design a state feedback of the form $F = -(k_1 x_1 + k_2 \dot{x}_1)$. The closed-loop transfer function P_{cl} of interest is the transfer function from the force on the second mass w to the position of the first mass x_1 (see Figure 10).

We will restrict the state feedback gains k_1 and k_2 to satisfy

$$1/2 \le k_1 \le 1, \qquad 1/2 \le k_2 \le 1, \tag{41}$$

The above parameter range has been chosen so that the system is stable for all values of k_1 and k_2 lying in it.

1. $\mathcal{D}_{\max}$-optimal incomplete state feedback

Our first design objective is to maximize $\mathcal{D}_{\max}$: We design an incomplete state feedback that maximizes the slowest decay rate of the system, with the gains restricted to lie in (41).

Figure 11 shows the result from the branch and bound algorithm applied to the computation of $\mathcal{D}_{\max}$. The solid lines correspond to the bounds in (22) and (24), and the dashed lines once again to the "improved" bounds[7] (in (23) and in (25)). We note that:

---

[7] The optimal scaling problem in (25) was again not solved exactly; instead, an upper bound to the optimum was used.

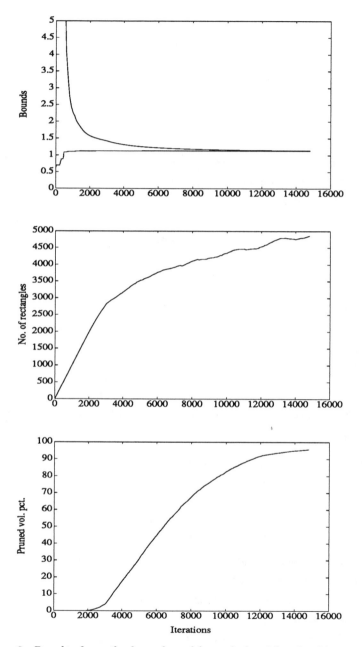

Fig. 9. Results from the branch and bound algorithm for $\mathcal{H}_{2,\max}$.

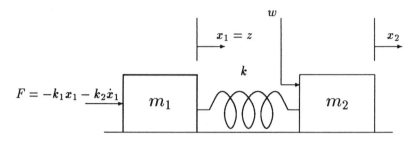

Fig. 10. The closed-loop system with incomplete state feedback
$F = -k_1 x_1 - k_2 \dot{x}_1$, where k_1 and k_2 are the design parameters,
each constrained to lie in $[1/2, 1]$. $[x_1 \; \dot{x}_1 \; x_2 \; \dot{x}_2]^T$ is the state
vector. $m_1 = 1$, $m_2 = 1$ and $k = 1$ are fixed, w is the force on
the second mass and z is the position of the first mass.

- Using the bounds from from (22) and (24), the branch and bound algorithm returns $0.2133 \leq \mathcal{D}_{\max} \leq 0.2141$ after 52 iterations. The algorithm takes 43 iterations to yield the same result using bounds (23)
  and (25).

- The upper bound from (25) performs only marginally better than the
  bound from (24). Thus, scaling does not help the upper bound much
  in this particular example.

- The algorithm returns the optimal gains $k_1 = 0.5$ and $k_2 = 1.0$, which
  corresponds to a vertex of the rectangle in (41).

2. $\mathcal{H}_{\infty,\min}$-optimal incomplete state feedback

We next compute the optimal values of the gains k_1 and k_2 that yield the
smallest possible $\mathbf{H}_\infty$ norm of the closed-loop transfer function between w,
the force on the second mass and z, the position of the first mass, with the
gains restricted as in (41).

Figure 12 shows the result from the branch and bound algorithm applied
to the computation of $\mathcal{H}_{\infty,\min}$. We note the following:

- After 275 iterations, the algorithm returns $2.5928 \leq \mathcal{H}_{2,\min} \leq 2.6006$.

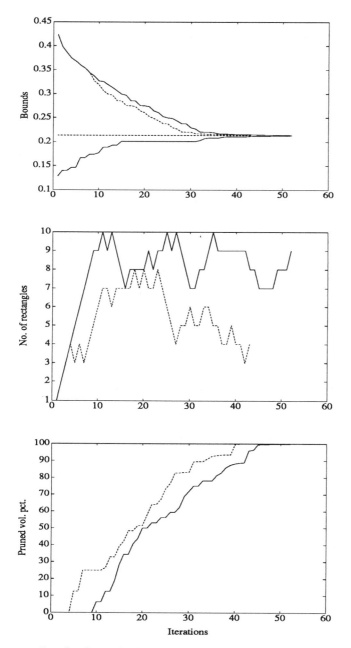

Fig. 11. Results from the branch and bound algorithm for $\mathcal{D}_{\max}$.

- The algorithm returns optimal gains $k_1 = 0.831$, $k_2 = 0.999$, which does *not* correspond to a vertex of the parameter rectangle. In fact, the minimum value among the $\mathbf{H}_\infty$ norms at the vertices is 2.6225, which is slightly worse than $\mathcal{H}_{\infty,\min}$ returned by the algorithm.

- Not only does the algorithm return a set of parameters that achieves the upper bound (2.6006) on $\mathcal{H}_{\infty,\min}$, but it also can *prove* that the smallest achievable $\mathcal{H}_{\infty,\min}$ must be at least 2.5928.

3. $\mathcal{H}_{2,\min}$-optimal incomplete state feedback

We finally compute the optimal feedback gains k_1 and k_2 that minimize the $\mathbf{H}_2$ norm of the transfer function from w to z with the gains restricted as in (41).

The results from the branch and bound algorithm are shown in Figure 13. We note that:

- After 17,500 iterations, the algorithm returns $0.9900 \leq \mathcal{H}_{2,\min} \leq 1.0002$. Thus, the performance of the algorithm, as with the case of $\mathcal{H}_{2,\max}$, is rather slow. This can be traced once again to the poor quality of the bounds (39) and (38) for $\mathcal{H}_{2,\min}$.

- The optimal set of gains for $\mathcal{H}_{2,\min}$ is $k_1 = 1$ and $k_2 = 1$, which is different than the gains returned for either $\mathcal{D}_{\max}$ or $\mathcal{H}_{\infty,\min}$. Almost all of the work done by the algorithm has gone towards establishing the lower bound of 0.9900. The algorithm can *guarantee* that for every choice of parameters in (41), the $\mathbf{H}_2$ norm from w to z is at least 0.9900.

C.   Simultaneous maximization of multiple objectives

Finally, we present an example that illustrates the branch and bound algorithm applied towards maximization of several objectives at the same time. We consider a robustness analysis problem and will use the same setup as in subsection A (see Figure 5). For this system, we consider the simultaneous maximization of the $\mathbf{H}_\infty$ norms from w to x_1 and w to $\dot{x}_1$.

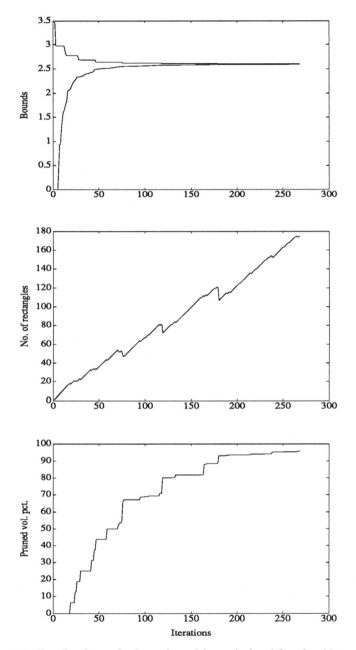

Fig. 12. Results from the branch and bound algorithm for $\mathcal{H}_{\infty,min}$

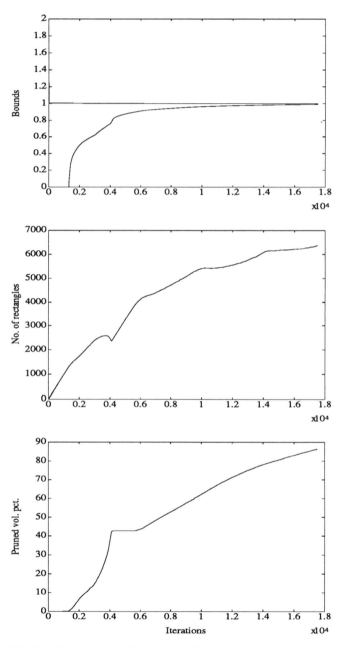

Fig. 13. Results from the branch and bound algorithm for $\mathcal{H}_{2,\min}$.

Figure 14 shows the result from the branch and bound algorithm applied to this problem using the bounds (26) and (29). The solid lines show the bounds $\mathcal{H}_{\infty,\max}$ between w and x_1 and the dashed lines, the bounds between w and $\dot{x}_1$. It is clear that the algorithm takes quite a few iterations more compared to the maximization of just $\mathcal{H}_{\infty,\max}$ between w and x_1 (see Figure 8). The reason for this is clear from Figure 15, which shows the parameter region under consideration. The two objective functions that we seek to maximize achieve their maxima at two opposite corners of the parameter region. Thus, in effect, the simultaneous maximization algorithm has to do two separate individual branch and bound maximizations, interlaced.

VI. CONCLUSION

We have described some parameter problems in control systems analysis and design which may be cast as global optimization problems, and how a branch and bound algorithm may be used to solve them. *Our main point is that we may combine recent (and continuing) gains in computing power with advances in theory to answer questions that were previously unanswerable.* We know of no existing methods that compute exactly any of the six quantities that we have described in this chapter.

We must emphasize that branch and bound algorithm can be computation intensive (worst-case combinatoric), and requires much more computation than most local optimization methods, which, in many instances, do yield the global optimum. The main strength of the branch and bound algorithm, however, is in yielding *guaranteed* bounds for the global optimum. These bounds may be used, in turn, to inspire confidence in corresponding local optimization methods. For instance, if for a certain class of problems, a local minimization procedure consistently returns objective values that are not much larger than the lower bound on the minimum returned by the branch and bound algorithm, we may conclude that the local minimization method is "good enough", especially if it involves far less computation.

Clearly, there is a trade-off between the computational effort spent on the branch and bound iterations and that spent on computing upper and lower bounds during each iteration: If the bounds are very good, they most likely require some computational effort; however, fewer branch and bound

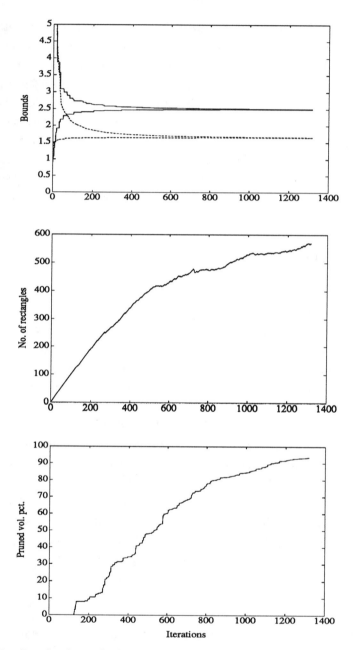

Fig. 14. Results from the branch and bound algorithm for multiple $\mathcal{H}_{\infty,max}$ computations. The solid lines show bounds on $\mathcal{H}_{\infty,max}$ between w and x_1 and the dashed lines, the bounds between w and $\dot{x}_1$.

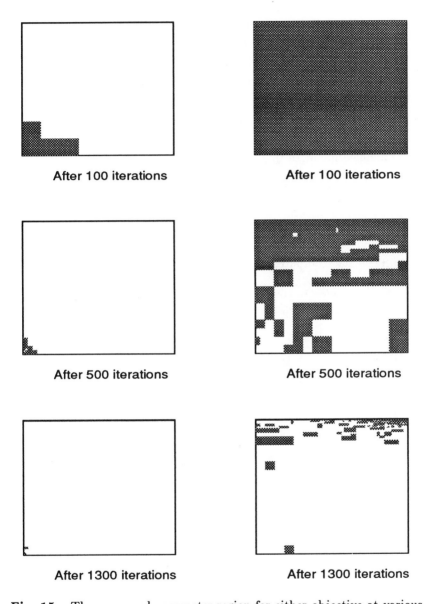

Fig. 15. The unpruned parameter region for either objective at various stages of the algorithm during the computation of $\mathcal{H}_{\infty,\text{max}}$. The left hand side corresponds to $\mathcal{H}_{\infty,\text{max}}$ between w and x_1 and the right hand side to $\mathcal{H}_{\infty,\text{max}}$ between w and $\dot{x}_1$. The x- and y-coordinates are k and $1/m_2$ respectively.

iterations will be needed. On the other hand, if the bounds are loose, the branch and bound algorithm may take a very long time to converge.

The basic branch and bound algorithm is easily extended to other problems, with the bound computation being the problem-specific task. We must mention, however, that the robust synthesis problem, which involves finding the design parameters that minimize the maximum of the objective over the uncertain parameters — the so-called "minimax" problem — cannot be handled in our setting. We refer the reader to [49] for work on this topic.

In conclusion,

- A number of global optimization problems in control systems analysis and design can be solved within a reasonable amount of time on present day computers.

- With advances in computing power and theory, the list of problems that can be solved using these methods is likely to grow.

- There will always be problems which will take inordinate amounts of time to solve with these methods.

VII.   ACKNOWLEDGEMENTS

We would like to thank Silvano Balemi, coauthor of [17] and [49], with whom major parts of this work were completed. This research supported in part by NSF under ECS-85-52465 and by AFOSR under 89-0228.

VIII.   REFERENCES

1.   A. Nemirovsky and D. Yudin, *Problem Complexity and Method Efficiency in Optimization*, John Wiley & Sons, (1983).

2.   C. H. Papadimitriou and K. Steiglitz, *Combinatorial Optimization: Algorithms and Complexity*, Prentice Hall, (1982).

3.   P. M. Pardalos and J. B. Rosen, *Constrained Global Optimization: Algorithms and Applications*, Springer-Verlag, (1987).

4. R. H. J. M. Otten and L. P. P. P. van Ginneken, *The Annealing Algorithm*, Kluwer Academic Publishers, (1989).

5. Y. Ye, "Interior-point algorithms for global optimization", *Annals of Operations Research*, **25**, pp. 59–74, (1990).

6. A. H. Land and A. G. Doig, "An automatic method for solving discrete programming problems", *Econometrica*, **28**, pp. 497–520, (1960).

7. R. J. Dakin, "A tree search algorithm for mixed integer programming problems", *The Computer Journal*, **8**, pp. 250–255, (1965).

8. E. L. Lawler and D.E. Wood, "Branch-and-bound methods: A survey", *Operations Research*, **14**, pp. 699–719, (1966).

9. K. Spielberg, "Enumerative methods in integer programming", *Annals of Discrete Mathematics*, **5**, pp. 139–183, (1979).

10. A. Schrijver, *Theory of Linear and Integer Programming*, Wiley-Interscience series in discrete mathematics. John Wiley & Sons, (1986).

11. M. Grötschel, L. Lovász, and A. Schrijver, *Geometric Algorithms and Combinatorial Optimization*, volume 2, Springer-Verlag, (1988).

12. R. R. E. De Gaston and M. G. Safonov, "Exact calculation of the multiloop stability margin", *IEEE Trans. Aut. Control*, **33**(2), pp. 156–171, (1988).

13. A. Sideris and R. S. S. Peña, "Fast computation of the multivariable stability margin for real interrelated uncertain parameters", *IEEE Trans. Aut. Control*, **34**(12), pp. 1272–1276, (1989).

14. B. C. Chang, O. Ekdal, H. H. Yeh, and S. S. Banda, "Computation of the real structured singular value via polytopic polynomials", *J. of Guidance*, **14**(1), pp. 140–147, (1991).

15. A. Vicino, A. Tesi, and M. Milanese, "Computation of nonconservative stability perturbation bounds for systems with nonlinearly correlated uncertainties", *IEEE Trans. Aut. Control*, **35**(7), pp. 835–841, (1990).

16. C. DeMarco, V. Balakrishnan, and S. Boyd, "A branch and bound methodology for matrix polytope stability problems arising in power systems", In *Proc. IEEE Conf. on Decision and Control*, , pp. 3022–3027, Honolulu, Hawaii, (1990).

17. V. Balakrishnan, S. Boyd, and S. Balemi, "Branch and bound algorithm for computing the minimum stability degree of parameter-dependent linear systems", *To appear, International Journal of Robust and Nonlinear Control*, (1992).

18. T. S. Parker and L. O. Chua, *Practical Numerical Algorithms for Chaotic Systems*, Springer-Verlag, (1989).

19. V. Balakrishnan and S. Boyd, "Computation of the worst-case covariance for linear systems with uncertain parameters", In *Proc. IEEE Conf. on Decision and Control*, Brighton, U. K, (1991).

20. J. Rohn, "Systems of linear interval equations", *Linear Algebra and its Applications*, **126**, pp. 39–78, (1989).

21. J. Rohn, "Nonsingularity under data rounding", *Linear Algebra and its Applications*, **139**, pp. 171–174, (1990).

22. J. Rohn and S. Poljak, "Radius of nonsingularity", *Mathematics of Control, Signals, and Systems*, (1991-92), To appear.

23. J. Demmel, "The componentwise distance to the nearest singular matrix", *SIAM J. on Matrix Analysis and Applications*, (1991-92), To appear.

24. V. L. Kharitonov, "Asymptotic stability of an equilibrium position of a family of systems of linear differential equations", *Differential'nye Uraveniya*, **14**(11), pp. 1483–1485, (1978).

25. B. R. Barmish, "Invariance of the strict Hurwitz property for polynomials with perturbed coefficients", *IEEE Trans. Aut. Control*, **AC-29**, pp. 935–936, (1984).

26. A. C. Bartlett, C. V. Hollot, and H. Lin, "Root locations of an entire polytope of polynomials: it suffices to check the edges", *Mathematics of Control, Signals, and Systems*, **1**(1), pp. 61–71, (1989).

27. M. Fu and B. R. Barmish, "Polytopes of polynomials with zeros in a prescribed region", *IEEE Trans. Aut. Control*, **34**(5), pp. 544–546, (1989).

28. S. Bialas, "A necessary and sufficient condition for stability of interval matrices", *Int. J. Control*, **37**(4), pp. 717–722, (1983).

29. B. R. Barmish and C. V. Hollot, "Counterexamples to a recent result on the stability of interval matrices by S. Bialas", *Int. J. Control*, **39**(5), pp. 1103–1104, (1984).

30. B. Ross Barmish, M. Fu, and S. Saleh, "Stability of a polytope of matrices: Counterexamples", *IEEE Trans. Aut. Control*, **33**(6), pp. 569–572, (1988).

31. B. D. Anderson, N. K. Bose, and E. I. Jury, "Output feedback stabilization and related problems—Solution via decision methods", *IEEE Trans. Aut. Control*, **AC-20**, pp. 53–66, (1975).

32. E. Zeheb, "Necessary and sufficient conditions for root clustering of a polytope of polynomials in a simply connected domain", *IEEE Trans. Aut. Control*, **34**, pp. 986–990, (1989).

33. C. A. Desoer and M. Vidyasagar, *Feedback Systems: Input-Output Properties*, Academic Press, New York, (1975).

34. D. G. Luenberger, *Linear and Nonlinear Programming*, Addison-Wesley, Reading, Mass., 2nd edition, (1984).

35. S. Boyd and C. Barratt, *Linear Controller Design: Limits of Performance*, Prentice-Hall, (1991).

36. S. Boyd, V. Balakrishnan, and P. Kabamba, "A bisection method for computing the $\mathbf{H}_\infty$ norm of a transfer matrix and related problems", *Mathematics of Control, Signals, and Systems*, **2**(3), pp. 207–219, (1989).

37. S. Boyd and V. Balakrishnan, "A regularity result for the singular values of a transfer matrix and a quadratically convergent algorithm for computing its $\mathbf{L}_\infty$-norm", *Syst. Control Letters*, **15**, pp. 1–7, (1990).

38. B. Anderson and J. B. Moore, "Linear system optimization with prescribed degree of stability", *Proc. IEEE*, **116**(12), pp. 2083–2087, (1969).

39. J. Doyle, "Analysis of feedback systems with structured uncertainties", *IEE Proc.*, **129-D**(6), pp. 242–250, (1982).

40. M. G. Safonov, "Stability margins of diagonally perturbed multivariable feedback systems", *IEE Proc.*, **129-D**, pp. 251–256, (1982).

41. J. Doyle, J. E. Wall, and G. Stein, "Performance and robustness analysis for structured uncertainties", In *Proc. IEEE Conf. on Decision and Control*, , pp. 629–636, (1982).

42. M. K. Fan and A. L. Tits, "Characterization and efficient computation of the structured singular value", *IEEE Trans. Aut. Control*, **AC-31**(8), pp. 734–743, (1986).

43. M. K. Fan and A. L. Tits, "*m*-form numerical range and the computation of the structured singular value", *IEEE Trans. Aut. Control*, **33**(3), pp. 284–289, (1988).

44. M. G. Safonov, "Exact calculation of the multivariable structured-singular-value stability margin", In *Proc. IEEE Conf. on Decision and Control*, , pp. 1224–1225, Las Vegas, NV, (1984).

45. M. G. Safonov, "Optimal diagonal scaling for infinity-norm optimization", *Syst. Control Letters*, **7**, pp. 257–260, (1986).

46. M. G. Safonov and J. Doyle, "Optimal scaling for multivariable stability margin singular value computation", In *Proceedings of MECO/EES 1983 Symposium*, (1983).

47. M. Saeki, "A method of robust stability analysis with highly structured uncertainties", *IEEE Trans. Aut. Control*, **31**(10), pp. 935–940, (1986).

48. M. K. H. Fan, A. L. Tits, and J. C. Doyle, "Robustness in the presence of mixed parametric uncertainty and unmodeled dynamics", *IEEE Trans. Aut. Control*, **36**(1), pp. 25–38, (1991).

49. S. Balemi and V. Balakrishnan, "Global optimization of H_∞-norm of parameter-dependent linear systems", Technical Report 91-15, Automatic Control Laboratory, ETH Zentrum, 8092 Zürich, Switzerland, (1991).

50. H. T. Toivonen and P. M. Mäkilä, "Computer-aided design procedure for multiobjective LQG control problems", *Int. J. Control*, **49**(2), pp. 655–666, (1989).

51. R. A. Horn and C. A. Johnson, *Matrix Analysis*, Cambridge University Press, (1985).

Robust Techniques for Combined Filtering and Parameter Estimation

Wolfgang J. Runggaldier
Michele Bonollo

Dipartimento di Matematica Pura ed Applicata,
Universitá di Padova, Via Belzoni 7, 35131 - Padova , Italy

Abstract

We present robust solution methods for combined filtering and parameter estimation problems in discrete time. The robustness is with respect to small variations in the a-priori distributions both of the noise disturbances as well as of the parameter. From the same point of view we also study linear discrete time filtering problems where the coefficients in the observations depend on an unobsrervable random process that is independent of the state and has a given a-priori distribution.

I. INTRODUCTION

One of the issues faced when solving practical problems is the following : A solution method can only be derived if first one constructs a model and the method then gives the solution for the model. Models however do in general not completely match reality, although they may come close to it. The question then arises as to how good is the solution corresponding to the model, but obtained on the basis of data coming from the real world ?

The answer to this question is outside the realm of mathematics, but one can cope with the problem by designing solution methods that are robust in the sense that they perform satisfactorily also when applied to a real situation that differs not too strongly from the model.

In stochastic problems such as filtering and stochastic control, one of the basic data needed to construct a model are the distributions of the noise disturbances. For mathematical convenience, these disturbances are usually taken as white Gaussian, which practically never corresponds to reality. What one often has in practice are disturbances with distributions that are close to Gaussian in the sense of weak convergence (convergence in distribution) and have fatter tails than a Gaussian. It is therefore of interest to have filtering and control algorithms that are robust with respect to small variations in the distributions of the noise disturbances. This issue has e.g. been approached in [1], [2] in the case of continuous-time models that do not depend on an unknown parameter (in this context see also [3]). The purpose of this paper is to extend the results of [1] and [2] to parameter-dependent models in discrete-time. Besides their practical importance (sampled observations), one of the main reasons why we consider discrete-time models stems from the fact that (see [4]) in discrete time an explicitly computable method is available to solve combined filtering and parameter estimation problems and this method is robust with respect to small variations in the data of the model wich are the white Gaussian noise distributions and the prior distribution of the parameter.

Noise distributions with heavy tails often produce what by analogy to Statistics we may call "outliers". Such outliers may cause poor performance of standard methods and, to robustify the methods, in the field of Statistics one often performs a preliminary trimming of the data (see e.g.[5]) with the purpose of cutting off the outliers. A related technique is known from engineering practice, namely the use of a limiter in the observations (see e.g. [6], see also [7]). In the present paper we shall also take into account the possibility of using such limiters to enhance robustness in the case of observation noise distributions with heavy tails.

In Section II.A.1 we describe some of the results in [1] and [2], which in Section II.A.2 are then extended to the case of models that depend on an unknown parameter. Section II.B contains the main results from [4] that allow the construction of a robust and explicitly computable method to solve combined filtering and parameter estimation problems in discrete time. In Section III we extend results from [1] and [2] to discrete-time models: More precisely, in subsection III.A we consider combined filtering and parameter estimation problems for models with heavy-tailed observation noise distributions and provide robust solution methods that are based on the results of [4] recalled in Section II.B. In subsection III.B we consider the (nonlinear) filtering problem in a linear discrete time model, where the coefficients in the observations depend on an unobservable random process that is independent of the state, and provide robust solution methods based on the use of the Kalman-Bucy filter (see e.g. [8]). Results of simulations, intended to test the actual performance of the methods, are reported in Section IV together with their discussion.

We finally remark that the results of this paper allow for extensions also to discrete-time stochastic and adaptive control; this extension is again possible due to the existence of an explicit method to determine nearly optimal controls for discrete-time adaptive control problems over a finite horizon (see [9]).

CONTENTS

Abstract

I. INTRODUCTION

II. REVIEW AND EXTENSION OF PREVIOUS RESULTS

    II.A. Robust methods for continuous-time filtering

        II.A.1. Models without unknown parameters

        II.A.2. Models depending on an unknown parameter

    II.B. An explicit method for combined filtering and parameter

        estimation im discrete time.

III. ROBUST METHODS IN DISCRETE TIME

    III.A. Robust filtering and parameter estimation for discrete time

        models with heavy-tailed observation noise distribution.

        III.A.1. The case without limiter

            III.A.1.a. Theoretical considerations

            III.A.1.b. Robust solution methods

        III.A.2. The case with a limiter

            III.A.2.a. Theoretical considerations

            III.A.2.b. Robust solution methods

    III.B. Robust filtering for linear discrete time models with random

        coefficients and heavy-tailed observation noise distribution

        III.B.1. The case without limiter

            III.B.1.a. Theoretical considerations

            III.B.1.b. Robust solution methods

        III.B.2. The case with a limiter

            III.B.2.a. Theoretical considerations

            III.B.2.b. Robust solution methods

IV. SIMULATION RESULTS AND THEIR DISCUSSION

Acknowledgments

References

II. REVIEW AND EXTENSIONS OF PREVIOUS RESULTS.

This section consists of two subsections : In II.A we describe and partly extend robustness results obtained in [1] and [2] for filtering problems in continuous time. In II.B we recall from [4] an explicitly computable method to solve combined filtering and parameter estimation problems in discrete time. These results are then used in the next section III to obtain robust and explicitly computable methods to solve filtering problems for discrete-time models with unknown parameters.

II.A. Robust methods for continuous-time filtering
II.A.1. Models without unknown parametes

In this subsection we recall results from [1] and [2] that partly motivated the investigations described in this paper.

Given two independent, stationary and ergodic stochastic processes ψ_t, ϕ_t with integrable covariance function, assume that for a given small positive parameter $\varepsilon > 0$, the state and observation processes x_t^ε and y_t^ε of a partially observable dynamical system satisfy

$$\dot{x}_t^\varepsilon = A\, x_t^\varepsilon \;+\; \frac{1}{\varepsilon}\, \psi_{t/\varepsilon^2}$$

$$\dot{y}_t^\varepsilon = C\, x_t^\varepsilon \;+\; \frac{1}{\varepsilon}\, \phi_{t/\varepsilon^2}$$

(II.1)

The filtering problem corresponding to model (II.1) then consists of estimating x_t^ε given y_s^ε, $s \le t$. Since model (II.1) is linear, it may be of interest to find a linear filter of x_t^ε given the observations.

For the purpose of constructing a linear filter, consider the family of filtering models obtained from (II.1) by letting ε vary over the set of small positive reals. Since ψ_t, ϕ_t were assumed to have an integrable covariance function, letting the symbol $\Rightarrow$ denote weak convergence (convergence in distribution), it can be shown that

$$w_t^\varepsilon = \int_0^t \frac{1}{\varepsilon} \psi_{t/\varepsilon^2} \, ds \Rightarrow B \, w_t$$

$$v_t^\varepsilon = \int_0^t \frac{1}{\varepsilon} \phi_{t/\varepsilon^2} \, ds \Rightarrow D \, v_t$$

(II.2)

where w_t, v_t are standard Brownian motions (Wiener processes) and

$$B^2 = 2 \int_0^\infty E \{\psi_t \psi_0\} \, dt$$

$$D^2 = 2 \int_0^\infty E \{\phi_t \phi_0\} \, dt$$

(II.3)

The following result is now shown in [1]

Proposition II.1 : We have

$$(x_t^\varepsilon, y_t^\varepsilon) \Rightarrow (x_t, y_t)$$

where x_t, y_t satisfy the linear Gaussian model

$$d \, x_t = A \, x_t \, dt + B \, dw_t$$

$$d \, y_t = C \, x_t \, dt + D \, dv_t$$

(II.4)

Given the above proposition, a natural way to construcy a linear filter estimate $\hat{x}_t^\varepsilon$ of x_t^ε given y_s^ε, $s \le t$, is to consider the Kalman-Bucy filter for the limit model (II.4) and to feed it with the real observations y_t^ε from (II.1), i.e.

$$\dot{\hat{x}}_t^\varepsilon = A \hat{x}_t^\varepsilon + K_t [\dot{y}_t^\varepsilon - C x_t^\varepsilon]$$

$$K_t = P_t C^T D^{-2}$$
\hfill (II.5)

$$\dot{P}_t = A P_t + P_t A^T + BB^T - P_t C^T D^{-2} C P_t$$

Notice also that, since $\hat{x}_t^{\wedge\varepsilon}$ is a linear functional of y_t^ε, from the weak convergence $(x_t^\varepsilon, y_t^\varepsilon) \Rightarrow (x_t, y_t)$ it follows $(x_t^\varepsilon, \hat{x}_t^{\wedge\varepsilon}, y_t^\varepsilon) \Rightarrow (x_t, \hat{x}_t, y_t)$ where $\hat{x}_t$ corresponds to the Kalman-Bucy filter for model (II.4).

The question now arises as to how good is the filter estimate $\hat{x}_t^{\wedge\varepsilon}$? For this purpose recall that the filter estimate $\hat{x}_t = E \{ x_t \mid y_s, s \le t \}$ for the Kalman-Bucy model (II.4) minimizes the filter variance $E \{ (x_t - \hat{x}_t)^2 \}$ among all possible estimates of x_t that are (measurable) functions of the observations $y_s, s \le t$. Denoting by $N (x; \hat{x}_t^{\wedge\varepsilon}, P_t)$ the Gaussian distribution with mean $\hat{x}_t^{\wedge\varepsilon}$ and variance P_t as determined from (II.5), in [1] the following theorem is shown

Theorem II.1 : If $f(x)$ is a continuous and bounded function and $F_t(y^\varepsilon)$ is a continuous and bounded functional of $y_s^\varepsilon, s \le t$, then

$$\lim_{\varepsilon \downarrow 0} E \{ [f(x_t^\varepsilon) - F_t(y^\varepsilon)]^2 \} \ge \lim_{\varepsilon \downarrow 0} E \{ [f(x_t^\varepsilon) - \int f(x) \, dN(x; \hat{x}_t^{\wedge\varepsilon}, P_t)]^2 \}$$

This theorem states the following : Given a continuous and bounded function $f(x_t^\varepsilon)$ of the state x_t^ε, among all estimators $F_t(y^\varepsilon)$ of $f(x_t^\varepsilon)$ that are continuous and bounded functions of the past observations, the estimator $\int f(x) \, dN(x; \hat{x}_t^{\wedge\varepsilon}, P_t)$ minimizes the estimation variance in the limit when $\varepsilon \downarrow 0$.

It is to be noted that the same result holds for all families of processes $x_t^\varepsilon, y_t^\varepsilon$ satisfying a "prelimit" model such that one can obtain the weak convergence $(x_t^\varepsilon, y_t^\varepsilon) \Rightarrow (x_t, y_t)$ with (x_t, y_t) satisfying the Kalman-Bucy model (II.4). It is shown in [1] that the above weak convergence can also be obtained for nonlinear "prelimit" models of the form

$$\dot{x}_t^\varepsilon = A \, x_t^\varepsilon + \frac{1}{\varepsilon} \psi_{t/\varepsilon^2}$$

$$\dot{y}_t^\varepsilon = \frac{1}{\varepsilon} H (\varepsilon \, C \, x_t^\varepsilon + \phi_{t/\varepsilon^2})$$

(II.6)

where the function H(.) plays the role of a "limiter". In particular, if H is a "hard limiter", namely $H(z) = \text{sign}(z)$ and the process ϕ_t is such that $E\{\phi_t^2\} = 1$ and $E\{\phi_t\phi_0\} = e^{-t/2}$, it is shown in [1] that the limit model (II.4) then takes the form

$$d \, x_t = A \, x_t \, dt + B \, dw_t$$

$$d \, y_t = \sqrt{\frac{2}{\pi}} C \, x_t \, dt + 2 \sqrt{\ln 2} \, dv_t$$

(II.7)

with B as in (II.3).

The above results were further generalized in [2] to the case when the processes x_t^ε, y_t^ε satisfy the following generalization of the prelimit models (II.1) and (II.6)

$$\dot{x}_t^\varepsilon = A \, (\xi_{t/\varepsilon^2}) \, x_t^\varepsilon + \frac{1}{\varepsilon} \psi_{t/\varepsilon^2}$$

$$\dot{y}_t^\varepsilon = \frac{1}{\varepsilon} H (\varepsilon \, C \, (\eta_{t/\varepsilon^2}) \, x_t^\varepsilon + \phi_{t/\varepsilon^2})$$

(II.8)

where $\xi_t, \psi_t, \eta_t, \phi_t$ are strictly stationary and ergodic processes satisfying some additional appropriate assumptions that are essentially variants of the requirement of the integrability of the covaraince function. The generalization is thus in the sense that the coefficients A and C in (II.1) and (II.6) may be rapidly oscillating in correspondence with the properties of the stationary processes ξ_t and η_t. In [2] the following is shown

Proposition II.2 : Under the abovementioned appropriate assumptions we have

$$(x_t^\varepsilon, y_t^\varepsilon) \Rightarrow (x_t, y_t)$$

with (x_t, y_t) satisfying the linear-Gaussian model

$$d\,x_t = E\{A(\xi_0)\}\, x_t\, dt + B\, dw_t$$

$$\text{(II.9)}$$

$$d\,y_t = C_H\, x_t\, dt + D_H\, dv_t$$

where B, C_H, D_H can be computed on the basis of the statistical properties of the processes ψ_t, η_t, ϕ_t and where C_H, D_H depend also on the form of the "limiter" function H.

Again, from $(x_t^\varepsilon, y_t^\varepsilon) \Rightarrow (x_t, y_t)$ it follows that $(x_t^\varepsilon, \hat{x}_t^\varepsilon, y_t^\varepsilon) \Rightarrow (x_t, \hat{x}_t, y_t)$ where, as before, $\hat{x}_t$ corresponds to the Kalman-Bucy filter for the limit model (II.9) and $\hat{x}_t^\varepsilon$ is the linear estimate for x_t^ε obtained by feeding the Kalman-Bucy filter for (II.9) with the observations y^ε from (II.8). Furthermore, an analogue of Theorem II.1 can be shown also in this case.

The results described so far can be seen as results concerning the robustness of the Kalman-Bucy filter for the limit models (II.4), (II.7), (II.9) with respect to small variations in the a-priori distributions in these models. We can however use the above results also for the following considerations that we report here for the case of the more general prelimit model (II.8). Notice in fact that from the weak convergence $(x_t^\varepsilon, y_t^\varepsilon) \Rightarrow (x_t, y_t)$ and under a uniform integrability assumption (in practice one may simply use a truncation), we have

$$\lim_{\varepsilon \downarrow 0} E\{ [x_t^\varepsilon - \hat{x}_t^\varepsilon]^2\} = E\{ [x_t - \hat{x}_t]^2\} = P_t^H \qquad \text{(II.10)}$$

where the filter variance P_t^H satisfies (assume for simplicity the scalar case) the Riccati equation

$$\dot{P}_t^H = 2\,P_t^H\,E\{A(\xi_0)\} + B^2 - \frac{C_H^2}{D_H^2}\,[\,P_t^H\,]^2 \qquad (II.11)$$

From this equation it is immediately clear that the magnitude of P_t^H depends on the "signal-to-noise ratio" C_H^2/D_H^2 in the sense that, if for two limiter functions H_1 and H_2 we have $C_{H_1}^2/D_{H_1}^2 > C_{H_2}^2/D_{H_2}^2$, then $P_t^{H_1} < P_t^{H_2}$ for all t. This result gives a possibility to choose appropriately the limiter function $H(z)$ in order to improve the estimation. To this effect, a theoretical investigation is made in [10]. As Example II.1 below is intended to show, it is however generally true that, analogously to the situation in Statistics where trimming of the data is used to eliminate outliers and thus to improve estimation, the use of a suitable limiter $H(z)$ in the observations improves the filter performance when the observation-noise distributions have heavy tails.

Example II.1 :   (taken from [2]) Consider the following family of prelimit models

$$\dot{x}_t^\varepsilon = A\,x_t^\varepsilon + \frac{1}{\varepsilon}\,\psi_{t/\varepsilon^2}$$

$$\qquad\qquad\qquad\qquad\qquad\qquad\qquad (II.12)$$

$$\dot{y}_t^\varepsilon = \frac{1}{\varepsilon}\,H\,(\,\varepsilon\,C\,x_t^\varepsilon + \phi_{t/\varepsilon}^q\,)$$

for the two situations :

a) $H(z) = H_L(z) = z$; i.e. there is no limiter in the observations

b) $H(z) = H_S(z) = \text{sign}\,(z)$; i.e. there is a "hard limiter" in the observations

The stationary process ϕ_t^q is defined by

$$\phi_t^q = \alpha\,\gamma_t + (\,1 - \alpha\,)\,\gamma_t^{2q-1} \qquad (II.13)$$

where $\alpha \in (0,1)$, q is a positive integer, and γ_t is the stationary Gaussian process given by

$$\gamma_t = \gamma_0 \, e^{-t/2} + \int_0^t e^{-(t-u)/2} \, d\beta_u \, , \quad \gamma_0 \sim N(0,1), \qquad (II.14)$$

with β_t a standard Wiener process independent of ψ_t, so that $\gamma_t \sim N(0,1)$ and $E\{\gamma_t\gamma_0\} = e^{-t/2}$. In other words, ϕ_t^q is a process whose distribution has tails

that, as q increases, become more and more heavy than those of a Gaussian. In [2] it is shown that for $(x_t^\varepsilon, y_t^\varepsilon)$ from (II.12) we have $(x_t^\varepsilon, y_t^\varepsilon) \Rightarrow (x_t, y_t)$ with (x_t, y_t) satisfying the model

$$d\,x_t = A\,x_t\,dt + B\,dw_t$$

$$(II.15)$$

$$d\,y_t = C_H\,x_t\,dt + D_H\,dv_t$$

where B is as in (II.3) and the values of C_H, D_H are such that we have

$$C_{H_S}^2 / D_{H_S}^2 \, < \, C_{H_L}^2 / D_{H_L}^2 \qquad \text{for } q = 1$$

$$(II.16)$$

$$C_{H_S}^2 / D_{H_S}^2 \, > \, C_{H_L}^2 / D_{H_L}^2 \qquad \text{for } q > q(\alpha)$$

Using the Kalman-Bucy filter corresponding to model (II.15), but fed with the observations from (II.12), by (II.10) and the considerations following (II.11),we can thus reduce the filter variance $E\{[x_t^\varepsilon - \hat{x}_t^{\wedge \varepsilon}]^2\}$ for small ε by introducing a hard limiter in the observations whenever we suspect that the observation-noise distribution has heavy tails. In other words, in the latter situation we can improve the robustness of the Kalman-Bucy filter for the limit model with the additional use of a limiter in the observations.

II.A.2. Models depending on an unknown parameter.

In this subsection we show that the previous results can be extended to the case when the models depend on an unknown parameter θ, which

(Bayesian point of view) is considered to be a realization of a random variable Θ with given prior distribution. More specifically, consider the situation when model (II.6) (particular case of model (II.8)) depends on a parameter θ, i.e.

$$\dot{x}_t^\varepsilon = A \, \theta \, x_t^\varepsilon + \frac{1}{\varepsilon} \, \psi_{t/\varepsilon^2}$$

$$\dot{y}_t^\varepsilon = \frac{1}{\varepsilon} \, H \, (\varepsilon \, C \, \theta \, x_t^\varepsilon + \phi_{t/\varepsilon^2})$$

(II.17)

As a consequence of Proposition II.2 we then have

Corollary II.1 : For $(x_t^\varepsilon, y_t^\varepsilon)$ from (II.17) we have

$$(x_t^\varepsilon, y_t^\varepsilon) \implies (x_t, y_t) \quad \text{for } \varepsilon \downarrow 0$$

where (x_t, y_t) satisfies the model

$$d\,x_t = A \, \theta \, x_t \, dt + B \, dw_t$$

$$d\,y_t = C_H \, \theta \, x_t \, dt + D_H \, dv_t$$

(II.18)

with the values of B, C_H, D_H according to Proposition II.2.

Proof. It suffices to show that, for a continuous and bounded functional g, we have

$$E \, \{ \, g \, (x^\varepsilon, y^\varepsilon) \, \} \, \to \, E \, \{ \, g \, (x, y) \, \} \quad \text{for } \varepsilon \downarrow 0$$

From the results of Proposition II.2 we know that, pointwise, for each θ we have

$$G^\varepsilon(\theta) := E \, \{ \, g \, (x^\varepsilon, y^\varepsilon) \, | \, \theta \} \, \to \, E \, \{ g \, (x, y) \, | \, \theta \} := G(\theta) \quad \text{for } \varepsilon \downarrow 0$$

Furthermore, since $g(x,y)$ is bounded, $G^\varepsilon(\theta)$ is bounded uniformly in ε, θ. From the bounded convergence theorem we then have

$$E \, \{ \, g \, (x^\varepsilon, y^\varepsilon) \, \} = E \, \{ \, G^\varepsilon(\theta) \, \} \, \to \, E \, \{ \, G(\theta) \, \} = E \, \{ \, g(x,y) \, \} \quad \text{for } \varepsilon \downarrow 0.$$

Assume now we have an explicit method to solve the combined filtering and parameter estimation problem corresponding to (II.18) and that this method gives estimates $\hat{x}_t$ and $\hat{\theta}_t$ that are continuous functions of the observations y_s, $s \le t$. Defining, consistently with previous terminology, $\hat{x}_t^\varepsilon$, $\hat{\theta}_t^\varepsilon$ as the above estimates when fed with the observations y_s^ε, $s \le t$, from (II.17), Corollary II.1 then allows to conclude that

$$(x_t^\varepsilon, \hat{x}_t^\varepsilon, \hat{\theta}_t^\varepsilon, y_t^\varepsilon) \;\Rightarrow\; (x_t, \hat{x}_t, \hat{\theta}_t, y_t) \quad \text{for } \varepsilon \downarrow 0 \tag{II.19}$$

so that one can adapt the previous robustness results also to the present parameter-dependent case.

The point is now that, to the best of our knowledge, in continuous time there exists no explicit method to solve the combined filtering and parameter estimation problem corresponding to model (II.18) which, in fact, is a nonlinear filtering problem. On the other hand, such an explicit method can be worked out in discrete time and the purpose of the next section is to describe it.

II.B. An explicit method for combined filtering and parameter estimation in discrete time.

In this section we review some of the results in [4] (see also [11]). Consider a process triple (x_n, θ_n, y_n) satisfying over a finite horizon $n=0,1,2,....,N$ the model

$$x_n = A_{n-1}(\theta_{n-1})\, x_{n-1} + Q_{n-1}(\theta_{n-1})\, w_n \quad , \quad x_0 = Q_0(\theta_0)\, w_0 \tag{II.20}$$

$$y_n = C_n(\theta_n)\, x_n + R_n(\theta_n)\, v_n$$

where the process $\{\theta_n\}$ takes only a finite number of values and is defined through its joint a-priori distribution $p(\theta_0,....,\theta_N)$, equivalent to assigning

$p_0(\theta_0)$ and $p(\theta_n \mid \theta_{n-1},....,\theta_0)$ for $n = 1,....,N$. The processes $\{w_n\}$ and $\{v_n\}$ are standard white Gaussian sequences, independent among themselves and from $\{\theta_n\}$ and the matrices $Q_n(\theta_n)$ and $R_n(\theta_n)$ are positive definite. We may view model (II.20) as follows : Considering the pair (x_n,θ_n) as the state of a dynamical system, partially observed through the process $\{y_n\}$, we have that the component θ_n of the state evolves as a nonlinear, non-necessarily Markov process; conditionally on θ_n however, the pair (x_n,y_n) satisfies a linear Gaussian model. If, in particular, θ_n is constant $(\theta_n \equiv \theta)$, then we may think of θ as an unknown parameter in the linear model (II.20) for (x_n,y_n) for which an a-priori distribution $p(\theta)$ is given. We shall also interpret model (II.20) in the sense that y_n is being observed starting from $n=1$ and that, at a generic period n, first a transition $(x_{n-1},\theta_{n-1}) \to (x_n,\theta_n)$ takes place and then an observation y_n is generated according to (II.20).

The problem here consists of computing recursively for $n=1,....,N$ the conditional distribution $p_n(x_n,\theta_n \mid y^n)$ of (x_n,θ_n) given the observations $y^n :=$ $\{y_1,...,y_n\}$. In the particular case of $\theta_n \equiv \theta$ this solves the combined filtering and parameter estimation problem corresponding to (II.20).

The complete recursive solution to compute $p_n(x_n,\theta_n \mid y^n)$ is described in [4]. For simplicity, in Theorem II.2 below we report it for the particular case when $\theta_n \equiv \theta$, so that it coincides with the one already derived in [11]. In this theorem, the symbol $\propto$ stands for "proportional to" in the sense that the quantities on the left and on the right of $\propto$ are equal modulo a normalization factor which, in the case of conditional densities, implies that its integral is equal to one.

Theorem II.2 : We have

$$p_n(x_n, \theta \mid y^n) \propto q_n(x_n, \theta, y^n) =$$

$$p(\theta)\, \varphi_n(\theta)\, \exp\left[-\frac{1}{2} x_n^T\, M_n(\theta)\, x_n + x_n^T\, h_n(\theta, y^n) + k_n(\theta, y^n) \right] \quad \text{(II.21)}$$

where, for each of the finite number of possible values of θ, the quantities $\varphi_n(\theta)$, $M_n(\theta)$, $h_n(\theta, y^n)$, $k_n(\theta, y^n)$ can be computed recursively as follows :

Letting

$$N_n(\theta) := A_{n-1}^T (\theta) \, Q_{n-1}^{-2}(\theta) \, A_{n-1}(\theta) + M_{n-1}(\theta) \qquad \text{(II.22)}$$

we have

$$\varphi_n(\theta) = \varphi_{n-1}(\theta) \, [\det (Q_{n-1}(\theta) \, R_n(\theta))]^{-1} \, [\det N_n(\theta)]^{-1/2}$$

$$\varphi_0(\theta) = [\det (Q_0(\theta))]^{-1} \qquad \text{(II.23)}$$

$$M_n(\theta) = C_n^T (\theta) \, R_n^{-2} (\theta) \, C_n(\theta) + Q_{n-1}^{-2}(\theta) \, -$$

$$Q_{n-1}^{-2}(\theta) \, A_{n-1}(\theta) \, N_n^{-1} (\theta) \, A_{n-1}^T (\theta) \, Q_{n-1}^{-2} (\theta)$$

$$M_0(\theta) = Q_0^{-2} (\theta) \qquad \text{(II.24)}$$

$$h_n(\theta, y^n) = C_n^T (\theta) \, R_n^{-2}(\theta) \, y_n + Q_{n-1}^{-2} (\theta) \, A_{n-1}(\theta) \, N_n^{-1}(\theta) \, h_{n-1}(\theta, y^{n-1})$$

$$h_0(\theta) = 0 \qquad \text{(II.25)}$$

$$k_n(\theta, y^n) = k_{n-1}(\theta, y^{n-1}) + \frac{1}{2} \, h_{n-1}^T(\theta, y^{n-1}) \, N_n^{-1} (\theta) \, h_{n-1}(\theta, y^{n-1}) \, -$$

$$- \frac{1}{2} \, y_n^T \, R_n^{-2}(\theta) \, y_n \; ; \; k_0(\theta) = 0 \qquad \text{(II.26)}$$

Furthermore, the matrices $M_n(\theta)$ are positive definite.

Recalling that we had used the symbol $N (x; m, V)$ to denote the Gaussian distribution with mean m and variance V, from Theorem II.2 we immediately obtain

Corollary II.2 : The marginal conditional distributions for the parameter θ and the state x_n are given by

$$p_n(\theta \mid y^n) \propto p(\theta) \; \varphi_n(\theta) \; M_n^{-1/2}(\theta)$$

$$\exp[\; k_n(\theta,y^n) + \frac{1}{2} \; h_n^T(\theta,y^n) \; M_n^{-1}(\theta) \; h_n(\theta,y^n) \;] \qquad (\text{II.27})$$

$$p_n(x_n \mid y^n) = \sum_\theta p_n(x_n, \theta \mid y^n) =$$

$$\sum_\theta p_n(\theta \mid y^n) \; N(\; x_n; \; h_n^T(\theta,y^n) \; M_n^{-1}(\theta), \; M_n^{-1}(\theta) \;) \qquad (\text{II.28})$$

i.e. $p_n(x_n \mid y^n)$ is a combination of Gaussian distributions with the weights given by the conditional distributions of the various values of θ.

In [4] it is furthermore shown that, also in the general case when θ_n is non-constant, the explicit method to compute recursively $p_n(x_n, \theta_n \mid y^n)$ is robust with respect to small variations in the a-priori distributions in model (II.20) in the following sense : Consider the family, indexed by $\varepsilon > 0$, of "prelimit" perturbed models

$$x_n^\varepsilon = A_{n-1}(\theta_{n-1}^\varepsilon) \, x_{n-1}^\varepsilon + Q_{n-1}(\theta_{n-1}^\varepsilon) \, w_n^\varepsilon \quad , \quad x_0^\varepsilon = Q_0(\theta_0^\varepsilon) \, w_0^\varepsilon$$

$$\qquad\qquad\qquad\qquad\qquad\qquad\qquad\qquad\qquad\qquad\qquad (\text{II.29})$$

$$y_n^\varepsilon = C_n(\theta_n^\varepsilon) \, x_n^\varepsilon + R_n(\theta_n^\varepsilon) \, v_n^\varepsilon$$

where the processes $\{\theta_n^\varepsilon\}$, $\{w_n^\varepsilon\}$, $\{v_n^\varepsilon\}$ are such that, for $\varepsilon \downarrow 0$, we have the weak convergence (convergence in distribution)

$$(\, \{w_n^\varepsilon\}, \{v_n^\varepsilon\}, \{\theta_n^\varepsilon\} \,) \; \Rightarrow \; (\, \{w_n\}, \{v_n\}, \{\theta_n\} \,)$$

with $\{w_n\}$, $\{v_n\}$, $\{\theta_n\}$ satisfying the same properties as the corresponding processes in model (II.20).

Let $p_n(x_n, \theta_n \mid (y^\varepsilon)^n)$ be obtained from the same relations as $p_n(x_n, \theta_n \mid y^n)$ except for replacing the observations y_n from (II.20) by the actual observations y_n^ε corresponding to a model of the form (II.29). By analogy to the robustness statement of Theorem II.1, and with the same meaning of the symbols, in [4] the following is shown

Theorem II.3 : Under the above assumptions we have for all continuous and bounded functions $f(x,\theta)$

$$\lim_{\varepsilon \downarrow 0} E \left\{ \left[f(x_n^\varepsilon, \theta_n^\varepsilon) - F_n(y^\varepsilon) \right]^2 \right\} \geq$$

$$\lim_{\varepsilon \downarrow 0} E \left\{ \left[f(x_n^\varepsilon, \theta_n^\varepsilon) - \sum_\theta \int f(x,\theta) \, p_n(x, \theta \mid (y^\varepsilon)^n) \, dx \right]^2 \right\}$$

III. ROBUST METHODS IN DISCRETE TIME

Taking advantage of the explicit method to solve the combined filtering and parameter estimation problems in discrete time, which was described in section II.B, in the present section we present robust methods to solve combined filtering and parameter estimation problems in discrete time models that are analogous to those in continuous time that we mentioned in section II.A.2. We shall also present robust methods to solve filtering problems for discrete-time models that depend on an unobservable parameter process and are analogous to those in continuous time recalled in section II.A.1.. A discussion with simulation results follows in section IV.

III.A. Robust filtering and parameter estimation for discrete time models with heavy-tailed observation noise distribution

In this subsection we study robust methods to solve the combined filtering and parameter estimation problem for a discrete-time model that corresponds to one instance of the continuous-time model (II.17). We recall that (II.17) is the parameter-dependent extension of model (II.6) that was considered in [1] and is a particular case of model (II.8) considered in [2]. We proceed along two further subsections : In the first one we consider the case without limiter in the observations, i.e. when the function H(.) in (II.17) is the identity function; in the second we then study the case when there is a limiter in the observations. For simplicity, in the sequel we shall only consider the scalar case.

III.A.1. The case without limiter

III.A.1.a. Theoretical considerations

As mentioned above, we start from one instance of the continuous-time model (II.17) for which we take the following

$$dx_t = A\, x_t\, dt\; +\; d\, w_t$$

$$\dot{y}_t^\varepsilon = C\,\theta\, x_t\; +\; \frac{1}{\varepsilon}\; \phi_{t/\varepsilon^2}^q \tag{III.1}$$

i.e. we assume for simplicity of exposition that the state process x_t satisfies an ideal white noise equation, which does not depend on any unknown parameter. As will become apparent below, this restriction does not however limit the generality of the approach to be presented. We assume furthermore that the process ϕ_t^q determining the observation noise in (III.1) is given by (II.13), i.e. is non-Gaussian with distribution that, for $q > 1$, has heavier tails than the Gaussian.

Given a time-discretization step $\Delta > 0$, consider then the following discrete-time version of (III.1)

$$x_n = (\, A\, \Delta + 1\,)\, x_{n-1} + \Delta\, w_n$$

$$y_n^\varepsilon = C\,\Delta\,\theta\, x_n\; +\; v_n^\varepsilon \tag{III.2}$$

where Δw_n are independent, identically distributed according to a Gaussian law with mean zero and variance Δ. To specify the observation noise process

v_n^ε, notice that y_n^ε corresponds to the increment, over the time interval $[n\Delta,(n+1)\Delta)$, of y_t^ε when x_t remains constant and equal to its value at $n\Delta$. It is then natural to choose

$$v_n^\varepsilon = \frac{1}{\varepsilon} \int_{n\Delta}^{(n+1)\Delta} \phi_{u/\varepsilon^2}^q \, du =$$

$$\frac{1}{\varepsilon} \left\{ \alpha \int_{n\Delta}^{(n+1)\Delta} \gamma_{u/\varepsilon^2} \, du + (1-\alpha) \int_{n\Delta}^{(n+1)\Delta} \gamma_{u/\varepsilon^2}^{2q-1} \, du \right\} =$$

$$\varepsilon \left\{ \alpha \int_{n\Delta/\varepsilon^2}^{(n+1)\Delta/\varepsilon^2} \gamma_u \, du + (1-\alpha) \int_{n\Delta/\varepsilon^2}^{(n+1)\Delta/\varepsilon^2} \gamma_u^{2q-1} \, du \right\} \quad \text{(III.3)}$$

where the second equality follows from the definition of ϕ_t^q in (II.13) and γ_t is given by (II.14). For simulation purposes it will be convenient to replace the process γ_t, given by (II.14), by a right-continuous process γ_t^Δ that is constant over intervals of length Δ and whose values $\gamma_{n\Delta}^\Delta$ at the time discretization points $n\Delta$ are given by

$$\gamma_{n\Delta}^\Delta = \gamma_0 \, e^{-n\Delta/2} + e^{-n\Delta/2} \sum_{h=0}^{n-1} e^{h\Delta/2} \, \Delta\beta_h \quad \text{(III.4)}$$

with $\Delta\beta_n$ independent and identically distributed according to a Gaussian law with mean zero and variance Δ, so that $\gamma_{n\Delta}^\Delta = \gamma_{n\Delta}$ for all n. Instead of (III.3) we then consider

$$v_n^\varepsilon = \varepsilon\Delta \left\{ \alpha \sum_{h=0}^{1/\varepsilon^2-1} \gamma_{(n/\varepsilon^2+h)\Delta} + (1-\alpha) \sum_{h=0}^{1/\varepsilon^2-1} \gamma_{(n/\varepsilon^2+h)\Delta}^{2q-1} \right\} \quad \text{(III.5)}$$

Notice that, by its construction, we have $\lim_{\Delta\to 0} \gamma_t^\Delta = \gamma_t$ a.s. uniformly in t on compacts so that

$$\lim_{\Delta \to 0} \sum_{n=0}^{N} v_n^\varepsilon = \frac{1}{\varepsilon} \int_0^{(N+1)\Delta} \phi_{u/\varepsilon 2}^q \, du \qquad (III.6)$$

a.s. for any given N. On the other hand, for a given $T > 0$, the process

$$x_t^\Delta = \sum_{n=0}^{[T/\Delta]} x_n \, 1_{[n\Delta,(n+1)\Delta)}(t) \qquad (t \leq T) \qquad (III.7)$$

with x_n as in (III.2), converges, for $\Delta \downarrow 0$, in distribution to x_t in (III.1). Letting finally

$$y_t^{\Delta,\varepsilon} := \sum_{n=0}^{[t/\Delta]} y_n^\varepsilon = \int_0^t C \, \theta \, x_u^\Delta \, du + \sum_{n=0}^{[t/\Delta]} v_n^\varepsilon \qquad (III.8)$$

we immediately obtain the following

Proposition III.1 : The discrete time model (III.2) with v_n^ε given by (III.4) and (III.5) is such that for the processes x_t^Δ and $y_t^{\Delta,\varepsilon}$, defined in (III.7) and (III.8) respectively, we have the weak convergence (convergence in distribution)

$$(x_t^\Delta, y_t^{\Delta,\varepsilon}) \Rightarrow (x_t, y_t^\varepsilon) \quad \text{for } \Delta \downarrow 0$$

with x_t and y_t^ε satisfying the continuous-time model (III.1).

Notice now that from Corollary II.1, combined with the weak convergence recalled in Example II.1 for the particular case of the process ϕ_t^q, we obtain, for $\varepsilon \downarrow 0$, the weak convergence $(x_t, y_t^\varepsilon) \Rightarrow (x_t, y_t)$ with (x_t, y_t) satisfying

$$d\,x_t = A\,x_t\,dt + dw_t$$

$$\qquad\qquad\qquad\qquad\qquad\qquad (III.9)$$

$$d\,y_t = C\,\theta\,x_t\,dt + D\,dv_t$$

for a suitable value of D according to Proposition II.2. It follows from [2] that a good approximation to the value of D is given by

$$D \cong \sqrt{4 \alpha^2 + 4 \alpha (1-\alpha) 2q (2q-3)!!} \qquad \text{(III.10)}$$

where the values of α and q are those used in (III.5) provided $1-\alpha$ is small.

On the other hand, considering the Euler-type approximation of (III.9), namely

$$x_n = (A \Delta + 1) x_{n-1} + \Delta w_n$$

$$\text{(III.11)}$$

$$y_n = C \Delta \theta x_n + D \Delta v_n$$

where, again, y_n is to be considered as an approximation to the increment of y_t over the interval $[n\Delta,(n+1)\Delta)$, and defining x_t^Δ as in (III.7) and y_t^Δ as

$$y_t^\Delta := \sum_{n=0}^{[t/\Delta]} y_n \qquad \text{(III.12)}$$

it is well known that, for $\Delta \downarrow 0$ we have the weak convergence $(x_t^\Delta, y_t^\Delta) \Rightarrow (x_t, y_t)$. Combining these facts with Proposition III.1 we obtain

Corollary III.1 : For small Δ and ε, the process pairs $(x_t^\Delta, y_t^{\Delta,\varepsilon})$ and (x_t^Δ, y_t^Δ), where x_t^Δ is given by (III.7), $y_t^{\Delta,\varepsilon}$ by (III.8), and y_t^Δ by (III.12), are close in the sense of convergence in distribution.

III.A.1.b. Robust solution methods.

Combining the results of subsection III.A.1.a with those of Section II.B, we now present a robust method to solve the combined filtering and parameter estimation problem for model (III.2). Notice in fact that the prelimit model (III.2) is a particular case of (II.29) and that the discrete time limit model (III.11) is a particular case of (II.20) with $A_{n-1}(\theta_{n-1}) = A \Delta + 1$, $Q_{n-1}(\theta_{n-1}) = \sqrt{\Delta}$, $C_n(\theta_n) = C \Delta \theta$, $R_n(\theta_n) = D \sqrt{\Delta}$. Furthermore, from

Corollary III.1 it also follows that, for small values of Δ and ε, v_n^ε is close in distribution to D Δv_n. Applying then the method described in Theorem II.2 to model (III.11) and running it with the real observations y_n^ε from (III.2), we obtain an algorithm to solve the combined filering and parameter estimation problem for model (III.2) and this algorithm is, by the results of Section II.B, robust in the sense specified by Theorem II.3. In this context notice furthermore that, by applying a practically insignificant truncation, in Theorem II.3 we can ignore the boundedness requirement for the function $f(x,\theta)$, so that the inequality of Theorem II.3 applies also to the estimation error variance.

III.A.2. The case with a limiter

III.A.2.a. Theoretical considerations

This time we start from the following instance of model (II.17)

$$dx_t = A \, x_t \, dt + d \, w_t$$

$$\dot{y}_t^\varepsilon = \frac{1}{\varepsilon} \, H \, (\varepsilon \, C \, \theta \, x_t + \phi_{t/\varepsilon^2}^q)$$

(III.13)

which is model (III.1) with a limiter in the observations. We shall assume that the limiter function H(.) is continuous which is e.g. the case when H is a "soft limiter" of the form

$$H_k(z) = \begin{cases} -1 & \text{for} & x < -k \\ x/k & \text{for} & -k \le x \le k \\ +1 & \text{for} & x > k \end{cases}$$

(III.14)

Given a time-discretization step Δ, we now consider the following discrete-time version of (III.13), namely

$$x_n = (A \Delta + 1) x_{n-1} + \Delta w_n$$

(III.15)

$$y_n^\varepsilon = \frac{\Delta}{\varepsilon} H (\varepsilon C \theta x_n + \frac{\varepsilon}{\Delta} v_n^\varepsilon)$$

where v_n^ε is as defined in (III.5). We again have that x_t^Δ, as defined in (III.7), converges for $\Delta \downarrow 0$ in distribution to x_t. Via Skorokhod imbedding (see [12]) we can then consider x_t^Δ to converge also uniformly in t on compacts a.s. on a suitable probability space. Furthermore, from the continuity of the trajectories of the process ϕ_t^q , from the uniform on compacts convergence a.s. of γ_t^Δ to γ_t, and from the first equality in (III.3) we have that a.s. uniformly in t on compacts

$$\lim_{\Delta \downarrow 0} \frac{\varepsilon}{\Delta} v_{[t/\Delta]}^\varepsilon = \phi_{t/\varepsilon^2}^q$$

(III.16)

Analogously to (III.8) let

$$y_t^{\Delta,\varepsilon} := \sum_{n=0}^{[t/\Delta]} y_n^\varepsilon = \frac{1}{\varepsilon} \sum_{n=0}^{[t/\Delta]} \Delta H [\varepsilon C \theta x_n + \frac{\varepsilon}{\Delta} v_n^\varepsilon] =$$

$$\frac{1}{\varepsilon} \sum_{n=0}^{[t/\Delta]} \Delta H [\varepsilon C \theta x_{n\Delta}^\Delta + \frac{\varepsilon}{\Delta} v_n^\varepsilon]$$

(III.17)

The a.s. convergence of x_t^Δ, (III.16) and the continuity of H(.) then lead to the following analogue of Proposition III.1, namely

Proposition III.2 : The discrete-time model (III.15) with v_n^ε given by (III.4) and (III.5) is such that for the processes x_t^Δ and $y_t^{\Delta,\varepsilon}$, defined in (III.7) and (III.17) respectively, we have the weak convergence

$$(x_t^\Delta, y_t^{\Delta,\varepsilon}) \Rightarrow (x_t, y_t^\varepsilon) \quad \text{for } \Delta \downarrow 0$$

with x_t and y_t^ε satisfying the continuous-time model (III.13).

Analogously to the previous subsection III.A.1, from Corollary II.1 (together with Example II.1) we also obtain, for $\varepsilon \downarrow 0$, the weak convergence $(x_t, y_t^\varepsilon) \Rightarrow (x_t, y_t)$ with (x_t, y_t) satisfying

$$d\, x_t = A\, x_t\, dt + dw_t$$

$$d\, y_t = C_H\, \theta\, x_t\, dt + D_H\, dv_t \tag{III.18}$$

for suitable values of C_H, D_H according to Proposition II.2. It follows from [2] (see also [7]) that in the case of a "hard limiter" i.e. $H(z) = \text{sign}\,(z)$, one has, with α as used in (III.5) and independently of the value of q there,

$$C_H = \sqrt{\frac{2}{\pi}}\frac{C}{\alpha} \; ; \quad D_H = 2\sqrt{\ln 2} \tag{III.19}$$

The computation of these values for a "soft limiter" is more involved, but in the case of $H(.)$ given by (III.14) with a small value of k, one may keep the above values with a good approximation. Finally, we have for $\Delta \downarrow 0$ the weak convergence $(x_t^\Delta, y_t^\Delta) \Rightarrow (x_t, y_t)$, where, this time, x_t^Δ, y_t^Δ correspond to (III.11) with C_H, D_H in place of C, D. Combining again these facts with the weak convergence in Proposition III.2 we obtain

Corollary III.2 : Consider the process pairs $(x_t^\Delta, y_t^{\Delta,\varepsilon})$ and (x_t^Δ, y_t^Δ), where x_t^Δ is given by (III.7), $y_t^{\Delta,\varepsilon}$ by (III.17), and y_t^Δ by (III.12) with x_n, y_n according to (III.11) in which C and D are replaced by C_H and D_H respectively. Then, for small values of Δ and ε, the above pairs are close in the sense of convergence in distribution.

III.A.2.b. Robust solution methods

Contrary to the previous subsection III.A.1, here we do not anymore have that (III.15) is a particular case of (II.29) ; we still have however that the discrete-time limit model (III.11), in which C and D are now C_H and D_H, is a particular case of model (II.20). We can thus again apply the method of Theorem II.2 to this particular case of model (II.20) and run it with the real observations y_n^ε from (III.15).

The so obtained algorithm for the combined filtering and parameter estimation in (III.15) is again robust in the sense of Theorem II.3, but this time we cannot directly apply this theorem since it is stated for model (II.29) which does not include our perturbed prelimit model (III.15) as a special case. With the use of Corollary III.2 we can however show directly that the statement of Theorem II.3 remains valid also in the present case. We have in fact the following corollary, where $O(\Delta,\varepsilon)$ denotes a quantity that goes to zero with Δ and ε.

Corollary III.3 : Let $f(x,\theta)$ be continuous and bounded and $F_t(y)$ any continuous and bounded functional of y_s for $s \le t$. Letting

$$\Phi_t(y) = \sum_\theta \int f(x,\theta)\, p_t(x,\theta \mid y^t)\, dx \qquad (III.20)$$

where, for $t \in [n\Delta,(n+1)\Delta)$, $p_t(x,\theta \mid y^t)$ is given by the $p_n(x,\theta \mid y^n)$ of Theorem II.2 applied to the limit model (III.11), we have

$$E\{ [f(x_t^\Delta,\theta) - F_t(y^{\Delta,\varepsilon})]^2 \} \ge$$

$$E\{ [f(x_t^\Delta,\theta) - \Phi_t(y^{\Delta,\varepsilon})]^2 \} - O(\Delta,\varepsilon) \qquad (III.21)$$

Proof. By the continuity and boundedness of $f(.)$ and $F_t(.)$ we have from Corollary III.2

$$E\{ [f(x_t^\Delta,\theta) - F_t(y^{\Delta,\varepsilon})]^2 \} = E\{ [f(x_t^\Delta,\theta) - F_t(y^\Delta)]^2 \} - O(\Delta,\varepsilon) \quad (III.22)$$

On the other hand, it follows immediately from (II.21)-(II.26) that $\Phi_t(y)$ is a continuous and bounded function of y_s for $s \le t$. Therefore

$$E\{ [f(x_t^\Delta,\theta) - \Phi_t(y^{\Delta,\varepsilon})]^2 \} = E\{ [f(x_t^\Delta,\theta) - \Phi_t(y^\Delta)]^2 \} + O(\Delta,\varepsilon) \quad (III.23)$$

Finally, $\Phi_t(y^\Delta)$ is by definition the conditional expectation of $f(x_t, \theta)$ given y_s^Δ for $s \leq t$ and as such minimizes the mean square estimation error among all other estimators that are (measurable) functions of y_s^Δ for $s \leq t$, i.e. we have

$$E \{ [f(x_t^\Delta, \theta) - \Phi_t(y^\Delta)]^2 \} \leq E \{ [f(x_t^\Delta, \theta) - F_t(y^\Delta)]^2 \} \qquad \text{(III.24)}$$

The result follows by combining (III.22)-(III.24),

Analogously to the considerations made after Proposition II.2, in the present case with a limiter we can say more about robustness. Let

$$\hat{\hat{x}}_t^\Delta = \int x \, p_t(x \mid (y^\Delta)^t) \, dx \quad ; \quad \hat{\hat{x}}_t^{\Delta,\varepsilon} = \int x \, p_t(x \mid (y^{\Delta,\varepsilon})^t) \, dx \qquad \text{(III.25)}$$

i.e. $\hat{\hat{x}}_t^\Delta$ is the estimate for x_t^Δ obtained with the method of Theortem II.2 applied to the limit model (III.11) and corresponding to the ideal observations y_n of that model ; $\hat{\hat{x}}_t^{\Delta,\varepsilon}$ is this same estimate corresponding to the real observations y_n^ε from (III.15). Assuming uniform integrability (in practice this can be achieved by means of a truncation), from (III.23) we obtain in particular

$$E \{ [x_t^\Delta - \hat{\hat{x}}_t^{\Delta,\varepsilon}]^2 \} = E \{ [x_t^\Delta - \hat{\hat{x}}_t^\Delta]^2 \} + O(\Delta,\varepsilon) \qquad \text{(III.26)}$$

i.e. the mean square estimation errors incurred when estimating x_t^Δ by $\hat{\hat{x}}_t^\Delta$ or $\hat{\hat{x}}_t^{\Delta,\varepsilon}$ are close for small Δ and ε.

On the other hand, from the weak convergence $(x_t^\Delta, y_t^\Delta) \Rightarrow (x_t, y_t)$ we have

$$E \{ [x_t^\Delta - \hat{\hat{x}}_t^\Delta]^2 \} = E \{ [x_t - \hat{x}_t]^2 \} + O(\Delta) \qquad \text{(III.27)}$$

where $\hat{x}_t$ is the filter estimate for the continuous-time limit model (III.18). Combining (III.26) and (III.27) we thus have

$$E \{ [x_t^\Delta - \hat{\hat{x}}_t^{\Delta,\varepsilon}]^2 \} = E \{ [x_t - \hat{x}_t]^2 \} + O(\Delta,\varepsilon) \qquad \text{(III.28)}$$

Recall now that $P_t := E \{ [x_t - \hat{x}_t]^2 \}$ satisfies the Riccati equation (II.11) with $E \{ A(\xi_0) \} = A$ and $B = 1$ and its magnitude depends on the ratio C_H^2 / D_H^2. By analogy to the considerations after formula (II.11) and in Example II.1, this fact allows us now to enhance the robustness of the filter estimate $\hat{\hat{x}}_t^{\Delta, \varepsilon}$ by using an appropriate limiter in the observations whenever the observation noise disturbances have heavy tails.

III.B. Robust filtering for linear discrete-time models with random coefficients and heavy-tailed observation noise distribution

In this section we study robust methods to solve the filtering problem for a discrete-time model that corresponds to one instance of the more general continuous-time model (II.8). Again, we consider the two cases without and with a limiter in the observations.

III.B.1. The case without limiter

III.B.1.a. Theoretical considerations

We consider here the following instance of model (II.8)

$$dx_t = A x_t dt + d w_t$$

$$\dot{y}_t^\varepsilon = C \eta_{t/\varepsilon^2} x_t + \frac{1}{\varepsilon} \phi_{t/\varepsilon^2}^q$$

(III.29)

where ϕ_t^q is given by (II.13) and the process η_t is independent of x_t. For a time-discretization step Δ we then consider the following discrete-time version of (III.29)

$$x_n = (A \Delta + 1) x_{n-1} + \Delta w_n$$

$$y_n^\varepsilon = C \Delta \eta_n^\varepsilon x_n + v_n^\varepsilon \tag{III.30}$$

where v_n^ε is as in (III.5). Concerning η_n^ε , we take again into account that y_n^ε represents an increment so that , analogously to the definition of v_n^ε in (III.3), we may take

$$\eta_n^\varepsilon = \frac{1}{\Delta} \int_{n\Delta}^{(n+1)\Delta} \eta_{u/\varepsilon^2} \, du = \frac{\varepsilon^2}{\Delta} \int_{n\Delta/\varepsilon^2}^{(n+1)\Delta/\varepsilon^2} \eta_u \, du \tag{III.31}$$

For simulation as well as algorithmic purposes it will be convenient to have η_t a right-continuous process that is constant over the intervals $[h\Delta,(h+1)\Delta)$, h=0,1,....If η_t is not of such form, assuming from now on that it is a.s. piecewise continuous, we approximate it with the process

$$\eta_t^\Delta = \sum_{h=0}^{\infty} \eta_{h\Delta} \, 1_{[h\Delta,(h+1)\Delta)}(t) \tag{III.32}$$

so that $\lim_{\Delta \to 0} \eta_t^\Delta = \eta_t$ a.s. uniformly in t on compacts. Analogously to (III.5), we then consider instead of (III.31) the following

$$\eta_n^\varepsilon = \frac{\varepsilon^2}{\Delta} \int_{n\Delta/\varepsilon^2}^{(n+1)\Delta/\varepsilon^2} \eta_u^\Delta \, du = \varepsilon^2 \sum_{h=0}^{1/\varepsilon^2-1} \eta_{(n/\varepsilon^2+h)\Delta} \tag{III.33}$$

Finally, corresponding to (III.8) let

$$y_t^{\Delta,\varepsilon} := \sum_{n=0}^{[t/\Delta]} y_n^\varepsilon = C \sum_{n=0}^{[t/\Delta]} \Delta \, \eta_n^\varepsilon x_n + \sum_{n=0}^{[t/\Delta]} v_n^\varepsilon =$$

$$C \sum_{n=0}^{[t/\Delta]} \Delta \left[\varepsilon^2 \sum_{h=n/\varepsilon^2}^{(n+1)/\varepsilon^2-1} \eta_{h\Delta} \right] x_n + \sum_{n=0}^{[t/\Delta]} v_n^\varepsilon =$$

$$\int_0^t C \, \eta^\Delta_{u/\varepsilon^2} \, x^\Delta_u \, du \; + \; \sum_{n=0}^{[t/\Delta]} v^\varepsilon_n \tag{III.34}$$

Taking into account (III.6) as well as the a.s. uniform convergence on compacts of η^Δ_t to η_t and of x^Δ_t to x_t (see comment after (III.15)), we immediately obtain the following analogue of Proposition III.1

Proposition III.3   The discrete-time model (III.30), with v^ε_n given by (III.4) and (III.5) and with η^ε_n given by (III.33), is such that for the processes x^Δ_t and $y^{\Delta,\varepsilon}_t$ defined in (III.7) and (III.34) respectively, we have the weak convergence

$$(x^\Delta_t, y^{\Delta,\varepsilon}_t) \; \Rightarrow \; (x_t, y^\varepsilon_t) \quad \text{for } \Delta \downarrow 0$$

with x_t and y^ε_t satisfying the continuous-time model (III.29).

Proceeding by analogy to subsection III.A.1 and using the results in [2] mentioned in Proposition II.2, it is possible to show that we also have the weak convergence $(x_t, y^\varepsilon_t) \Rightarrow (x_t, y_t)$ with (x_t, y_t) satisfying

$$d \, x_t = A \, x_t \, dt \; + \; dw_t$$

$$\tag{III.35}$$

$$d \, y_t = C \, E\{\eta_0\} \, x_t \, dt \; + \; D \, dv_t$$

and where, by the considerations after formula (III.9), the value of D can be approximated by the expression on the right hand side of (III.10).

Finally, considering the Euler-type approximation of (III.35), namely

$$x_n = (A \, \Delta + 1) \, x_{n-1} + \Delta \, w_n$$

$$\tag{III.36}$$

$$y_n = C \, \Delta \, E\{\eta_0\} \, x_n \; + \; D \, \Delta \, v_n$$

we have, for $\Delta \downarrow 0$, the weak convergence $(x_t^\Delta, y_t^\Delta) \Rightarrow (x_t, y_t)$ where x_t^Δ and y_t^Δ are given by (III.7) and (III.12) respectively with x_n and y_n from (III.36). Combining these facts with Proposition III.3 we obtain

Corollary III.4 : Consider the process pairs $(x_t^\Delta, y_t^{\Delta,\varepsilon})$ and (x_t^Δ, y_t^Δ), where x_t^Δ is given by (III.7), $y_t^{\Delta,\varepsilon}$ by (III.34), and y_t^Δ by (III.12) with x_n, y_n according to (III.36). Then, for small values of Δ and ε, the above pairs are close in the sense of convergence in distribution.

III.B.1.b. Robust solution methods

We shall use the results of III.B.1.a to obtain a robust method to solve the filtering problem for model (III.30). For this purpose notice that, if $\eta_{k\Delta}$ is a finite-state Markov chain, also η_n^ε in (III.33) is, so that (III.30) is a particular case of model (II.29), where the process θ_n^ε there is here replaced by the process η_n^ε. Proceeding then by analogy to subsection III.A.1.b, we can use the method of Theorem II.2 for the limit model (II.20), with the coefficients suitably chosen to correspond to the prelimit model (III.30), and run it with the real observations y_n^ε from (III.30). This algorithm is then robust in the sense specified by Theorem II.3.

If the interest is only in estimating the state x_n, then in the present case we can obtain also an alternative robust method for model (III.30). Notice in fact that the discrete-time model (III.36) is of the form of a linear-Gaussian model, so that a Kalman-Bucy filter can be used to compute the conditional distribution $p_n(x_n \mid y^n)$ $(y^n := y_0, y_1, ..., y_n)$ which is Gaussian and depends continuously on the data y_h with $h \le n$. If we now use the Kalman-Bucy filter for (III.36) with the real observations y_n^ε from (III.30), we obtain a robust method to estimate the state in (III.30). With a proof completely analogous to that of Corollary III.3 we obtain in fact

Corollary III.5 : Let f(x) be continuous and bounded and $F_t(y)$ any continuous and bounded functional of y_s for $s \leq t$. Letting

$$\Phi_t(y) = \int f(x) \, p_t(x \mid y^t) \, dx \tag{III.37}$$

where $p_t(x \mid y^t)$ is given by the $p_n(x \mid y^n)$, obtained from the Kalman-Bucy filter for (III.36) whenever $t \in [n\Delta, (n+1)\Delta)$, we have

$$E \{ [f(x_t^\Delta) - F_t(y^{\Delta,\varepsilon})]^2 \} \geq$$

$$E \{ [f(x_t^\Delta) - \Phi_t(y^{\Delta,\varepsilon})]^2 \} - O(\Delta,\varepsilon) \tag{III.38}$$

III.B.2 The case with a limiter

III.B.2.a Theoretical considerations

The starting point here is the particular case of model (II.8), when in (III.29) we have a limiter in the observations, namely

$$dx_t = A \, x_t \, dt + d \, w_t$$

$$\dot{y}_t^\varepsilon = \frac{1}{\varepsilon} H (\varepsilon \, C \, \eta_{t/\varepsilon 2} \, x_t + \phi_{t/\varepsilon 2}^q) \tag{III.39}$$

As in (III.13), here too we assume H(.) continuous. Given a time-discretization step $\Delta > 0$, we consider the following discrete-time version of (III.39)

$$x_n = (A \, \Delta + 1) \, x_{n-1} + \Delta \, w_n$$

$$y_n^\varepsilon = \frac{\Delta}{\varepsilon} H (\varepsilon \, C \, \eta_n^\varepsilon \, x_n + \frac{\varepsilon}{\Delta} v_n^\varepsilon) \tag{III.40}$$

with η_n^ε and v_n^ε as in (III.30). Proceeding along the lines of subsection III.A.2, and using the first equality in (III.3) as well as (III.33), let (see (III.17))

$$y_t^{\Delta,\varepsilon} := \sum_{n=0}^{[t/\Delta]} y_n^\varepsilon = \frac{1}{\varepsilon} \sum_{n=0}^{[t/\Delta]} \Delta H \left[\varepsilon C \eta_n^\varepsilon x_n + \frac{\varepsilon}{\Delta} v_n^\varepsilon \right] =$$

$$\frac{1}{\varepsilon} \sum_{n=0}^{[t/\Delta]} \Delta H \left(\varepsilon C x_n \left[\varepsilon^2 \sum_{h=n/\varepsilon^2}^{(n+1)/\varepsilon^2 - 1} \eta_{h\Delta} \right] + \frac{\varepsilon}{\Delta} v_n^\varepsilon \right) =$$

$$\frac{1}{\varepsilon} \sum_{n=0}^{[t/\Delta]} \Delta H \left(\varepsilon C x_{n\Delta}^\Delta \frac{1}{\Delta} \int_{n\Delta}^{(n+1)\Delta} \eta_{u/\varepsilon^2}^\Delta \, du + \frac{1}{\Delta} \int_{n\Delta}^{(n+1)\Delta} \phi_{u/\varepsilon^2}^q \, du \right) \quad \text{(III.41)}$$

The a.s. uniform convergence on compacts of η_t^Δ and x_t^Δ as well as the continuity of $H(.)$ then lead to the following analogue of Proposition III.2

Proposition III.4 : The discrete-time model (III.40) is such that for the processes x_t^Δ and $y_t^{\Delta,\varepsilon}$, defined in (III.7) and (III.41) respectively, we have the weak convergence

$$(x_t^\Delta, y_t^{\Delta,\varepsilon}) \Rightarrow (x_t, y_t^\varepsilon) \quad \text{for } \Delta \downarrow 0$$

with x_t and y_t^ε satisfying the continuous-time model (III.39).

Proceeding by analogy to subsection III.A.2.a and using results from [2] mentioned in Proposition II.2, it is possible to show that, for $\varepsilon \downarrow 0$, we also have the weak convergence $(x_t, y_t^\varepsilon) \Rightarrow (x_t, y_t)$ with (x_t, y_t) satisfying

$$d x_t = A x_t \, dt + dw_t$$

$$\text{(III.42)}$$

$$d y_t = \bar{C}_H x_t \, dt + D_H \, dv_t$$

for suitable values of $\bar{C}_H$ and D_H, where D_H is the same as in (III.18). Using the independence of the processes η_t and ϕ_t^q , it follows immediately from formulas (2.27) as well as (2.13) and (2.19) in [2] that in the case when H(.) is either a hard limiter, i.e. H(z) = sign (z), or a soft limiter as in (III.14), then $\bar{C}_H = C_H \, E \, \{\eta_0\}$ where C_H is the same as in (III.18). The values of C_H and D_H for H(z) = sign (z) are given in (III.19) and can be used with a good approximation also when H is given by (III.14) with a small value of k.

Considering then the following Euler-type approximation of (III.42), namely

$$x_n = (A \, \Delta + 1) \, x_{n-1} + \Delta \, w_n$$

(III.43)

$$y_n = \bar{C}_H \, \Delta \, x_n + D_H \, \Delta \, v_n$$

we have, with the usual definitions of x_t^Δ, y_t^Δ, the weak convergence (x_t^Δ, y_t^Δ) $\Rightarrow (x_t, y_t)$ for $\Delta \downarrow 0$. Combining these facts with Proposition III.4 we obtain

Corollary III.6 : Consider the process pairs $(x_t^\Delta, y_t^{\Delta,\varepsilon})$ and (x_t^Δ, y_t^Δ), where x_t^Δ is given by (III.7), $y_t^{\Delta,\varepsilon}$ by (III.41), and y_t^Δ by (III.12) with x_n, y_n according to (III.43). Then, for small values of Δ and ε, the above pairs are close in the sense of convergence in distribution.

III.B.2.b. Robust solution methods

As was already the case in subsection III.A.2.b, here the discrete-time prelimit model (III.40) is not a particular case of (II.29), so that the analogue of the first robust method discussed in subsection III.B.1.b is not applicable here. We can however apply the analogue of the alternative robust method discussed in that subsection, since the discrete-time limiting model (III.43) is linear-Gaussian. Interpreting here $p_n(x_n \mid y^n)$ as the conditional distribution obtained from the Kalman-Bucy filter applied to (III.43) and noticing that $\Phi_t(y)$ in (III.37) is a continuous and bounded functional of y_s for $s \leq t$, we then have that Corollary III.5 holds in the same form also in the present case, thus specifying the sense in which this alternative method is robust. Finally, the possibility of using in the present case a limiter in the observations, gives us an additional means to enhance the robustness and for this we can repeat almost identically the last part of subsection III.A.2.b, provided we replace the method of Theorem II.2 with the Kalman-Bucy filter for the discrete-time limit model (III.43).

IV. SIMULATION RESULTS AND THEIR DISCUSSION

In order to test the actual performance of the various robust methods, that are proposed in this paper, some simulations have been carried out. More precisely, for each of the prelimit models a certain number of sequences (x_n), (y_n^ε) were generated and one of the algorithms proposed for that model was applied to produce for each generated sequence a corresponding estimate $\hat{x}_n^\varepsilon$ of x_n. The limiter function H was in all cases chosen to be a "soft limiter" of the form (III.14). The algorithms that were applied are more precisely the following :

1) For model (III.2) : Method of Theorem II.2 corresponding to the limit model (III.11) with D given by the approximate expression (III.10)

2) For model (III.15) : Method of Theorem II.2 corresponding to the limit model (III.11) with $C = C_H$, $D = D_H$, where the values of C_H and D_H are approximated by those corresponding to a hard limiter and given by (III.19).

3) For model (III.30) : Kalman-Bucy filter corresponding to the limit model (III.36) with the value of D again approximated by (III.10).

4) For model (III.40) : Kalman-Bucy filter corresponding to (III.43) with $\bar{C}_H = C_H E\{\eta_0\}$ and C_H, D_H again approximated by (III.19).

The process $\{\eta_{h\Delta}\}$ was simulated as a two-state ergodic Markov chain, independent of x_n, with states $\{\ 1\ ,\ 2\ \}$ and transition matrix $\begin{vmatrix} 1/2 & 1/2 \\ 1/3 & 2/3 \end{vmatrix}$ and starting from its invariant measure. For the generation of (v_n^ε) we selected $\alpha = .8$.

The choice of the other data was based taking into account the following facts :

i) The performance of all the algorithms is strongly dependent on the signal-to-noise ratio. In order to prevent this ratio from varying considerably over successive time periods, we chose the coefficient A so as to have the value of $A \Delta + 1$ close to 1.

ii) The limiter in the observations fulfills its purpose only if it cuts off just the "outliers" that are due to noise distributions with heavier tails than the Gaussian (recall that the algorithms are those corresponding to the limit models where the observation noise distribution is Gaussian)

Therefore, when using a soft limiter of the form of (III.14), the smaller we take the value of k in (III.14), the smaller we should take the value of ε. Since on the one hand we are obliged to take the value of k sufficiently small so that the approximation of the values of C_H, D_H by those in (III.20) is acceptable, on the other hand a too small value of ε increases considerably the simulation burden (see (III.5) and (III.33) where the number of variates to be generated is of the order of $1 / \varepsilon^2$), we selected the values $k = 1$, $\varepsilon = 1 / 12$.

Finally, in order to affect neither the signal-to-noise ratio in the limiting models, nor the balance between the choice of k and ε, we put the coefficient $C = 2.5$ and (for the tables shown below) performed the simulations with $\theta = 1$. The value of the time-discretization parameter was chosen to be $\Delta = .2$.

The criteria, according to which to evaluate the performance of the algorithms, were chosen as :

a) The empirical filter variance, namely $V_n = \dfrac{1}{M} \sum\limits_{i=1}^{M} [\, x_n(i) - \hat{x}_n^{\varepsilon}(i)\,]^2$

where $x_n(i)$ and $\hat{x}_n^{\varepsilon}(i)$ are the values of x_n and its estimate respectively, corresponding to the i-th simulation run, with M denoting the number of these runs.

b) The mean relative estimation error for x_n, namely

$$E_n = \frac{1}{M} \sum_{i=1}^{M} \frac{|x_n(i) - \hat{x}_n^{\varepsilon}(i)|}{|x_n(i)|}$$

This criterion appears more appropriate than V_n , but is less robust to "outliers" caused by simulation runs leading to small values of $|x_n(i)|$, which therefore have to be cut off.

Figures 1 and 2 below plot the values of V_n and E_n respectively for the algorithms 1) and 2) as functions of the parameter q in (III.5), which determines the fatness of the tails of the observation noise distribution (Combined filtering and parameter estimation without and with limiter in the observations). These values were computed for the time period n = 10 performing M = 50 simulation runs. The other data were chosen as specified above, in particular A Δ + 1 = 1.05. Figures 3 and 4 serve the same purpose, but for the algorithms 3) and 4). (Filtering in models with random coefficients, without and with limiter in the observations).

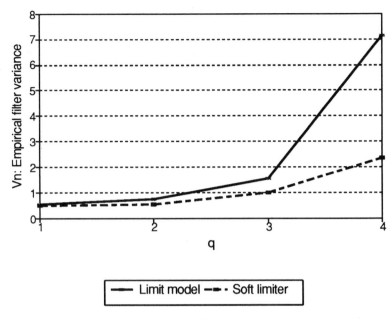

Fig.1. Plotting V_n for combined filtering and parameter estimation

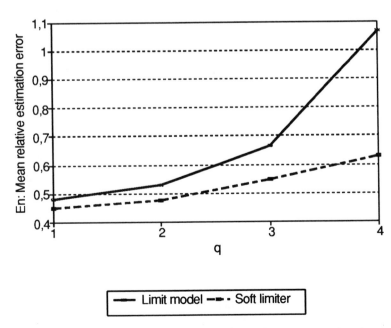

Fig.2. Plotting E_n for combined filtering and parameter estimation

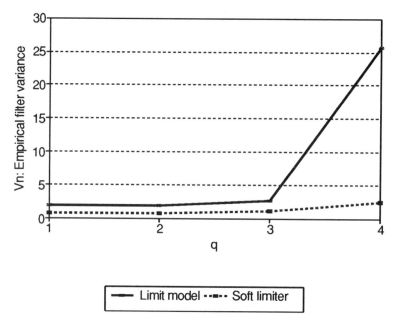

Fig.3. Plotting V_n for filtering with random coefficients

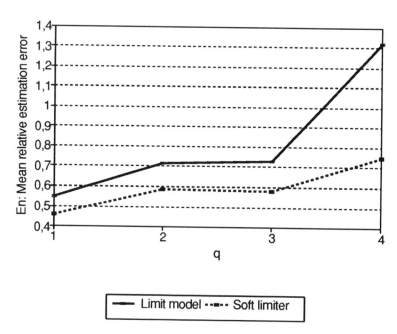

Fig.4. Plotting E_n for filtering with random coefficients

The Figures show, as one would expect,that the algorithms without the use of a limiter continue to perform well also with heavy-tailed noise-distributions that correspond to values of q up to q = 3; thereafter, in order to have the algorithm still performing satisfactorily, i.e. in order to ensure robustness, the use of a limiter becomes unavoidable.

We conclude this section with a couple of further remarks due to additional simulations that are not reported here.

It turned out that the algorithms performed better when passing from a Gaussian noise distribution to a uniform noise distribution, rather than when fattening its tails ; for example, the performance corresponding to a uniform distribution with support [- 1.73 , 1.73] (variance = 1) was close to that for a standard Gaussian. This could lead to the conclusion that the robustness is stronger with respect to the shape of the distribution (as long as its symmetry is not changed) than with respect to its support.

In the limit models considered in Section III, the parameter enters only the observation equation and, furthermore, linearly. This implies that in the formulas corresponding to the method of Theorem II.2 the parameter θ enters only as a square, so that the algorithm may well estimate quite accurately the absolute value of θ, but is unable to distinguish its sign. This problem disappears when the parameter enters also the state equation and it turned out that the absence of the parameter in the state equation deteriorates the quality of the state estimation. Furthermore, with heavy tails and no limiter, the parameter tends to be overestimated ; it is as though an increase in the absolute value of the observations due to heavier noise is attributed by the algorithm to an increase in the absolute value of the parameter. In such a situation the use of a limiter in the observations obviously improves the performance of the filter.

Acknowledgment : The authors wish to acknowledge valuable suggestions made by Dr. Diego Bricio Hernandez concerning the simulations and ECOMATICA s.a.s. of Padova for aiding us with computer graphics.

References

1. H. J. Kushner and W. J. Runggaldier, "Filtering and Control for Wide Bandwidth Noise Driven Systems", *IEEE Transactions* **AC-32**,123-133 (1987).

2. R. Sh. Liptser and W. J. Runggaldier, "On Diffusion Approximations for Filtering", *Stochastic Processes and Their Applications* **38** (1991) to appear.

3. I. C. Schick, "Robust Recursive Estimation of the State of a Discrete-Time Stochastic Linear Dynamic System in the Presence of Heavy-Tailed Observation Noise" , Ph.D. Thesis, Massachusetts Institute of Technology (1990).

4. W. J. Runggaldier and C. Visentin, "Combined Filtering and Parameter Estimation : Approximations and Robustness", *Automatica* **26**, 401-404 (1990).

5. P. J. Huber, "Robust Statistics", J.Wiley, New York, 1981.

6. W. B. Davenport, "Signal to Noise Ratios in Band Pass Limiters", *J. Appl. Phys.* **24**, 720-727 (1953).

7. H. J. Kushner, "Approximation and Weak Convergence Methods for Random Processes with Applications to Stochastic System Theory", MIT Press, Cambridge, 1984.

8. B. D. O. Anderson and J. B. Moore, "Optimal Filtering", Prentice-Hall, 1979.

9. W. J. Runggaldier and O. Zane, "Approximations for Discrete-Time Adaptive Control : Construction of ε-Optimal Controls", *Math.Control Signals Systems* **4**, 269-291 (1991).

10. R. Sh. Liptser and S. V. Lototski, "Diffusion Approximation and Robust Kalman Filter", Preprint.

11. G. B. Di Masi and W. J. Runggaldier, "On Measure Transformations for Combined Filtering and Parameter Estimation in Discrete Time", *Syst. Control Lett.* **2**, 57-62 (1982).

12. A. V. Skorokhod, "Limit Theorems for Stochastic Processes", *Theory of Probability and its Applications* **1**, 262-290 (1956).

Stability and Robustness of Multivariable Model Reference Adaptive Control Schemes

Gang Tao
Electrical and Computer Engineering Department
Washington State University, Pullman, WA 99164

Petros A. Ioannou
Department of Electrical Engineering-Systems
University of Southern California, Los Angeles, CA 90089

Abstract

The design and analysis of model reference adaptive control schemes for both continuous and discrete time multivariable plants with additive and multiplicative unmodeled dynamics are presented in a unified framework. Issues such as plant and controller parametrizations, design of adaptive laws, stability, robustness and performance are clarified.

1 Introduction

Starting with the early 70's considerably progress has been made in adaptive control that continued through the 80's and 90's. Most of the recent work and that of the 70's and 80's in adaptive control concentrated on single-input, single-output (SISO) plants. Work for multi-input, multi-output (MIMO) plants received some attention in (Dion et al. 1988; Elliott and Wolovich 1982, 1984; Goodwin and Sin 1984; Monopoli and Hsing 1975; Morse 1980; Narendra and Annaswamy 1989; Sastry and Bodson 1989; Singh and Narendra 1984; Tao and Ioannou 1988, 1989; Tsiligiannis and Svoronos 1986) but

did not progress as much as in the case of SISO plants due to various difficulties that arise in the MIMO case. These difficulties include problems on
parametrization, a priori knowledge, commutativity, stability and robustness
that are not as easy to resolve as in the SISO case. In this paper we consider
the design and analysis of model reference adaptive control (MRAC) schemes
for multivariable plants with additive and multiplicative unmodeled dynamics. It is the goal of this paper to clarify the class of unmodeled dynamics
admissible to robust multivariable adaptive control, to develop the adaptive
laws for updating the controller parameters and to analyze the stability, robustness and tracking performance of the multivariable adaptive control system in a unified framework for both the continuous and discrete time plants.
In addition issues such as plant and controller parametrizations and a priori
knowledge are also addressed and discussed.

The paper is organized as follows. In Section 2 we present and discuss the
plant assumptions, control objective and model reference control structure.
In Section 3 we develop the error model and adaptive law to estimate the
controller parameters on line. In Section 4 we show that the closed-loop plant
with the developed adaptive controller is globally stable in the presence of
unmodeled dynamics and the tracking error between the plant output and the
reference signal is small, i.e., of the order of the plant unmodeled dynamics,
in the mean sense. In Section 5 we discuss the a priori plant knowledge and
performance improvement for multivariable adaptive controllers.

To give a unified presentation for both the continuous and discrete time
adaptive control systems, we first introduce the following definitions and notations.

(1) The dummy variable D is used as the Laplace transform variable or the
differential operator $D[x](t) = \dot{x}(t)$, $t \in [0, \infty)$, in the continuous time (CT)
case, and as the z-transform variable or the advance operator $D[x](t) = x(t +
1)$, $t \in \{0, 1, 2, 3, \cdots\}$, in the discrete time (DT) case, respectively, whenever
it applies.

(2) A transfer matrix (resp. a polynomial, a square matrix) is called stable

(resp. Hurwitz) if all its poles (resp. zeros, eigenvalues) are in the open left half complex plane for the CT case, and on the open unit disk of the complex plane for the DT case.

(3) A vector signal $x(t)$ is uniformly bounded if $x(t) \in L_\infty$ for the CT case or $x(t) \in l_\infty$ for the DT case.

(4) $\|x(t)\|_1 \triangleq \int_0^\infty \|x(t)\| dt$ in the CT case; $\|x(t)\|_1 \triangleq \sum_{t=0}^\infty \|x(t)\|$ in the DT case.

(5) $a \in R^+$ means that a is a finite positive constant.

(6) A linear operator $T(D,t)$ with impulse response matrix $T(\tau,t)$ is quasi-time-invariant, stable and strictly proper with finite gain if $\|T(\tau,t)\| \leq c_1 e^{-c_2 \tau}$, $\tau \geq 0$, for some c_1, $c_2 \in R^+$ and any $t \geq 0$; $T(D,t)$ is quasi-time-invariant, stable and proper with finite gain if $T_\infty(t) = \lim_{D\to\infty} T(D,t)$ is finite for $t \geq 0$, and $T(D,T) - T_\infty(t)$ is quasi-time-invariant, stable and strictly proper with finite gain. We will simply call a quasi-time-invariant, stable operator a stable operator.

(7) Equations whose numbers have the letter "a" (resp. "b") are used only for the CT (resp. DT) case; otherwise equations are used for both the CT and DT cases.

2   Plant and Controller Structure

Consider the following linear time-invariant plant with N inputs, N outputs:

$$y(t) = G(D)[u](t) \tag{2.1}$$

where $y(t)$ is the plant output, $u(t)$ is the plant input, with $t \in [0, \infty)$ for the CT case and $t \in \{0, 1, 2, 3, \cdots\}$ for the DT case,

$$G(D) = G_0(D)[I + \mu\Delta_m(D)] + \mu\Delta_a(D) \tag{2.2}$$

is the plant transfer matrix, $G_0(D) = Z_0(D)R_0^{-1}(D)$ is the modeled or nominal part of $G(D)$, $Z_0(D)$, $R_0(D)$ are some $N \times N$ right coprime polynomial matrices with $R_0(D)$ being column proper (Elliott and Wolovich 1984), and

$\mu\Delta_m(D)$, $\mu\Delta_a(D)$ are the multiplicative, additive plant unmodeled dynamics respectively with $\mu \geq 0$.

The objective of MRAC can be stated as follows: given a reference model of the form:

$$y_m(t) = W_m(D)[r](t) \qquad (2.3)$$

where $W_m(D)$ is an $N \times N$ rational transfer matrix, $y_m(t)$ is the reference output and $r(t)$ is an external input signal, find the feedback control $u(t)$ for the plant (2.1) with unknown $G(D)$ such that $y(t)$ tracks $y_m(t)$ as close as possible and the closed-loop plant is globally stable in the sense that all signals in the system are uniformly bounded for any bounded initial conditions, in spite of the presence of the unmodeled dynamics $\mu\Delta_m(D)$, $\mu\Delta_a(D)$.

An important concept used in designing multivariable model reference control (MRC) schemes is the zero structure at infinity of the nominal plant. This structure is characterized by the Hermite normal form (Morse 1980; Singh and Narendra 1984) or the interactor matrix (Wolovich and Falb 1976; Elliott and Wolovich 1982). We present the modified interactor matrix which can be used in the design of MRC schemes in both the CT and DT cases.

Lemma 2.1 *(Tao and Ioannou 1988, 1989) For any $N \times N$ strictly proper rational full rank transfer matrix $G_0(D)$ there exists a (non-unique) lower triangular polynomial matrix $\xi_m(D)$, defined as the modified left interactor (MLI) matrix of $G_0(D)$, of the form:*

$$\xi_m(D) = \begin{pmatrix} d_1(D) & 0 & \cdots & \cdots & 0 \\ h_{21}(D) & d_2(D) & 0 & \cdots & 0 \\ & \cdots & \cdots & \cdots & \\ h_{N1}(D) & \cdots & \cdots & h_{N\,N-1}(D) & d_N(D) \end{pmatrix} \qquad (2.4)$$

where $h_{ij}(D)$, $j = 1, \cdots, N-1$, $i = 2, \cdots, N$, are some polynomials, and $d_i(D)$, $i = 1, \cdots, N$, are arbitrary monic Hurwitz polynomials of certain degrees $l_i > 0$, such that $\lim_{D\to\infty} \xi_m(D)G_0(D) = K_p$, the high frequency gain matrix of $G_0(D)$, is finite and nonsingular.

In order to design MRAC schemes to achieve the control objective we make the following assumptions about the plant (2.1) and the reference model (2.3):

(A1) $G_0(D)$ is strictly proper, of full rank and has a known MLI matrix $\xi_m(D)$;

(A2) an upper bound $\bar{\nu}_0$ on the observability index ν_0 of $G_0(D)$ is known;

(A3) all zeros of $G_0(D)$ are stable;

(A4) all poles of $W_m(D)$ are stable and the zero structure at infinity of $W_m(D)$ is the same as that of $G_0(D)$, i.e., $\lim_{D\to\infty} \xi_m(D)W_m(D)$ is finite and nonsingular;

(A5) $r(t)$ is uniformly bounded;

(A6C) S_p is known such that $K_pS_p = (K_pS_p)^T > 0$ in the CT case;

(A6D) S_p is known such that $2I > K_pS_p = (K_pS_p)^T > 0$ in the DT case;

(A7) $\Delta_m(D)$, $\Delta_a(D)$ are rational transfer matrices such that

$$D_m = \lim_{D\to\infty} W_m(D)K_p\Delta_m(D), \; D_a = \lim_{D\to\infty} \Delta_a(D) \qquad (2.5)$$

are finite, and $(W_m(D-q)K_p\Delta_m(D-q) - D_m)D$, $(\Delta_a(D-q) - D_a)D$ are stable and proper with finite gain for some known $q \in R^+$ in the CT case, or $(W_m(qD)K_p\Delta_m(qD) - D_m)(D+1)$, $(\Delta_a(qD) - D_a)(D+1)$ are stable and proper with finite gain for some known q such that $0 < q < 1$ in the DT case.

From (A1), (A4), without loss of generality, we can take $W_m(D) = \xi_m^{-1}(D)$.

From (A2), we choose the controller structure as:

$$u(t) = \Theta_1^T\omega_1(t) + \Theta_2^T\omega_2(t) + \Theta_3 r(t) \qquad (2.6)$$

where $\omega_1(t) = F(D)[u](t)$, $\omega_2(t) = F(D)[y](t)$, $F(D) = \frac{A_0(D)}{n(D)}$, $A_0(D) = (I, DI, \cdots, D^{\bar{\nu}_0-1}I)^T$, $\Theta_1 = (\Theta_{11}, \cdots, \Theta_{1\bar{\nu}_0})^T$, $\Theta_2 = (\Theta_{21}, \cdots, \Theta_{2\bar{\nu}_0})^T$, Θ_3, $\Theta_{ij} \in R^{N\times N}$, $i = 1,2$, $j = 1, \cdots, \bar{\nu}_0$, and $n(D)$ is any monic polynomial of degree $\bar{\nu}_0$ such that $n(D - q_0)$ is Hurwitz for some $q_0 \in R^+$ in the CT case or $n(q_0 D)$ is Hurwitz for some $0 < q_0 < 1$ in the DT case.

With the specification of $n(D)$, $\xi_m(D)$, $R_0(D)$, $Z_0(D)$, it follows that there exist Θ_1^*, Θ_2^*, $\Theta_3^* = K_p^{-1}$ (Elliott and Wolovich 1982) such that

$$\Theta_1^{*T}A_0(D)R_0(D) + \Theta_2^{*T}A_0(D)Z_0(D) = n(D)(R_0(D) - \Theta_3^*\xi_m(D)Z_0(D)) \; (2.7)$$

So we have the following plant-model transfer matrix matching equation:

$$I - \Theta_1^{*T}F(D) - \Theta_2^{*T}F(D)G_0(D) = \Theta_3^*W_m^{-1}(D)G_0(D) \qquad (2.8)$$

From (2.1) - (2.7) it follows (Tao and Ioannou 1988; Tao 1989) that for $\Theta_i = \Theta_i^*$, $i = 1, 2, 3$, in (2.5), there exists $\mu_0 > 0$ such that for any $\mu \in [0, \mu_0)$, the closed-loop plant is internally stable and

$$y(t) = W_m(D)[r](t) + \mu\Delta(D)[r](t) + \epsilon_0(t) \tag{2.9}$$

where $\Delta(D)$ is stable and proper with finite gain, and $\epsilon_0(t)$ is an exponentially decaying term due to initial conditions of the control system.

When $G_0(D)$ in (2.1) is known, we can calculate the parameter matrices Θ_i^*, $i = 1, 2, 3$, by solving (2.6) and use them in the controller (2.5). If $G_0(D)$ is unknown, then we need to develop an adaptive law to update the controller parameters $\Theta_i(t)$, $i = 1, 2, 3$, on line so that the closed-loop plant is globally stable and has the desired tracking performance, i.e., the control objective stated above is achieved.

3   Error Model and Adaptive Law

To develop an adaptive law for the controller (2.5), we need to express the closed-loop system in terms of the tracking error $e(t) = y(t) - y_m(t)$ and the parameter error $\tilde{\Theta}(t) = \Theta(t) - \Theta^*$ where $\Theta(t)$ is the estimate of $\Theta^* = (\Theta_1^{*T}, \Theta_2^{*T}, \Theta_3^{*T})^T$ at time t.

From (2.1) - (2.7), by ignoring the effect of initial conditions, we have

$$\xi_m(D)[y - y_m](t) = \Theta_3^{*-1}(\tilde{\Theta}^T(t)\omega(t)$$

$$+ \mu(I - \Theta_1^{*T}F(D))(\Delta_m(D) + G_0^{-1}(D)\Delta_a(D))[u](t)) \tag{3.1}$$

where $\omega(t) = (\omega_1^T(t), \omega_2^T(t), r^T(t))^T$.

Let d_m be the maximum degree of $\xi_m(D)$, $\Psi^* = \Theta_3^{*-1} = K_p$, $f(D)$ be any polynomial of degree d_m such that $f(D - q_0)$ is Hurwitz the CT case or $f(q_0D)$ is Hurwitz in the DT case, and $h(D) = \frac{1}{f(D)}$, from (3.1) we have

$$\xi_m(D)h(D)[y - y_m](t) = \Psi^*(h(D)[u](t) - \Theta^{*T}h(D)[\omega](t)) + \mu\eta_1(t) \tag{3.2}$$

where

$$\eta_1(t) = \Delta_1(D)[u](t) \tag{3.3}$$

$$\Delta_1(D) = \Theta_3^{*-1}h(D)(I - \Theta_1^{*T}F(D))(\Delta_m(D) + G_0^{-1}(D)\Delta_a(D)) \qquad (3.4)$$

Let $\Psi(t)$ be the estimate of Ψ^* at time t, we define the estimation error as:

$$\epsilon(t) = \frac{\xi_m(D)h(D)[y - y_m](t) + \Psi(t)\xi(t)}{1 + \beta(\zeta^T(t)\zeta(t) + \xi^T(t)\xi(t)) + \alpha m^2(t)} \qquad (3.5)$$

where $\alpha \in R^+$ is arbitrary, $0 \leq \beta \in R^+$ for the CT case and $1 \leq \beta \in R^+$ for the DT case,

$$\xi(t) = \Theta^T(t)\zeta(t) - \frac{1}{f(D)}[\Theta^T\omega](t), \; \zeta(t) = \frac{1}{f(D)}[\omega](t) \qquad (3.6)$$

and $m(t)$ is the normalizing signal defined as:

$$(D - \sigma)[m](t) = -\delta_0 m(t) + \delta_1(\|u(t)\| + \|y(t)\| + 1)) \qquad (3.7)$$

where $\sigma = 0$, $0 < \delta_0 < min\{q, q_0\}$ for the CT case, $\sigma = 1$, $max\{q, q_0\} < 1 - \delta_0 < 1$ for the DT case, $\delta_1 \in R^+$, and $m(0) > \frac{\delta_1}{\delta_0}$.

The adaptive laws for updating the controller parameters are chosen as:

$$\dot{\Theta}^T(t) = -S_p\epsilon(t)\zeta^T(t) - G_1^T(t) \qquad (3.8a)$$

$$\dot{\Psi}(t) = -\Gamma\epsilon(t)\xi^T(t) - G_2(t) \qquad (3.9a)$$

for the CT case, and

$$\Theta^T(t + 1) = \Theta^T(t) - S_p\epsilon(t)\zeta^T(t) - G_1^T(t) \qquad (3.8b)$$

$$\Psi(t + 1) = \Psi(t) - \Gamma\epsilon(t)\xi^T(t) - G_2(t) \qquad (3.9b)$$

for the DT case, where $G_1(t)$, $G_2(t)$ are the modifying terms for robustness, and Γ, S_p are constant matrices such that $\Gamma = \Gamma^T > 0$ and S_p satisfies the assumptions (A6C) and (A6D).

The switching σ-modification (Ioannou and Tsakalis 1986(a), (b)) are extended to the multivariable case as:

$$G_1^T(t) = \sigma_1(t)\Theta^T(t), \; G_2(t) = \sigma_2(t)\Gamma\Psi(t) \qquad (3.10)$$

where, for the CT case,

$$\sigma_1(t) = \begin{cases} 0 & \text{if } \|\Theta(t)\| < M_1 \\ \sigma_{10}(\frac{\|\Theta(t)\|}{M_1} - 1) & \text{if } M_1 \leq \|\Theta\| < 2M_1 \\ \sigma_{10} & \text{if } \|\Theta(t)\| \geq 2M_1 \end{cases} \qquad (3.11a)$$

$$\sigma_2(t) = \begin{cases} 0 & \text{if } \|\Psi(t)\| < M_2 \\ \sigma_{20}(\frac{\|\Psi(t)\|}{M_2} - 1) & \text{if } M_2 \leq \|\Psi(t)\| < 2M_2 \\ \sigma_{20} & \text{if } \|\Psi(t)\| \geq 2M_2 \end{cases} \tag{3.12a}$$

$M_1, M_2 \in R^+$ such that

$$\frac{\|\Theta^*\|\|\Gamma_p\|}{\lambda_1} < M_1, \; \|\Psi^*\| < M_2 \tag{3.13a}$$

$$\Gamma_p = K_p^T S_p^{-1} = (S_p^{-1})^T (K_p S_p)^T S_p^{-1} \tag{3.14a}$$

λ_1 is the minimum eigenvalue of Γ_p, and $\sigma_{10}, \sigma_{20} \in R^+$ are arbitrary; and for the DT case,

$$\sigma_1(t) = \begin{cases} 0 & \text{if } \|\Theta(t)\| < M_1 \\ \sigma_{10} & \text{if } \|\Theta(t)\| \geq M_1 \end{cases} \tag{3.11b}$$

$$\sigma_2(t) = \begin{cases} 0 & \text{if } \|\Psi(t)\| < M_2 \\ \sigma_{20} & \text{if } \|\Psi(t)\| \geq M_2 \end{cases} \tag{3.12b}$$

$M_1, M_2, \sigma_{10}, \sigma_{20} \in R^+$ such that

$$\frac{3\|\Theta^*\|\|\Gamma_p\|}{2\lambda_1} < M_1, \; \frac{3\|\Psi^*\|}{2} < M_2 \tag{3.13b}$$

$$\sigma_{10} \leq \frac{\lambda_1 \lambda_2}{2(\lambda_2\|\Gamma_p\| + \|S_p\|^2)}, \; \sigma_{20} \leq \frac{\lambda_3}{2(\lambda_3 + 1)} \tag{3.14b}$$

and $\lambda_1, \lambda_2, \lambda_3$ are the minimum eigenvalues of Γ_p, $2\beta I - S_p^T K_p^T$, $2\beta I - \Gamma$ respectively with Γ_p being given in (3.14a).

Define $\tilde{\Psi}(t) = \Psi(t) - \Psi^*$, from (3.2), (3.5), we obtain the error model as:

$$\epsilon(t) = \frac{K_p \tilde{\Theta}^T(t)\zeta(t) + \tilde{\Psi}(t)\xi(t) + \mu\eta_1(t)}{1 + \beta(\zeta^T(t)\zeta(t) + \xi^T(t)\xi(t)) + \alpha m^2(t)} \tag{3.15}$$

From (2.6), (3.3), (3.4), it follows that

$$\Delta_1(D) = \Theta_3^{*-1}(I - \Theta_1^{*T} F(D))K_p^{-1}h(D)\xi_m(D)W_m(D)K_p\Delta_m(D)$$

$$+ h(D)\xi_m(D)(I + W_m(D)\Theta_3^{*-1}\Theta_2^{*T} F(D))\Delta_a(D) \tag{3.16}$$

Using (A7), (3.16), we can express $\Delta_1(D)$ as:

$$\Delta_1(D) = \bar{\Delta}_1(D)\frac{1}{D + q_1} + D_1 \tag{3.17}$$

for some $q_1 > \delta_0$ such that $\bar{\Delta}_1(D - q_1)$ is stable and proper with finite gain in the CT case, or some $0 < q_1 < 1 - \delta_0$ such that $\bar{\Delta}_1(q_1 D)$ is stable and proper with finite gain in the DT case.

Hence it follow from (A7), (3.17), (3.7) that for both the CT and DT cases there exist k_1, $k_2 \in R^+$ such that

$$\frac{\|\eta_1(t)\|}{m(t)} \le k_1 + k_2\|\Theta(t)\| \tag{3.18}$$

Here we used the fact that $\frac{\|\omega(t)\|}{m(t)} \le k_0$ for some $k_0 \in R^+$.

Remark 3.1 For the CT case, we can define an alternate estimation error from the following equation:

$$\epsilon(t) = \xi_m(D)h(D)[y - y_m](t) + h(D)f(D)[\Psi\xi - \alpha\epsilon m^2](t) \tag{3.19a}$$

where $f(D)$ is a polynomial of degree $d_m - 1$ such that $f(D - q_0)$ is Hurwitz, $h(D) = \frac{1}{f(D)(D+a_m)}$, a_m, $\alpha \in R^+$ are arbitrary, and

$$\xi(t) = \Theta^T(t)\zeta(t) - \frac{1}{f(D)}[\Theta^T\omega](t), \; \zeta(t) = \frac{1}{f(D)}[\omega](t) \tag{3.20a}$$

In this case, $h(D)f(D) = \frac{1}{D+a_m}$ and the error model is of the form:

$$\dot{\epsilon}(t) = -a_m\epsilon(t) + K_p\tilde{\Theta}^T(t)\zeta(t) + \tilde{\Psi}(t)\xi(t) - \alpha\,\epsilon(t)\,m^2(t) + \mu\eta_2(t) \tag{3.21a}$$

where

$$\eta_2(t) = \Delta_2(D)[u](t) \tag{3.22a}$$

$$\Delta_2(D) = \frac{D + a_m}{D + q}(h(D)W_m^{-1}(D)(I + W_m(D)\Theta_3^{*-1}\Theta_2^{*T}F(D))(D+q)\Delta_a(D)$$

$$+ \Theta_3^{*-1}(I - \Theta_1^{*T}F(D))(h(D)K_p^{-1}W_m^{-1}(D))(D+q)W_m(D)K_p\Delta_m(D)) \tag{3.23a}$$

For this choice of the estimation error, instead of (A7), we use:

(A7a) $\Delta_m(D)$, $\Delta_a(D)$ are rational transfer matrices such that

$$D_m = \lim_{D\to\infty} DW_m(D)K_p\Delta_m(D), \; D_a = \lim_{D\to\infty} D\Delta_a(D) \tag{3.24a}$$

are finite, and $(DW_m(D - q)K_p\Delta_m(D - q) - D_m)D$, $(D\Delta_a(D - q) - D_a)D$ are stable and proper with finite gain for some known $q \in R^+$.

It follows from (A7a), (3.22a), (3.23a), (3.7) that (3.18) also holds.

We should note that the class of unmodeled dynamics satisfying (A7a) is a subset of those satisfying (A7). However if the estimation error $\epsilon(t)$ (3.19a) is used then the reduction of the order of the adaptive control system could be up to $3N$ where N is the dimension of the plant output vector.

4   Stability and Robustness Analysis

In this section we analyze the stability, robustness and the tracking performance of the proposed adaptive control schemes.

We first establish the stability properties of the adaptive laws (3.8) - (3.14).

Lemma 4.1 *The adaptive law (3.8a) - (3.14a) guarantees that $\Theta(t)$, $\Psi(t)$, $\epsilon(t) \in L_\infty$, and*

$$\int_{t_1}^{t_2} x^2(t)dt \leq \alpha_c \mu^2(t_2 - t_1) + a_c \qquad (4.1a)$$

for some α_c, $a_c \in R^+$, any $t_2 > t_1 \geq 0$, where $x(t) = \|\dot{\Theta}(t)\| + \|\epsilon(t)\|m(t)$; the adaptive law (3.8b) - (3.14b) guarantees that $\Theta(t)$, $\Psi(t)$, $\epsilon(t) \in l_\infty$, and

$$\sum_{t=t_1}^{t_2-1} x^2(t) \leq \alpha_d \mu^2(t_2 - t_1) + a_d \qquad (4.1b)$$

for some α_d, $a_d \in R^+$, any $t_2 > t_1 \geq 0$, where $x(t) = \|\Theta(t+1) - \Theta(t)\| + \|\epsilon(t)\|m(t)$.

Proof: (1) For the CT case, we consider the following positive definite function:

$$V(t) = tr[\tilde{\Theta}(t)\Gamma_p\tilde{\Theta}^T(t)] + tr[\tilde{\Psi}^T(t)\Gamma^{-1}\tilde{\Psi}(t)] \qquad (4.2a)$$

where $t \in [0, \infty)$. Use (3.8a) - (3.14a), (3.15), (3.18), (4.2a), we have

$$\dot{V}(t) = -2\epsilon^T(t)\epsilon(t) - 2\alpha\epsilon^T(t)\epsilon(t)m^2(t) + 2\mu\epsilon^T(t)\eta_1(t)$$

$$-2\beta\epsilon^T(t)\epsilon(t)(\zeta^T(t)\zeta(t) + \xi^T(t)\xi(t)) - 2\sigma_1(t)tr[\tilde{\Theta}(t)\Gamma_p\Theta^T(t)]$$

$$-2\sigma_2(t)tr[\tilde{\Psi}^T(t)\Psi(t)] \leq -2\epsilon^T(t)\epsilon(t) - \alpha\epsilon^T(t)\epsilon(t)m^2(t)$$

$$+ k_3\mu^2 + k_4\mu^2\|\Theta(t)\|^2 - 2\sigma_1(t)tr[\tilde{\Theta}(t)\Gamma_p\Theta^T(t)] - 2\sigma_2(t)tr[\tilde{\Psi}^T(t)\Psi(t)] \quad (4.3a)$$

for some k_3, $k_4 \in R^+$. From the definitions of λ_1, σ_1, $\sigma_2(t)$, we have that

$$\sigma_1(t)tr[\tilde{\Theta}(t)\Gamma_p\Theta^T(t)] + \sigma_2(t)tr[\tilde{\Psi}^T(t)\Psi(t)] \geq \sigma_1(t)(\lambda_1\|\Theta(t)\|^2$$

$$- \|\Gamma_p\|\|\Theta(t)\|\|\Theta^*\|) + \sigma_2(t)(\|\Psi(t)\|^2 - \|\Psi(t)\|\|\Psi^*\|) \geq 0 \qquad (4.4a)$$

It follows from (4.3a), (4.4a) that there exists $\mu_1^* > 0$ such that for any $\mu \in [0, \mu_1^*)$ the existence of $k_5 \in R^+$ is guaranteed such that $\dot{V}(t) \leq 0$ whenever $\|\Theta(t)\| + \|\Psi(t)\| \geq k_5$, so $\Theta(t)$, $\Psi(t) \in L_\infty$, and from (3.7), (3.15), (3.18) we have that $\epsilon(t) \in L_\infty$.

From (3.11a), (3.12a) it follows that there exists k_6, $k_7 \in R^+$ such that

$$\|G_1(t)\| \leq k_6\sigma_1(t)tr[\tilde{\Theta}(t)\Gamma_p\Theta^T(t)], \quad \|G_2(t)\| \leq k_7\sigma_2(t)tr[\tilde{\Psi}^T(t)\Psi(t)] \quad (4.5a)$$

From (4.3a) it follows that there exists $k_8 \in R^+$ such that

$$2\sigma_1(t)tr[\tilde{\Theta}(t)\Gamma_p\Theta^T(t)] + 2\sigma_2(t)tr[\tilde{\Psi}^T(t)\Psi(t)] \leq -\dot{V}(t) + k_8\mu^2 \qquad (4.6a)$$

Therefore by using (3.8a), (4.3a) - (4.6a), we have (4.1a).

(2) For the DT case, we consider the following positive definite function:

$$V(t) = tr[\tilde{\Theta}(t)\Gamma_p\tilde{\Theta}^T(t)] + tr[\tilde{\Psi}^T(t)\Gamma^{-1}\tilde{\Psi}(t)] \qquad (4.2b)$$

where $t \in \{0, 1, 2, \cdots\}$. Using (3.8b) - (3.14b), (3.15), (3.18), (4.2b) and the definitions of Γ_p (3.14a), λ_i, $i = 1, 2, 3$, we have that

$$V(t+1) - V(t) = -2\epsilon^T(t)\epsilon(t) - 2\alpha\epsilon^T(t)\epsilon(t)m^2(t) + 2\mu\epsilon^T(t)\eta_1(t)$$

$$-\zeta^T(t)\zeta(t)(2\beta I - S_p^T K_p^T)\epsilon(t) + 2\sigma_1(t)\epsilon^T(t)S_p^T\Theta(t)\zeta(t) + \sigma_1^2(t)tr[\Theta(t)\Gamma_p\Theta^T(t)]$$

$$-\frac{1}{2}\sigma_1(t)tr[\Theta(t)\Gamma_p\Theta^T(t)] - \frac{1}{2}\sigma_1(t)tr[\tilde{\Theta}(t)\Gamma_p\Theta^T(t)] - \frac{1}{2}\sigma_1(t)tr[(2\Theta(t)-$$

$$3\Theta^*)\Gamma_p\Theta^T(t)] - \xi^T(t)\xi(t)(2\beta I - \Gamma)\epsilon(t) + 2\sigma_2(t)\epsilon^T(t)\Gamma\Psi(t)\xi(t)$$

$$+\sigma_2^2(t)tr[\Psi^T(t)\Gamma\Psi(t)] - \frac{1}{2}\sigma_2(t)tr[\Psi^T(t)\Psi(t)] - \frac{1}{2}\sigma_2(t)tr[\tilde{\Psi}^T(t)\Psi(t)]$$

$$-\frac{1}{2}\sigma_2(t)tr[(2\Psi(t) - 3\Psi^*)^T\Psi(t)] \leq -2\epsilon^T(t)\epsilon(t) - \alpha\epsilon^T(t)\epsilon(t)m^2(t)$$

$$+ k_3\mu^2 + k_4\mu^2\|\Theta(t)\|^2 - \frac{1}{2}\sigma_1(t)tr[\tilde{\Theta}(t)\Gamma_p\Theta^T(t)] - \frac{1}{2}\sigma_2(t)tr[\tilde{\Psi}^T(t)\Psi(t)] \quad (4.3b)$$

for some k_3, $k_4 \in R^+$. Similarly we have

$$\sigma_1(t)tr[\tilde{\Theta}(t)\Gamma_p\Theta^T(t)] + \sigma_2(t)tr[\tilde{\Psi}^T(t)\Psi(t)] \geq \sigma_1(t)(2\lambda_1\|\Theta(t)\|^2$$

$$- 3\|\Gamma_p\|\|\Theta(t)\|\|\Theta^*\|) + \sigma_2(t)(2\|\Psi(t)\|^2 - 3\|\Psi(t)\|\|\Psi^*\|) \geq 0 \quad (4.4b)$$

It follows from (4.3b), (4.4b) that there exists $\mu_1^* > 0$ such that for any $\mu \in [0, \mu_1^*)$ the existence of $k_5 \in R^+$ is guaranteed such that $V(t+1) - V(t) \leq 0$ whenever $\|\Theta(t)\| + \|\Psi(t)\| \geq k_5$, so $\Theta(t)$, $\Psi(t) \in l_\infty$, and from (3.7), (3.15), (3.18) we have that $\epsilon(t) \in l_\infty$.

From (3.11b), (3.12b) it follows that there exists k_6, $k_7 \in R^+$ such that

$$\|G_1(t)\| \leq k_6\sigma_1(t)tr[\tilde{\Theta}(t)\Gamma_p\Theta^T(t)], \quad \|G_2(t)\| \leq k_7\sigma_2(t)tr[\tilde{\Psi}^T(t)\Psi(t)] \quad (4.5b)$$

From (4.3b) it follows that there exists $k_8 \in R^+$ such that

$$\frac{\sigma_1(t)}{2}tr[\tilde{\Theta}(t)\Gamma_p\Theta^T(t)] + \frac{\sigma_2(t)}{2}tr[\tilde{\Psi}^T(t)\Psi(t)] \leq -V(t+1) + V(t) + k_8\mu^2 \quad (4.6b)$$

Therefore from (3.8), (4.3b) - (4.6b), we have (4.2b). ∇

Remark 4.1 For the alternate choice of the estimation error $\epsilon(t)$ (3.19a) in the CT case, we consider the positive definite function:

$$V(t) = \epsilon^T(t)\epsilon(t) + tr[\tilde{\Theta}(t)\Gamma_p\tilde{\Theta}^T(t)] + tr[\tilde{\Psi}^T(t)\Gamma^{-1}\tilde{\Psi}(t)] \quad (4.7a)$$

and use (3.8a), (3.9a), (3.18), (3.21a) to obtain the following:

$$\dot{V}(t) = -2a_m\epsilon^T(t)\epsilon(t) - 2\alpha\epsilon^T(t)\epsilon(t)m^2(t) + 2\mu\epsilon^T(t)\eta_2(t)$$

$$-2\sigma_1(t)tr[\tilde{\Theta}(t)\Gamma_p\Theta^T(t)] - 2\sigma_2(t)tr[\tilde{\Psi}^T(t)\Psi(t)] \leq -(2a_m + \alpha m^2(t))\epsilon^T(t)\epsilon(t)$$

$$+ k_3\mu^2 + k_4\mu^2\|\Theta(t)\|^2 - 2\sigma_1(t)tr[\tilde{\Theta}(t)\Gamma_p\Theta^T(t)] - 2\sigma_2(t)tr[\tilde{\Psi}^T(t)\Psi(t)] \quad (4.8a)$$

for some k_3, $k_4 \in R^+$. Hence for some $\mu_1^* > 0$ and any $\mu \in [0, \mu_1^*)$, there exists $k_5 \in R^+$ such that $\dot{V}(t) \leq 0$ whenever $\|\Theta(t)\| + \|\Psi(t)\| \geq k_5$, so $\Theta(t)$, $\Psi(t)$, $\epsilon(t) \in L_\infty$. Similarly we have (4.1a).

The stability and robustness properties of the adaptive control schemes (2.5), (3.8) - (3.9) are given in the following theorem.

Theorem 4.1 *There exists $\mu^* > 0$ such that for any $\mu \in [0, \mu^*)$, all signals in the closed-loop plant (2.1), (2.3), (2.5), (3.8) - (3.9) are uniformly bounded and the tracking error $e(t) = y(t) - y_m(t)$ satisfies that*

$$\int_{t_1}^{t_2} \bar{e}^T(t)\bar{e}(t)dt \leq \beta_c \mu^2 (t_2 - t_1) + b_c \qquad (4.9a)$$

for some β_c, $b_c \in R^+$ in the CT case, and

$$\sum_{t=t_1}^{t_2-1} \bar{e}^T(t)\bar{e}(t) \leq \beta_d \mu^2 (t_2 - t_1) + b_d \qquad (4.9b)$$

for some β_d, $b_d \in R^+$ in the DT case, where $\bar{e}(t) = h(D)\xi_m(D)[e](t)$, and $t_2 \geq t_1$. Moreover $\lim_{t \to \infty} e(t) = 0$ if $\mu = 0$ in (2.1).

<u>Proof</u>: We use fictitious filters $H_i(D)$, $K_i(D)$ defined as:

$$DH_i(D) = 1 - K_i(D), \quad K_i(D) = \frac{a_i^{d_m}}{(D + a_i)^{d_m}}, \quad i = 1, 2, 3, \qquad (4.10a)$$

for the CT case, and

$$(D - 1)H_1(D) = 1 - K_1(D), \quad K_1(D) = \frac{(1 - \frac{1}{a_1})^{d_m} D^{d_m}}{(D - \frac{1}{a_1})^{d_m}} \qquad (4.10b)$$

for the DT case, where $a_i > 0$ is to be chosen to be sufficiently large but finite. Let $h_i(t)$ be the impulse response function of $H_i(D)$, we have that

$$\|h_i(t)\|_1 \leq \frac{c_0}{a_i} \qquad (4.11)$$

for some $c_0 \in R^+$ in both the CT and DT cases.

Using (2.1), (2.5), $H_1(D)$ and $K_1(D)$ defined in (4.10), we have

$$F(D)G_0^{-1}(D)[y](t) = K_1^{-1}(D)[\omega_1 - H_1(D)(D - \sigma)[\omega_1]](t)$$

$$+ \mu F(D)(\Delta_m(D) + G_0^{-1}(D)\Delta_a(D))[u](t) \qquad (4.12)$$

Let $\omega_1(t) = F(D)[u](t)$ (where $F(D) = \frac{A_0(D)}{n(D)}$) have a controllable realization (A, B), i.e.,

$$D[\omega_1](t) = (A + \sigma I)\omega_1(t) + Bu(t) \qquad (4.13)$$

where $A + \sigma I$ is a Hurwitz matrix, from (2.5), (4.12), (4.13), we have that

$$\omega_1(t) = K_1(D)F(D)G_0^{-1}(D)[y](t) + H_1(D)[A\omega_1](t)$$

$$+ H_1(D)B[\Theta_1^T\omega_1 + \Theta_2^T(\cdot)F(D)[y] + \Theta_3^T r](t)$$

$$- \mu K_1(D)F(D)(\Delta_m(D) + G_0^{-1}(D)\Delta_a(D))[u](t) \qquad (4.14)$$

For the CT case, since $H_1(D)$ satisfies (4.11) and $\Theta_1(t)$ is uniformly bounded, there exists $a_1^0 \in R^+$ such that $(I - H_1(D)(A + B\Theta_1^T(t)))^{-1}$ is a stable and proper operator with finite gain for any finite $a_1 > a_1^0$, where $a_1 > 0$ is defined in (4.10). For $a_1^0 < a_1 \in R^+$ it follows from (4.14) that

$$\omega_1(t) = G_1(D, \cdot)[y](t) + G_2(D, \cdot)[r](t) + \mu G_{\delta 1}(D, \cdot)[u](t) \qquad (4.15a)$$

where, for $T_1(D, t) \triangleq (I - H_1(D)(A + B\Theta_1^T(t)))^{-1}$,

$$G_1(D, t) = T_1(D, t)(K_1(D)F(D)G_0^{-1}(D) + H_1(D)B\Theta_2^T(t)F(D)) \qquad (4.16a)$$

$$G_2(D, t) = T_1(D, t)H_1(D)B\Theta_3(t) \qquad (4.17a)$$

$$G_{\delta 1}(D, t) = -T_1(D, t)K_1(D)F(D)(\Delta_m(D) + G_0^{-1}(D)\Delta_a(D)) \qquad (4.18a)$$

are stable and proper operators with finite gains. Hence it follows from (4.15a) that

$$\omega(t) = G_3(D, \cdot)[y](t) + G_4(D, \cdot)[r](t) + \mu G_{\delta 2}(D, \cdot)[u](t) \qquad (4.19a)$$

where

$$G_3(D, t) = \begin{pmatrix} G_1(D, t) \\ F(D) \\ 0 \end{pmatrix} \qquad (4.20a)$$

$$G_4(D, t) = \begin{pmatrix} G_2(D, t) \\ 0 \\ I \end{pmatrix} \qquad (4.21a)$$

$$G_{\delta 2}(D, t) = \begin{pmatrix} G_{\delta 1}(D, t) \\ 0 \\ 0 \end{pmatrix} \qquad (4.22a)$$

From (3.1), we have

$$D[y](t) = D[y_m](t) + DW_m(D)\Theta_3^{*-1}[\tilde{\Theta}^T\omega$$

$$+ \mu(I - \Theta_1^{*T}F(D))(\Delta_m(D) + G_0^{-1}(D)\Delta_a(D))[u]](t) \qquad (4.23a)$$

Defining $\bar{y}(t) = h(D)[y](t)$, using (2.6), (4.19a), (4.23a) and $H_2(D)$, $K_2(D)$ in (4.10), we have that

$$y(t) = K_2(D)h^{-1}(D)[\bar{y}](t) + H_2(D)DW_m(D)\Theta_3^{*-1}\tilde{\Theta}^T(\cdot)G_3(D,\cdot)[y](t)$$

$$+H_2(D)DW_m(D)\Theta_3^{*-1}\tilde{\Theta}^T(\cdot)G_4(D,\cdot)[r](t)$$

$$+ \mu H_2(D)D(W_m(D)\Theta_3^{*-1}\tilde{\Theta}^T(\cdot)G_{\delta 2}(D,\cdot) + \Delta_0(D))[u](t) \qquad (4.24a)$$

where

$$\Delta_0(D) = W_m(D)\Theta_3^{*-1}(I - \Theta_1^{*T}F(D))\Delta_m(D)$$

$$+ (W_m(D)\Theta_3^{*-1}\Theta_2^{*T}F(D) + I)\Delta_a(D) \qquad (4.25a)$$

with $\Theta_3^{*-1} = K_p$. From (A7), (2.7), (4.25a) we have that $\Delta_0(D)$ is stable and proper with finite gain.

Similarly $(1 - H_2(D)DW_m(D)\Theta_3^{*-1}\tilde{\Theta}^T(t)G_3(D,t))^{-1}$ is a stable and proper operator with finite gain for any finite $a_2 > a_2^0$ and some $a_2^0 \in R^+$. For $a_2^0 < a_2 \in R^+$, it follows from (4.24a) that

$$y(t) = G_5(D,\cdot)[\bar{y}](t) + G_6(D,\cdot)[r](t) + \mu G_{\delta 3}(D,\cdot)[u](t) \qquad (4.26a)$$

where, for $T_2(D,t) \triangleq (I - H_2(D)DW_m(D)\Theta_3^{*-1}\tilde{\Theta}^T(t)G_3(D,t))^{-1}$,

$$G_5(D,t) = T_2(D,t)K_2(D)h^{-1}(D) \qquad (4.27a)$$

$$G_6(D,t) = T_2(D,t)H_2(D)DW_m(D)(\Theta_3^{*-1}\tilde{\Theta}^T(t)G_4(D,t) + I) \qquad (4.28a)$$

$$G_{\delta 3}(D,t) = T_2(D,t)H_2(D)D(W_m(D)\Theta_3^{*-1}\tilde{\Theta}^T(t)G_{\delta 2}(D,t) + \Delta_0(D)) \qquad (4.29a)$$

are stable and proper operators with finite gains. Therefore it follows from (4.19a), (4.26a) that

$$\omega(t) = G_3(D,\cdot)G_5(D,\cdot)[\bar{y}](t) + (G_3(D,\cdot)G_6(D,\cdot) + G_4(D,\cdot))[r](t)$$

$$+ \mu(G_3(D,\cdot)G_{\delta 3}(D,\cdot) + G_{\delta 2}(D,\cdot))[u](t) \qquad (4.30a)$$

From (3.5) we have that

$$\bar{y}(t) = \bar{y}_m(t) + W_m(D)[\epsilon - \Psi\xi + \beta(\zeta^T\zeta + \xi^T\xi) + \alpha\epsilon m^2](t) \qquad (4.31a)$$

where $\bar{y}_m(D) = h(D)[y_m](t)$. From (2.5), (4.30a) we have

$$u(t) = \Theta^T(t)\omega(t) = \Theta^T(t)G_3(D,\cdot)G_5(D,\cdot)[\bar{y}](t) + \Theta^T(t)(G_3(D,\cdot)G_6(D,\cdot)+$$

$$G_4(D,\cdot))[r](t) + \mu\Theta^T(t)(G_3(D,\cdot)G_{\delta 3}(D,\cdot) + G_{\delta 2}(D,\cdot))[u](t) \qquad (4.32a)$$

From (4.32a) it follows that

$$u(t) = G_7(D,\cdot)\Theta^T(\cdot)G_3(D,\cdot)G_6(D,\cdot)[\bar{y}](t)$$

$$+ G_7(D,\cdot)\Theta^T(\cdot)(G_3(D,\cdot)G_6(D,\cdot) + G_4(D,\cdot))[r](t) \qquad (4.33a)$$

where $G_7(D,t) = (I - \mu\Theta^T(t)(G_3(D,t)G_{\delta 3}(D,t) + G_{\delta 2}(D,t)))^{-1}$ is stable and proper with finite gain for any $\mu \in [0, \mu_2^*)$ and some $\mu_2^* > 0$ because $\Theta(t)$ is uniformly bounded and $G_3(D,t)G_{\delta 3}(D,t) + G_{\delta 2}(D,t)$ is stable and proper with finite gain.

For the DT case, we define $\bar{\omega}(t) = (\omega_1^T(t), \omega_2^T(t))^T$, $\bar{\Theta}^* = (\Theta_1^{*T}, \Theta_2^{*T})^T$, $\bar{\Theta}(t) = (\Theta_1^T(t), \Theta_2^T(t))^T$, $\tilde{\bar{\Theta}}(t) = \bar{\Theta}(t) - \bar{\Theta}^*$, $\tilde{\Theta}_3(t) = \Theta_3(t) - \Theta_3^*$. It follows from (3.1) and the definition of $\bar{\omega}(t)$ that

$$D[\bar{\omega}](t) = P_1(D,\cdot)[\bar{\omega}](t) + b_1(t) \qquad (4.15b)$$

where

$$P_1(D,t) = \begin{pmatrix} DF(D)\bar{\Theta}^T(t) \\ DF(D)(W_m(D)\Theta_3^{*-1}\tilde{\bar{\Theta}}^T(t) + \mu\Delta_0(D)\bar{\Theta}(t)) \end{pmatrix} \qquad (4.16b)$$

is stable and proper with finite gain, and

$$b_1(t) = \begin{pmatrix} DF(D)[\Theta_3 r](t) \\ DF(D)(W_m(D)(\Theta_3^{*-1}\tilde{\Theta}_3(\cdot) + I) + \mu\Delta_0(D)\Theta_3(\cdot))[r](t) \end{pmatrix} (4.17b)$$

is uniformly bounded, with $\Delta_0(D)$ being of the same form as that in (4.25a) for the CT case.

Using the fact that the signals $r(t+k)$, $\Theta(t+k)$ are uniformly bounded for any finite $k > 0$, we have that

$$D^2[\bar{\omega}](t) = P_2(D,\cdot)[\bar{\omega}](t) + b_2(t) \qquad (4.18\text{b})$$

where $P_2(D,t) = P_1(D,t+1)P_1(D,t)$ is stable and proper with finite gain, and $b_2(t) = \bar{P}_1(D,\cdot)[b_1](t) + D[b_1](t)$ is uniformly bounded with $\bar{P}_1(D,t) = P_1(D,t+1)$. By deduction, we have that

$$D^{d_m}[\bar{\omega}](t) = P_0(D,\cdot)[\bar{\omega}](t) + b_0(t) \qquad (4.19\text{b})$$

where $P_0(D,t)$ is stable and proper with finite gain, and $b_0(t)$ is uniformly bounded.

Expressing $P_0(D,\cdot)[\bar{\omega}](t) = P_{01}(D,\cdot)[\omega_1](t) + P_{02}(D,\cdot)[\omega_2](t)$, using (2.5), (4.19b), we have that

$$D^{d_m}[u](t) = \bar{\Theta}^T(t+d_m)P_{01}(D,\cdot)[\omega_1](t) + \bar{\Theta}^T(t+d_m)P_{02}(D,\cdot)[\omega_2](t)$$

$$+ \bar{\Theta}^T(t+d_m)b_0(t) + D^{d_m}[\Theta_3 r](t) \qquad (4.20\text{b})$$

Since $H_1(D)$ satisfies (4.11) and $\Theta_1(t)$ is uniformly bounded, there exists $a_1^0 \in R^+$ such that

$$T_1(D,t) = (I - H_1(D)(A + B\Theta_1^T(t)) + \mu K_1(D)F(D)D^{-d_m}(\Delta_m(D)$$

$$+ G_0^{-1}(D)\Delta_a(D))\bar{\Theta}(t+d_m)P_{01}(D,t))^{-1} \qquad (4.21\text{b})$$

is a stable and proper operator with finite gain for any finite $a_1 > a_1^0$, any $\mu \in [0,\mu_2^*)$ and some $\mu_2^* = \mu_2^*(a_1) > 0$. For a finite and fixed $a_1 > a_1^0$, $\hat{y}(t) \triangleq \frac{\xi_m(D)}{D+a_0}[y](t)$ where $0 < a_0 < 1$ is arbitrary, it follows from (4.14), (4.20b) that

$$\omega_1(t) = G_1(D,\cdot)[\hat{y}](t) + b_3(t) \qquad (4.22\text{b})$$

where

$$G_1(D,t) = T_1(D,t)(K_1(D)F(D)(D+a_0)G_0^{-1}(D)\xi_m^{-1}(D)$$

$$+ (H_1(D)B\Theta_2^T(t) - \mu K_1(D)F(D)D^{-d_m}(\Delta_m(D) + G_0^{-1}(D)\Delta_a(D))$$

$$\bar{\Theta}(t + d_m)P_{02}(D,t))(D + a_0)F(D)\xi_m^{-1}(D)) \qquad (4.23b)$$

is stable and proper with finite gain, and

$$b_3(t) = T_1(D, \cdot)[H_1(D)B[\Theta_3 r] - \mu K_1(D)F(D)D^{-d_m}(\Delta_m(D)$$

$$+ G_0^{-1}(D)\Delta_a(D))[D^{d_m}[\Theta_3 r] + D^{d_m}[\bar{\Theta}^T]b_0]](t) \qquad (4.24b)$$

is uniformly bounded.

From (3.5) we have that

$$\hat{y}(t) = \frac{1}{D + a_0}[r](t) + \frac{1}{D + a_0}[\epsilon - \Psi\xi + \beta(\zeta^T\zeta + \xi^T\xi) + \alpha\epsilon m^2](t) \qquad (4.25b)$$

Let (A_1, B_1, C_1) be a minimal realization of $\frac{1}{\bar{f}(D)}$, i.e., $\frac{1}{\bar{f}(D)} = C_1(DI - A_1)^{-1}B_1$, express $\Theta(t) = (\theta_1(t), \cdots, \theta_N(t))$, where $\theta_i(t) \in R^{N(2\bar{\nu}+1)}$, $i = 1, \cdots, N$, define $W_c(D) = C_1(DI - A_1)^{-1}$, $W_b(D) = (DI - A_1)^{-1}B_1$, we can show that

$$\xi(t) = \begin{pmatrix} W_c(D)[W_b(D)[\omega^T]D[\theta_1]](t) \\ \cdot \\ \cdot \\ W_c(D)[W_b(D)[\omega^T]D[\theta_N]](t) \end{pmatrix} \qquad (4.34a)$$

for the CT case, and

$$\xi(t) = \begin{pmatrix} W_c(D)[W_b(D)D[\omega^T](D-1)[\theta_1]](t) \\ \cdot \\ \cdot \\ W_c(D)[W_b(D)D[\omega^T](D-1)[\theta_N]](t) \end{pmatrix} \qquad (4.34b)$$

for the DT case.

Hence for the CT case, from (3.7), Lemma 4.1, (4.26a), (4.32a) - (4.34a), we have that

$$\|\bar{y}(t)\| \leq x_0(t) + \beta_1 \int_0^t e^{-\alpha_1(t-\tau)}x(\tau)(\int_0^\tau e^{-\alpha_2(\tau-w)}\|\bar{y}(w)\|dw)d\tau \qquad (4.35a)$$

with $x(t) = \|\dot{\Theta}(t)\| + \|\epsilon(t)\|m(t)$, and for the DT case, from (3.7), Lemma 4.1, (4.22b), (4.25b), we have that

$$\|\hat{y}(t)\| \leq x_0(t) + \beta_1 \sum_{\tau=0}^{t-1} e^{-\alpha_1(t-\tau-1)}x(\tau)(\sum_{w=0}^{\tau-1} e^{-\alpha_2(\tau-w-1)}\|\hat{y}(w)\|) \qquad (4.35b)$$

where $x(t) = \|\Theta(t+1) - \Theta(t)\| + \|\epsilon(t)\|m(t)$, for some β_1, α_1, $\alpha_2 \in R^+$, $x_0(t)$ is uniformly bounded. Here we used the fact that $\frac{\|\zeta(t)\|}{m(t)}$, $\frac{\|\xi(t)\|}{m(t)}$ are uniformly bounded.

Applying the Gronwall lemma on (4.35), we conclude that there exists μ^*, $0 < \mu^* < min\{\mu_1^*, \mu_2^*\}$ such that $\bar{y}(t)$ or $\hat{y}(t)$ is uniformly bounded for any $\mu \in [0, \mu^*)$, so are $u(t)$, $y(t)$ and all signals in the closed-loop plant. For $\bar{e}(t) = \xi_m(D)h(D)[e](t)$, from (3.5), (4.1), (4.34), we have (4.9).

When $\mu = 0$, from (4.1), we have that $\|x^2(t)\|_1$ is finite. Hence it follows from (3.5), (4.34) that $\|\bar{e}^T(t)\bar{e}(t)\|_1$ is finite.

For the CT case, from (3.1) with $\mu = 0$, using $H_3(D)$ and $K_3(D)$ defined in (4.10), we have that

$$e(t) = W_m(D)\Theta_3^{*-1}[\tilde{\Theta}^T\omega](t) =$$

$$H_3(D)DW_m(D)\Theta_3^{*-1}[\tilde{\Theta}^T\omega](t) + W_m(D)K_3(D)h^{-1}(D)[\bar{e}](t) \qquad (4.36a)$$

where

$$\lim_{t\to\infty} W_m(D)K_3(D)h^{-1}(D)[\bar{e}](t) = 0 \qquad (4.37a)$$

for any finite $a_3 > 0$ in $K_3(D)$, and $DW_m(D)\Theta_3^{*-1}[\tilde{\Theta}^T\omega](t)$ is uniformly bounded. Hence from (4.36a), (4.37a) it follows that

$$\|e(t)\| \le c_3\|h_3(t)\|_1 + z_1(t) \le \frac{c_4}{a_3} + z_1(t) \qquad (4.38a)$$

where c_3, $c_4 \in R^+$, $\lim_{t\to\infty} z_1(t) = 0$. Since $a_3 > 0$ in $H_3(D)$ can be set to arbitrarily large, it follows from (4.38a) that $\lim_{t\to\infty} e(t) = 0$.

For the DT case the fact that $\|\bar{e}^T(t)\bar{e}(t)\|_1$ is finite implies that $\|e^T(t)e(t)\|_1$ is finite, so $\lim_{t\to\infty} e(t) = 0$. ∇

Some immediate comments are presented below for the developed adaptive control schemes.

Remark 4.2 The results and proofs have provided a new framework for analyzing the stability, robustness and tracking performance of the model reference adaptive control system. The adaptive law (3.8) - (3.14) guarantees that the estimation error $\epsilon(t)$ and the rate of the parameter adaptation, $\dot{\Theta}(t)$ in the CT case or $\Theta(t+1) - \Theta(t)$ in the DT case, are of the order of μ in a mean sense (see (4.1)), in addition to the boundedness of the estimation error and the estimated parameters. The latter results in that the closed-loop plant can be expressed as in (4.35) whose gain is ensured to be small from (4.1) if $\mu \geq 0$ is small so the stability of the closed-loop plant is guaranteed.

The filtered tracking error $h(D)\xi_m(D)[e](t)$ is of the order of the plant unmodeled dynamics in the mean sense (4.9). Recall that in generating the estimation error $\epsilon(t)$ (see (3.5)), we used the signal $\bar{e}(t) = h(D)\xi_m(D)[e](t)$, which is crucial for solving the non-commutativity problem in the design and analysis of multivariable adaptive controllers. The adaptive law (3.8) - (3.14) results in that $\epsilon(t)$ satisfies (4.1), so we have (4.9) for the tracking error $e(t)$. However it can be shown that in the CT case if $D^{d_m}W_m(D)$ is proper, then $\int_{t_1}^{t_2} e^T(t)e(t)dt \leq \bar{\beta}_c\mu^2(t_2 - t_1) + \bar{b}_c$, for some $\bar{\beta}_c, \bar{b}_c \in R^+$, which is the same as the SISO case when $N = 1$, and in the DT case (4.9b) implies that $\sum_{t=t_1}^{t_2-1} e^T(t)e(t) \leq \bar{\beta}_d\mu^2(t_2 - t_1) + \bar{b}_d$, for some $\bar{\beta}_d, \bar{b}_d \in R^+$.

Remark 4.3 In the absence of the plant unmodeled dynamics, i.e., $\mu = 0$ in the plant (2.1), we can design an adaptive controller by setting $G_1(t) = 0$, $G_2(t) = 0$ in (3.8), (3.9). The above unified theory can be used to show that $\|x^2(t)\|_1$ is finite where $x(t)$ is defined in (4.1) and $\Theta(t)$, $\Psi(t)$, $\epsilon(t)$ are uniformly bounded, which leads to (4.35) in which the gain is small for large t and in turn to the stability of the closed-loop plant. Moreover the tracking error $e(t)$ goes to zero asymptotically with time.

5   Discussions

In this section we discuss the reduction and substitution of a priori plant knowledge and the improvement of performance for the multivariable adaptive control schemes.

The MLI matrix $\xi_m(D)$ has been shown to play an important role in the design and analysis of multivariable MRAC schemes. In fact any $N \times N$ polynomial matrix $\xi_0(D)$ such that $\xi_0^{-1}(D)$ is stable and $\lim_{D\to\infty} \xi_0(D)G_0(D)$

is finite and non-singular can take the role of $\xi_m(D)$. The triangular form of $\xi_m(D)$ makes it possible to design a stable adaptive controller for multivariable plants with only the information about the degrees of the diagonal elements and the maximum degree of $\xi_m(D)$, which has been reported in Elliott and Wolovich (1984), Goodwin and Sin (1984) for the discrete time case where the concept of the interactor matrix $\xi(D)$ and an augmented estimator were used. The estimation on the maximum degree of $\xi(D)$ was given in Dion et al. (1988). In general $\xi_m(D)$ depends on the parameters of $G_0(D)$ which is unknown in the adaptive control problem. However $\xi_m(D)$ can be specified independent of the parameters of $G_0(D)$ if $\xi_m(D)$ is diagonal. The MLI matrix $\xi_m(D)$ can be constructed from the interactor matrix $\xi(D)$ (Tao and Ioannou 1988). $\xi_m(D)$ is diagonal if and only if $\xi(D)$ is diagonal. Designing an adaptive controller with less a priori plant knowledge has been an important topic in multivariable adaptive control. In Singh and Narendra (1984) a method for designing an input compensator for a multivariable plant so that the compensated plant has a diagonal Hermite normal form (equivalently the interactor matrix) was presented. This method needs the knowledge of the relative degrees of the elements of $G_0(D)$ and works generically. In Tao (1990) a stable adaptive control scheme was developed for plants with unknown interactor matrix and only the knowledge of an upper bound on the maximum degree of the interactor matrix is needed for implementation.

On the other hand, for many plants whose MLI matrix may be non-diagonal, the right interactor matrix (Tsiligiannis and Svoronos 1986) may be diagonal and thus can be specified without the knowledge of the plant parameters (Tao and Ioannou 1988). In this case a parametrization technique can be used to modify the plant zero structure at infinity so that the MLI matrix of the modified plant is diagonal and an adaptive controller can be designed with alternate information of the plant. Next we present such a design technique.

A dual concept to the MLI matrix defined in (2.4) is the modified right interactor (MRI) matrix which is defined by an upper triangular polynomial matrix $\xi_m^r(D)$ of the form:

$$\xi_m^r(D) = \begin{pmatrix} d_1(D) & h_{12}(D) & \cdots & \cdots & h_{1N}(D) \\ 0 & d_2(D) & h_{23}(D) & \cdots & h_{2N}(D) \\ & \cdots & \cdots & \cdots & \\ 0 & \cdots & & 0 & d_N(D) \end{pmatrix} \qquad (5.1)$$

where $h_{ij}(D)$, $i = 1, \cdots, N-1$, $j = 2, 3, \cdots, N$ are polynomials, and $d_i(D)$, $i = 1, \cdots, N$, are arbitrary monic Hurwitz polynomials of certain degrees r_i such that $\lim_{D \to \infty} G_0(D)\xi_m^r(D) = K_p^r$ is nonsingular and finite. The existence of $\xi_m^r(D)$ is guaranteed by Lemma 2.1 applied to $G_0^T(D)$.

If the MLI matrix is unknown but the MRI matrix of $G_0(D)$ is known, we can reparametrize the plant (2.1) so that the MLI matrix of the new nominal plant is diagonal and can be specified based on only the knowledge of the maximum degree of the MRI matrix $\xi_m^r(D)$ of $G_0(D)$.

The plant (2.1) may be represented as:

$$y(t) = G_r(D)[v](t), \; u(t) = g^{-1}(D)\xi_m^r(D)[v](t) \qquad (5.2)$$

where

$$G_r(D) = G_{0r}(D)[I + \mu\Delta_{mr}(D)] + \mu\Delta_{ar}(D) \qquad (5.3)$$

$$G_{0r}(D) = G_0(D)\xi_m^r(D)g^{-1}(D) \qquad (5.4)$$

$$\Delta_{mr}(D) = (\xi_m^r(D))^{-1}\Delta_m(D)\xi_m^r(D) \qquad (5.5)$$

$$\Delta_{ar}(D) = \Delta_a(D)g^{-1}(D)\xi_m^r(D) \qquad (5.6)$$

where $g(D)$ is any monic Hurwitz polynomial of degree d_r which is the maximum degree of $\xi_m^r(D)$. The MLI matrix $\bar{\xi}_m(D)$ of $G_{0r}(D)$ is $g(D)I$ and the associated high frequency gain matrix equals K_p^r. We note that any $N \times N$ polynomial matrix $\xi_0^r(D)$ such that $(\xi_0^r(D))^{-1}$ is stable and $\lim_{D \to \infty} G_0(D)\xi_0^r(D)$ is finite and non-singular can be used to take the role of $\xi_m^r(D)$ in this reparametrization for the design of multivariable adaptive controllers.

In this case we can design an adaptive controller to generate the signal $v(t)$ based on the remodeled plant (5.2), the controller structure (2.5) and the adaptive law (3.8) - (3.9). Then the control signal $u(t)$ is obtained through the compensator $u(t) = g^{-1}(D)\xi_m^r(D)[v](t)$. The assumptions on the remodeled nominal plant $G_{0r}(D)$, the unmodeled dynamics $\mu\Delta_{mr}(D)$, $\mu\Delta_{ar}(D)$ can be made similar to that in Section 2 for the plant (2.1). Our previous design and analysis hold for this case (Tao 1989). Moreover when defining the estimation error $\epsilon(t)$ we can choose $h(D) = g(D)$ so that $\bar{\xi}_m(D)h(D) = I$, which will result in an estimation error signal of the form:

$$\epsilon(t) = \frac{y(t) - y_m(t) + \Psi(t)\xi(t)}{1 + \beta(\zeta^T(t)\zeta(t) + \xi^T(t)\xi(t)) + \alpha m^2(t)} \qquad (5.7)$$

The use of (5.7) will result in that in the CT case the tracking error $e(t) = y(t) - y_m(t)$ satisfies that $\int_{t_1}^{t_2} e^T(t)e(t)dt \leq \beta_c\mu^2(t_2 - t_1) + b_c$, for some β_c, $b_c \in R^+$ in stead of (4.9a).

In the presence of plant unmodeled dynamics, the adaptive controllers developed and analyzed in previous sections can only guarantee that the tracking error $e(t) = y(t) - y_m(t)$ is small, i.e., of the order of the unmodeled dynamics which is characterized by μ in the mean (see (4.9), Remark 4.2 and the above discussions). Smallness in the mean does not imply smallness in the usual sense. The tracking error $e(t)$ may be large over short time intervals even in the steady state. However the external input $r(t)$ belongs the class of dominantly rich signals (Ioannou and Tao 1989) so that the vector signal $\zeta(t)$ is persistently exciting, then the tracking error $e(t)$ and the parameter error $\tilde{\Theta}(t)$ can be guaranteed to be small in the steady state. Therefore the performance of the adaptive control system is considerably improved, as shown in Ioannou and Tao (1989), Tao and Ioannou (1989).

6   Conclusions

In this paper we presented a unified procedure for designing MRAC schemes for MIMO continuous and discrete time plants. Issues such as parametrization of the plant, reference model and controller, design of adaptive laws for MIMO plants, stability, robustness, a priori plant knowledge and performance are addressed.

References

Dion, J. M., L. Dugard and J. Carrillo, Interactor and multivariable adaptive model matching, IEEE Trans. on Aut. Control, vol. 33, no. 4, pp. 399-401, 1988.

Elliott, H. and W. A. Wolovich, A parameter adaptive control structure for linear multivariable systems, IEEE Trans. on Aut. Control, vol. AC-27, no. 2, pp. 340-352, 1982.

Elliott, H. and W. A. Wolovich, Parametrization issues in multivariable adaptive control, Automatica, vol. 20, no. 5, pp. 533-545, 1984.

Goodwin, G. C. and K. S. Sin, Adaptive Filtering Prediction and Control, Prentice-Hall, Englewood Cliffs, N. J., 1984.

Ioannou, P. A. and G. Tao, Dominant richness and improvement of performance of robust adaptive control, Automatica, vol. 25, no. 2, pp. 287-291, 1989.

Ioannou, P. A. and K. Tsakalis, A robust direct adaptive controller, IEEE Trans. on Aut. Control, vol. AC-31, no. 11, pp. 1033-1043, 1986(a).

Ioannou, P. A. and K. Tsakalis, Robust discrete time adaptive control, in Adaptive and Learning Systems: Theory and Applications, Plenum Press, edited by K. S. Narendra, 1986(b).

Monopoli, R. V. and C. C. Hsing, Parameter adaptive control of multivariable systems, Int. J. Control, vol. 22, no. 3, pp. 313-327, 1975.

Morse, A. S., Parametrizations for multivariable adaptive control, Proc. of the 20th IEEE CDC, pp. 970-972, San Diego, CA, 1981.

Narendra, K. S. and A. M. Annaswamy, Stable Adaptive Systems, Prentice Hall, Englewood Cliffs, New Jersey, 1989.

Sastry, S. and M. Bodson, Adaptive Control Stability, Convergence, and Robustness, Prentice Hall, Englewood Cliffs, N. J., 1989.

Singh, R. P. and K. S. Narendra, Priori information in the design of multi-variable adaptive controllers, IEEE Trans. on Aut. Control, vol. AC-29, no. 12, pp. 1108-1111, 1984.

Tao, G., Model Reference Adaptive Control: Stability, Robustness and Performance Improvement, Ph. D. Dissertation, EE-Systems, University of Southern California, August 1989.

Tao, G., Model reference adaptive control of multivariable plants with unknown interactor matrix, Proc. of the 29th IEEE CDC, pp. 2730-2735, Honolulu, Hawaii, Dec. 1990.

Tao, G. and P. A. Ioannou, Robust model reference adaptive control for multivariable plants, Int. J. Adaptive Control and Signal Processing, vol. 2, no. 3, pp. 217-248, 1988.

Tao, G. and P. A. Ioannou, Robust stability and performance improvement of multivariable adaptive control systems, Int. J. of Control, vol. 50, no. 5, pp. 1835-1855, 1989.

Tsiligiannis, C. A. and S. A. Svoronos, Multivariable self-tuning control via right interactor matrix, IEEE Trans. on Aut. Control, AC-31. no. 4, pp. 987-989, 1986.

Wolovich, W. A and P. L. Falb, Invariants and canonical forms under dynamic compensation, SIAM J. Control and Opt., vol. 14, no. 6, pp. 996-1008, 1976.

Decentralized Adaptive Control *

A. Datta and P. Ioannou

Department of Electrical Engineering - Systems

University of Southern California

Los Angeles, CA 90089-0781

Abstract

A continuous time decentralized model reference adaptive controller is proposed for a large scale system composed of N interconnected linear subsystems with unknown parameters. Each local adaptive law utilizes a normalizing signal which is generated using the local input and the outputs of all the subsystems. Sufficient conditions for the closed loop stability of the adaptively controlled interconnected system are derived. These conditions provide bounds on the size of the allowable unmodelled interconnections and suggest ways to effect an improvement in the latter by appropriately choosing the free parameters of the adaptive controller.

*This work is supported in part by the National Science Foundation under Grant DMC-8452002 and in part by matching funds from the General Motors Foundation

1   Introduction

The control of large scale systems, which are composed of interconnected subsystems, is inherently accompanied by a poor knowledge of the plant parameters. As a result, the use of self-adjusting control systems, commonly referred to as adaptive schemes, is particularly appropriate in such a context. Even in the case where the plant parameters are accurately known, the design and implementation of a centralized controller for a large scale system may turn out to be a formidable task, both in terms of the degree of complexity as well as the associated expenditure. To reduce the design complexity, decentralized control schemes, whereby each subsystem is controlled independently on the basis of its own local performance criterion and locally available information, have been proposed in the literature. In order to handle parametric uncertainty and to simultaneously avoid the complexity of a centralized design, decentralized adaptive control schemes have been introduced [1],[2].

In [1], a decentralized model reference adaptive control scheme was proposed by Ioannou. A simple example was first used to demonstrate that without the use of "robustifying" modifications in the adaptive law, instability can result even in the presence of arbitrarily small interconnections. The adaptive laws were then modified for robustness using the so called "switching σ-modification" [3] and sufficient conditions were derived to guarantee the boundedness and exponential convergence of the state and parameter errors to bounded residual sets. The results of that paper are, however, valid only when the relative degree of the transfer function of each isolated subsystem is less than or equal to 2. The extension of these results to the case where the transfer functions of the decoupled subsystems have arbitrary, but known, relative degree is yet to be carried out.

Recently, Gavel et al. [2] proposed a decentralized model reference adaptive control scheme for large scale systems. Provided certain structural conditions are satisfied and assuming that the state of each subsystem is available for measurement, it was shown that the stability of the closed loop system could be guaranteed independent of the interconnection strengths. What essentially made such a result possible was the assumption about the unmodelled interconnections being in the range of the local control variables. Such an assumption is quite restrictive. Moreover, when only the local outputs and not the states are available for measurement, then the scheme in question is applicable only to the case where the transfer function of each decoupled subsystem has a relative degree of one.

In both [1] and [2] the designs are carried out without the exchange of any information between the subsystems. Such a restriction may be unnecessarily stringent, especially in cases where the subsystems involved are not separated by large distances. For example, instead of designing a multivariable adaptive controller for a predominantly diagonal multi-input multi-output system, considerable simplification may be achieved by treating the problem as one of decentralized adaptive control. In such a case, the exchange of signals between the local controllers can be easily carried out. Process control problems [4] typically fall into this category. Viewed against this background, the task of the designer becomes to retain the simplicity of a decentralized design by carrying out, if necessary, the partial exchange of information between the different subsystems. As shown in [1], no such exchange of subsystem information is required when the isolated subsystems have transfer functions of relative degree less than or equal to 2. As we will show in this paper, by allowing the exchange of output signals between the subsystems, the global stability result of [1] can be extended to the case where the isolated subsystems have transfer functions with arbitrary relative degrees. This, in itself, represents a significant development over the earlier results of [1] and [2].

In this paper, we present a direct decentralized adaptive control scheme for a linear time invariant large scale system composed of N interconnected subsystems. Each local subsystem is controlled by a local model reference adaptive controller. As already mentioned, the transfer function of each isolated subsystem is allowed to have arbitrary, but known, relative degree.

Due to the presence of unmodelled interconnections, instability may result from the use of adaptive controllers designed for systems without these unmodelled effects [1]. The adaptation algorithm used in this paper is structured after the switching σ-modification of Ioannou et al. [3] which was derived for a single input single output system and shown to be robust with respect to unmodelled dynamics. The robust design considered here requires the exchange of output signals between the different subsystems. As will be seen shortly, this exchange is necessary to ensure that the unmodelled interconnection terms enter the local adaptive laws as bounded disturbances.

The methodology used for carrying out the stability analysis for the decentralized adaptive control scheme of this paper differs from that in [3] and is structured along the lines of [5]. This new methodology approaches the whole problem from a quantitative point of view and, therefore, provides us with a quantitative feel for the size of the allowable unmodelled interconnections. Moreover, the dependence of the allowable interconnection strengths on the free parameters of the adaptive controller suggest ways to effect an improvement in the former by choosing the latter appropriately.

The paper is organized as follows. In Section 2, we introduce some notation and mathematical preliminaries. This is followed by the problem statement in Section 3. In Section 4, we deal with the problem of decentralized model reference control when the plant parameters are known. In Section 5, using the Certainty Equivalence Principle [6], we combine the control structure obtained in Section 4 with local robust adaptive laws to obtain a decentralized model reference adaptive control scheme. In Section 6, we carry out the stability analysis for the scheme proposed in Section 5. In Section 7, we summarize our conclusions.

2   Mathematical Preliminaries

The following notations will be extensively used in the subsequent sections:

$$\Sigma a \triangleq \{s \in \mathcal{C} | Re\ s \geq -a\}$$

$$[P_H](s) \triangleq \frac{1}{\sqrt{p}}(s+p)H(s)\ , p \in \mathcal{R}^+$$

$$[P_H](s-\delta) \triangleq \frac{1}{\sqrt{p}}(s+p-\delta)H(s-\delta)\ , p \in \mathcal{R}^+$$

$$H_1 H_2(s) \triangleq H_1(s)H_2(s)$$

In our derivations, we will make repeated use of the following identities:

$$(i)(x+y)^2 \leq (1+\epsilon)x^2 + (1+\frac{1}{\epsilon})y^2 \qquad (2.1)$$

where $\epsilon > 0$ is arbitrary.

$$(ii)\inf_{\epsilon>0} \{(1+\epsilon)x^2 + (1+\frac{1}{\epsilon})y^2\} = (|x|+|y|)^2 \qquad (2.2)$$

$$(iii)\inf_{\epsilon_1>0,\epsilon_2>0} (1+\epsilon_1)x^2 + (1+\frac{1}{\epsilon_1})(1+\epsilon_2)y^2 + (1+\frac{1}{\epsilon_1})(1+\frac{1}{\epsilon_2})z^2$$
$$= (|x|+|y|+|z|)^2 \quad (2.3)$$

The above identities allow us to avoid having to consider cross terms when squaring the sum of two or more quantities. They also permit the derivation of tighter bounds since the parameters $\epsilon_i > 0$ are arbitrary and can, therefore, be used as arguments of a minimization problem.

Our derivations will also make extensive use of the following results on exponentially weighted L^2 norms. The proofs are given in the Appendix.

Lemma 2.1: Let $y = H(s)[u]$ where $H(s)$ is a proper causal transfer matrix, analytic in $\Sigma \frac{\delta}{2}$ for some $\delta > 0$.

Then

$$\int_0^t e^{-\delta(t-\tau)}||y||^2 d\tau \leq ||H(s-\frac{\delta}{2})||_\infty^2 \int_0^t e^{-\delta(t-\tau)}||u||^2 d\tau$$

provided that u,y exist in $[0,t]$.[1]

Corollary 2.1: Assume that, in addition to the assumptions of Lemma 2.1, $u,y \in L_\infty^e$. Then, for any $t_0 \geq 0$ and $\forall t \geq t_0$

$$\int_{t_0}^t e^{-\delta(t-\tau)}||y||^2 d\tau \leq ||H(s-\frac{\delta}{2})||_\infty^2 \int_{t_0}^t e^{-\delta(t-\tau)}||u||^2 d\tau + c_{t_0}$$

---

[1]In this paper, the vector norm $||.||$ represents the Euclidean norm. The matrix norm is the corresponding induced norm.

where c_{t_0} is a finite constant depending on t_0 .

Lemma 2.2: Let $y = H(s)[u]$ where $H(s)$ is a strictly proper causal transfer matrix analytic in $\Sigma \frac{\delta}{2}$ for some $\delta > 0$. Then

$$||y||^2 \leq (1 + \lambda) \inf_{p \geq \delta} ||[P_H](s - \frac{\delta}{2})||_\infty^2 \int_0^t e^{-\delta(t-\tau)}||u||^2 \, d\tau + (1 + \frac{1}{\lambda})\epsilon_t^2$$

provided u, y exist in $[0, t]$. Here $\lambda > 0$ is arbitrary and ϵ_t^2 is an exponentially decaying to zero term that accounts for non zero initial conditions.

3 Problem Statement

Consider a system which is described as an interconnection of N subsystems

$$y_i = G_{Pi}(s)u_i + \sum_{j=1}^{N} \Delta_{ij}(s)y_j, i = 1, 2, 3, \ldots, N \qquad (3.1)$$

Here $u_i \in \mathcal{R}^1$ and $y_i \in \mathcal{R}^1$ are the input and output of the ith subsystem respectively, $G_{Pi}(s)$ is the transfer function of the ith isolated subsystem and $\Delta_{ij}(s)$ describes how the jth subsystem acts on the ith subsystem. It is assumed that $\Delta_{ij}(s), j = 1, 2, \ldots, N$ are strictly proper and analytic in $\Sigma \frac{\delta_{\Delta i}}{2}$ where $\delta_{\Delta i} > 0$ is known. The coefficients of $G_{Pi}(s)$, on the other hand, are assumed to be unknown.

The problem of decentralized MRAC can now be formulated as follows: design local MRAC for each of the isolated subsystems

$$y_i = G_{Pi}(s)u_i \qquad i = 1, 2, 3, \ldots, N \qquad (3.2)$$

so that when these controllers are applied to the interconnected system described by Eq. (3.1), global stability is still preserved.

As already mentioned in the introduction, the presence of the interconnection terms $\Delta_{ij}(s)$ may destabilize an adaptive scheme which has not been modified for robustness. To guarantee global stability in the presence of unmodelled interconnections, robust adaptive controllers have to be designed for each of the isolated subsystems. By using N such controllers, it is possible to guarantee global stability provided the interconnection strengths are "weak enough." In this respect, most of the available results, e.g., [1], are of a qualitative nature in the sense that they assert the existence of the bounds on the allowable interconnection strengths without providing effective tools for calculating the latter. In this paper, for the first time, we adopt a new approach in an attempt to address the problem of obtaining a quantitative measure of the allowable interconnection strengths. Our analysis yields H^∞ norm bounds on the latter.

4   Decentralized Model Reference Control: Known Parameters

In this section, we consider the design of a decentralized model reference controller for Eq. (3.1) in the case when the parameters of $G_{Pi}(s)$ happen to be known. As mentioned earlier, we will be designing local model reference controllers for each of the N isolated subsystems. Accordingly, we consider the ith isolated subsystem described by

$$y_i = G_{Pi}(s)u_i \tag{4.1}$$

The model reference control problem for Eq. (4.1) can now be stated as follows: design a control $u_i(t)$ so that the local output y_i of the i-th closed loop isolated subsystem tracks the output y_{mi} of the local reference model

$$y_{mi} = W_{mi}(s)r_i = \frac{k_{mi}}{D_{mi}(s)}\, r_i \qquad (4.2)$$

where $r_i(t)$ is any bounded piecewise continuous reference input signal,

$$k_{mi} > 0 \text{ is a known constant, and}$$

$$D_{mi}(s) \text{ is a known monic Hurwitz polynomial.}$$

To achieve such an objective, we make the following assumptions about

$$G_{Pi}(s) = k_{Pi}\, \frac{Z_i(s)}{R_i(s)} \;:$$

(A1) $R_i(s)$ is a monic polynomial of degree n_i .

(A2) $Z_i(s)$ is a monic Hurwitz polynomial of degree $m_i \le n_i - 1$.

(A3) k_{Pi} is a constant gain with known sign. Without any loss of generality, we assume $k_{Pi} > 0$.

(A4) The relative degree $n_i^* = n_i - m_i$ is known.

Choosing degree $[D_{mi}(s)] = n_i^*$, the following controller structure may be used

$$u_i = \theta_{1i}^T\, \frac{a_i(s)}{\Gamma_i(s)}\, u_i + \theta_{2i}^T\, \frac{a_i(s)}{\Gamma_i(s)}\, y_i + c_{oi}r_i \qquad (4.3)$$

where $\Gamma_i(s)$ is an arbitrary monic Hurwitz polynomial of degree n_i

$$a_i(s) \triangleq [s^{n_i-1}, s^{n_i-2}, \ldots, s, 1]^T$$

$\theta_{1i}, \theta_{2i} \in \mathcal{R}^{n_i}$ are controller parameters to be chosen

and c_{oi} is a feedforward scalar. For notational convenience, we define

$$\theta_i = [\theta_{1i}^T, \theta_{2i}^T, c_{oi}]^T$$

We now show that the above control structure meets the model reference control objective for Eq. (4.1) provided the controller parameters θ_{1i}, θ_{2i} and c_{oi} are properly chosen:

Lemma 4.1: There exists $\theta_i^* \in \mathcal{R}^{2n_i+1}$ such that for $\theta_i = \theta_i^*$, the local control law of Eq. (4.3) meets the model reference control objective for the local isolated subsystem described by Eq. (4.1).

Proof: From Eq. (4.3) we obtain

$$u_i = \frac{g_{1i}(s, \theta_{1i})}{\Gamma_i(s)} u_i + \frac{g_{2i}(s, \theta_{2i})}{\Gamma_i(s)} y_i + c_{oi} r_i$$

where

$$g_{1i}(s, \theta_{1i}) \triangleq \theta_{1i}^T a_i(s)$$

$$g_{2i}(s, \theta_{2i}) \triangleq \theta_{2i}^T a_i(s)$$

Thus

$$u_i = \frac{g_{2i}(s, \theta_{2i})}{[\Gamma_i(s) - g_{1i}(s, \theta_{1i})]} y_i + \frac{c_{oi}\Gamma_i(s)}{[\Gamma_i(s) - g_{1i}(s, \theta_{1i})]} r_i \qquad (4.4)$$

Substituting for u_i in Eq. (4.1) we obtain

$$R_i(s)y_i = k_{Pi} Z_i(s) \frac{g_{2i}(s, \theta_{2i})}{[\Gamma_i(s) - g_{1i}(s, \theta_{1i})]} y_i$$

$$+ k_{Pi} \frac{Z_i(s)c_{oi}\Gamma_i(s)}{[\Gamma_i(s) - g_{1i}(s, \theta_{1i})]} r_i$$

so that

$$\frac{y_i}{r_i} = \frac{k_{Pi} c_{oi} Z_i(s)\Gamma_i(s)}{R_i(s)[\Gamma_i(s) - g_{1i}(s, \theta_{1i})] - k_{Pi} Z_i(s)g_{2i}(s, \theta_{2i})} .$$

Now, in order to achieve model following for <u>all</u> piecewise continuous uniformly bounded signals $r_i(t)$, the above transfer function must be set equal to the reference model transfer function.

We thus have

$$\frac{k_{Pi}c_{oi}^*Z_i(s)\Gamma_i(s)}{R_i(s)[\Gamma_i(s) - g_{1i}(s,\theta_{1i}^*)] - k_{Pi}Z_i(s)g_{2i}(s,\theta_{2i}^*)} = W_{mi}(s) = \frac{k_{mi}}{D_{mi}(s)}$$

$$(4.5)$$

or

$$k_{Pi}c_{oi}^*Z_i(s)\Gamma_i(s)D_{mi}(s) = k_{mi}\{R_i(s)[\Gamma_i(s)-g_{1i}(s,\theta_{1i}^*)]-k_{Pi}Z_i(s)g_{2i}(s,\theta_{2i}^*)\}$$

Choosing $c_{oi}^* = \frac{k_{mi}}{k_{Pi}}$ we obtain

$$R_i(s)[\Gamma_i(s) - g_{1i}(s,\theta_{1i}^*)] - k_{Pi}Z_i(s)g_{2i}(s,\theta_{2i}^*) = Z_i(s)\Gamma_i(s)D_{mi}(s)$$

or

$$R_i(s)g_{1i}(s,\theta_{1i}^*) + k_{Pi}Z_i(s)g_{2i}(s,\theta_{2i}^*) = R_i(s)\Gamma_i(s) - Z_i(s)\Gamma_i(s)D_{mi}(s)$$

$$(4.6)$$

We note that degree $[R_i(s)] = n_i$, degree $[Z_i(s)] = m_i \leq n_i - 1$, degree $[R_i(s)\Gamma_i(s) - Z_i(s)\Gamma_i(s)D_{mi}(s)] = 2n_i - 1$, degree $[g_{1i}(s,\theta_{1i}^*)] = n_i - 1$, degree $[g_{2i}(s,\theta_{2i}^*)] = n_i - 1$. Now, if we assume that $R_i(s)$ and $Z_i(s)$ are coprime[2] then Eq. (4.6) can be solved for unique polynomials $g_{1i}(s,\theta_{1i}^*)$ and $g_{2i}(s,\theta_{2i}^*)$ each of degree $(n_i - 1)$ [7],[8].

Thus by setting $\theta_{1i} = \theta_{1i}^*, \theta_{2i} = \theta_{2i}^*$ and $c_{oi} = c_{oi}^*$ we are able to exactly meet the model reference control objective for Eq. (4.1). The corresponding control input is given by

$$u_i = \theta_{1i}^{*T}\frac{a_i(s)}{\Gamma_i(s)}u_i + \theta_{2i}^{*T}\frac{a_i(s)}{\Gamma_i(s)}y_i + c_{oi}^*r_i \qquad (4.7)$$

At this stage, it is also convenient to introduce some additional notation.

---

[2]If $R_i(s)$ and $Z_i(s)$ are not coprime, then any common factors are stable due to (A2) and cancel out in Eq. (4.6) so that the same conclusion continues to hold.

Define

$$w_{1i} = \frac{a_i(s)}{\Gamma_i(s)}\, u_i, w_{2i} = \frac{a_i(s)}{\Gamma_i(s)}\, y_i$$

$$\hat{G}_i(s) = \begin{bmatrix} \frac{a_i(s)}{\Gamma_i(s)} & 0 \\ 0 & \frac{a_i(s)}{\Gamma_i(s)} \end{bmatrix}$$

$$w_i = [w_{1i}^T, w_{2i}^T, r_i]^T$$

Thus

$$w_i = \left\{ \begin{array}{c} \hat{G}_i(s) \begin{bmatrix} u_i \\ y_i \end{bmatrix} \\ r_i \end{array} \right\}$$

where $\hat{G}_i(s)$ is a strictly proper transfer matrix.

If the coefficients of $R_i(s), Z_i(s)$ and the high frequency gain k_{Pi} are unknown, then the control law of Eq. (4.7) cannot be implemented since $\theta_{1i}^*, \theta_{2i}^*$ and c_{oi}^* are all unknown. In that case, instead of Eq. (4.7), Eq. (4.3) is used together with an appropriate adaptive law to update the time varying parameters θ_{1i}, θ_{2i} and c_{oi} . Such an adaptive law will be presented in the next section.

However, since the structure of the Model Reference controller is fixed by Eq. (4.3) irrespective of whether the parameters of $G_{Pi}(s)$ are known or not, it is important from the point of view of robustness to analyze the stability properties of Eq. (3.1) when subjected to the N local controls given by Eq. (4.7). For example, if the N local controls given by Eq. (4.7) are not robust with respect to the unmodelled interconnections, then there is little or no hope that Eq. (4.3) combined with a robust adaptive law would be able to perform any better. The robustness properties of the control law of Eq. (4.7) are given by the following theorem.

Theorem 4.1: Define $\Delta(s) = (\Delta_{ij}(s))_{i,j=1,2,\dots,N}$. Then $\exists\ M > 0$ such that for all $\Delta(s)$ satisfying $\|\Delta(s)\|_\infty < M$, all the signals of the closed loop plant described by Eqs. (3.1), (4.7) are bounded for any bounded initial conditions. Furthermore, the tracking error vector

$$e \stackrel{\Delta}{=} [y_1 - y_{m1}, y_2 - y_{m2}, \dots y_N - y_{mN}]^T$$

converges exponentially to the residual set

$D_e = \{e|\|e\| \le K(\Delta)r_o^2\}$ where $K(\Delta) > 0$ is a constant that depends on $\Delta(s)$ and r_o is a constant that depends on the upper bound for $\|[r_1, r_2, \dots r_N]^T\|$.

Proof: Now from Eqs. (3.1), (4.7) we obtain

$$y_i = k_{Pi} \frac{Z_i(s)}{R_i(s)} u_i + \sum_{j=1}^{N} \Delta_{ij}(s)\, y_j$$

$$= k_{Pi} \frac{Z_i(s)}{R_i(s)} \left\{ \frac{g_{2i}(s, \theta_{2i}^*)}{[\Gamma_i(s) - g_{1i}(s, \theta_{1i}^*)]} y_i + \frac{c_{oi}^* \Gamma_i(s) r_i}{[\Gamma_i(s) - g_{1i}(s, \theta_{1i}^*)]} \right\}$$

$$+ \sum_{j=1}^{N} \Delta_{ij}(s) y_j$$

or

$$R_i(s)[\Gamma_i(s) - g_{1i}(s, \theta_{1i}^*)]y_i = k_{Pi} Z_i(s) g_{2i}(s, \theta_{2i}^*) y_i + k_{Pi} c_{oi}^* Z_i(s)\Gamma_i(s) r_i$$

$$+ R_i(s)[\Gamma_i(s) - g_{1i}(s, \theta_{1i}^*)] \sum_{j=1}^{N} \Delta_{ij}(s) y_j$$

or

$$y_i = \frac{k_{Pi} c_{oi}^* Z_i(s)\Gamma_i(s)}{R_i(s)[\Gamma_i(s) - g_{1i}(s, \theta_{1i}^*)] - k_{Pi} Z_i(s) g_{2i}(s, \theta_{2i}^*)}\, r_i$$

$$+ \frac{R_i(s)[\Gamma_i(s) - g_{1i}(s, \theta_{1i}^*)]}{R_i(s)[\Gamma_i(s) - g_{1i}(s, \theta_{1i}^*)] - k_{Pi} Z_i(s) g_{2i}(s, \theta_{2i}^*)} \sum_{j=1}^{N} \Delta_{ij}(s) y_j$$

or

$$y_i = W_{mi}(s) r_i + [1 + \frac{W_{mi}(s) g_{2i}(s, \theta_{2i}^*)}{c_{oi}^* \Gamma_i(s)}] \sum_{j=1}^{N} \Delta_{ij}(s) y_j$$

where we made use of Eq. (4.5).

Now defining $y \triangleq [y_1, y_2, \ldots, y_N]^T$

$$W_m(s) \triangleq diag[W_{m1}(s), W_{m2}(s), \ldots, W_{mN}(s)]$$

$$r \triangleq [r_1, r_2, \ldots, r_N]^T$$

$$W(s) \triangleq diag[\frac{1 + W_{m1}(s)g_{21}(s, \theta_{21}^*)}{c_{o1}^* \Gamma_1(s)}, \frac{1 + W_{m2}(s)g_{22}(s, \theta_{22}^*)}{c_{o2}^* \Gamma_2(s)},$$
$$\ldots \frac{1 + W_{mN}(s)g_{2N}(s, \theta_{2N}^*)}{c_{oN}^* \Gamma_N(s)}]$$

and

$$\Delta(s) = (\Delta_{ij}(s))_{\substack{i=1,N \\ j=1,N}}$$

we obtain the multivariable system

$$y = W_m(s)r + W(s)\Delta(s)y \qquad (4.8)$$

It now follows from the multivariable small gain theorem [9] that all the closed loop signals are bounded provided

$$\|\Delta(s)\|_\infty < M$$

where

$$M \triangleq \frac{1}{\|W(s)\|_\infty}$$

Clearly, $M > 0$ since $W_{mi}(s), i = 1, 2, \ldots, N$ are stable and $\Gamma_i(s)$, $i = 1, 2, \ldots, N$ are all Hurwitz.

Further, from Eq. (4.8), with e as defined in the statement of the theorem, we obtain

$$e = [I - W(s)\Delta(s)]^{-1}W(s)\Delta(s)W_m(s)r.$$

Defining $E(s) \triangleq [I - W(s)\Delta(s)]^{-1}W(s)\Delta(s)W_m(s)$ and assuming that $E(s)$ is analytic in $\Sigma\frac{\delta^*}{2}$ for some $\delta^* > 0$ (follows from the first part of the proof), using Lemma 2.2, we obtain

$$\|e\|^2 \leq K_1 \inf_{p \geq \delta^*} \|[P_E](s - \frac{\delta^*}{2})\|_\infty^2 r_o^2 + \epsilon_t^2$$

where $K_1 > 0$ is a constant and ϵ_t^2 is an exponentially decaying to zero term.

Defining $K(\Delta) \triangleq K_1 \inf_{p \geq \delta^*} \|[P_E](s - \frac{\delta^*}{2})\|_\infty^2$, the expression for the residual set D_e follows and therefore the proof is complete.

5 Decentralized Model Reference Adaptive Control

In the last section, we proposed and analyzed a decentralized model reference control scheme. It was shown that when the coefficients of $G_{Pi}(s)$ are known, one can calculate $\theta_{1i}^*, \theta_{2i}^*$ and c_{oi}^* and use them in Eq. (4.7) to generate the control $u_i(t)$. It was also shown that the resulting decentralized scheme is robust with respect to the unmodelled interconnections.

In this section, we assume that the parameters of $G_{Pi}(s)$ are unknown. Hence Eq. (4.7) cannot be implemented. Instead, we will use a robust adaptive law to estimate $\theta_{1i}^*, \theta_{2i}^*$ and c_{oi}^* and use these estimates in Eq. (4.3) to calculate the control $u_i(t)$. This is in accordance with the Certainty Equivalence Principle [6].

The theory and design of robust adaptive laws has been extensively discussed in [6]. Our derivation of a robust adaptive law for estimating θ_i^* will be essentially based on the procedure outlined in that reference.

We will proceed in two steps. We first show that $\theta_{1i}^*, \theta_{2i}^*, c_{oi}^*$ satisfy a certain signal equation. Then we define an estimation error and obtain a robust adaptive law by examining the properties of the former. We now start with the first step.

Lemma 5.1: The desired parameter vector $\theta_i^* \triangleq [\theta_{1i}^{*T}, \theta_{2i}^{*T}, c_{oi}^*]^T$ satisfies the signal equation

$$e_i + \frac{1}{c_{oi}^*}[\theta_i^{*T}\zeta_i - W_{mi}(s)u_i] = \frac{1}{c_{oi}^*} W_{\Delta i}(s)y \qquad (5.1)$$

where

$$
\begin{aligned}
e_i &\triangleq y_i - y_{mi} \\
\zeta_{1i} &\triangleq \frac{a_i(s)}{\Gamma_i(s)} W_{mi}(s)u_i = W_{mi}(s)w_{1i} \\
\zeta_{2i} &\triangleq \frac{a_i(s)}{\Gamma_i(s)} W_{mi}(s)y_i = W_{mi}(s)w_{2i} \\
\zeta_i &\triangleq [\zeta_{1i}^T, \zeta_{2i}^T, y_{mi}]^T \\
\Delta_i(s) &\triangleq [\Delta_{i1}(s), \Delta_{i2}(s), \ldots, \Delta_{iN}(s)] \\
W_{\Delta i}(s) &\triangleq \frac{W_{mi}(s)[\Gamma_i(s) - g_{1i}(s, \theta_{1i}^*)]}{\Gamma_i(s)k_{Pi}Z_i(s)} R_i(s)\Delta_i(s) \\
y &\triangleq [y_1, y_2, \ldots, y_N]^T
\end{aligned}
$$

Proof: From Eq. (4.5), we obtain

$$\frac{k_{Pi}c_{oi}^*Z_i(s)\Gamma_i(s)D_{mi}(s)}{k_{mi}} = R_i(s)[\Gamma_i(s) - g_{1i}(s, \theta_{1i}^*)] - k_{Pi}Z_i(s)g_{2i}(s, \theta_{2i}^*)$$

Now operating both sides of the above identity on y_i, we obtain

$$
\begin{aligned}
\frac{k_{Pi}c_{oi}^*Z_i(s)\Gamma_i(s)D_{mi}(s)}{k_{mi}} y_i &= [\Gamma_i(s) - g_{1i}(s, \theta_{1i}^*)]R_i(s)y_i \\
&\quad - k_{Pi}Z_i(s)g_{2i}(s, \theta_{2i}^*)y_i
\end{aligned}
$$

From Eq. (3.1) we have

$$R_i(s)y_i = k_{Pi}Z_i(s)u_i + R_i(s)\sum_{j=1}^{N}\Delta_{ij}(s)y_j$$

Substituting this into the previous equation, we obtain

$$\frac{k_{Pi}c_{oi}^*Z_i(s)\Gamma_i(s)D_{mi}(s)}{k_{mi}}\,y_i \;=\; [\Gamma_i(s) - g_{1i}(s,\theta_{1i}^*)][k_{Pi}Z_i(s)u_i$$

$$+R_i(s)\sum_{j=1}^{N}\Delta_{ij}(s)y_j]$$

$$-k_{Pi}Z_i(s)g_{2i}(s,\theta_{2i}^*)y_i$$

Operating on both sides using the stable filter $\frac{W_{mi}(s)}{k_{Pi}Z_i(s)\Gamma_i(s)}$, we obtain

$$c_{oi}^*y_i \;=\; [W_{mi}(s) - \frac{g_{1i}(s,\theta_{1i}^*)W_{mi}(s)}{\Gamma_i(s)}][u_i + \frac{R_i(s)}{k_{Pi}Z_i(s)}\sum_{j=1}^{N}\Delta_{ij}(s)y_j]$$

$$-\frac{g_{2i}(s,\theta_{2i}^*)}{\Gamma_i(s)}\,W_{mi}(s)y_i$$

or

$$c_{oi}^*y_i - W_{mi}(s)u_i + \theta_{1i}^{*T}\frac{a_i(s)}{\Gamma_i(s)}\,W_{mi}(s)u_i + \theta_{2i}^{*T}\frac{a_i(s)}{\Gamma_i(s)}\,W_{mi}(s)y_i$$

$$=\;\frac{W_{mi}(s)[\Gamma_i(s) - g_{1i}(s,\theta_{1i}^*)]R_i(s)}{\Gamma_i(s)k_{Pi}Z_i(s)}\sum_{j=1}^{N}\Delta_{ij}(s)y_j$$

or

$$c_{oi}^*e_i - W_{mi}(s)u_i + \theta_{1i}^{*T}\frac{a_i(s)}{\Gamma_i(s)}\,W_{mi}(s)u_i + \theta_{2i}^{*T}\frac{a_i(s)}{\Gamma_i(s)}\,W_{mi}(s)y_i + c_{oi}^*W_{mi}(s)r_i$$

$$=\;\frac{W_{mi}(s)[\Gamma_i(s) - g_{1i}(s,\theta_{1i}^*)]}{\Gamma_i(s)k_{Pi}}\frac{R_i(s)}{Z_i(s)}\sum_{j=1}^{N}\Delta_{ij}(s)[y_j]$$

or

$$e_i + \frac{1}{c_{oi}^*}[\theta_i^{*T}\zeta_i - W_{mi}(s)u_i] = \frac{1}{c_{oi}^*}W_{\Delta_i}(s)[y]$$

where $e_i,\theta_i^*,\zeta_i,W_{\Delta_i}(s),y$ are as already defined in the statement of the Lemma and therefore the proof is complete.

Now define $\rho_{oi} = \frac{1}{c_{oi}^*}$. Let $\theta_i(t),\rho_i(t)$ be the estimates at time t of θ_i^* and ρ_{oi} respectively. Then we define the estimation error ϵ_{1i} as

$$\epsilon_{1i} = \rho_i[\theta_i^T\zeta_i - W_{mi}(s)u_i] + e_i \qquad (5.2)$$

From Eq. (5.2) and Lemma 5.1, it follows that

$$\epsilon_{1i} = \rho_{oi}\phi_i^T\zeta_i + \psi_i\xi_i + \rho_{oi}W_{\Delta_i}(s)[y] \tag{5.3}$$

where

$$\phi_i \overset{\Delta}{=} \theta_i - \theta_i^* , \quad \psi_i \overset{\Delta}{=} \rho_i - \rho_{oi}$$

and

$$\xi_i \overset{\Delta}{=} \theta_i^T\zeta_i - W_{mi}(s)u_i$$

If we now define

$$\begin{aligned}
\bar{\theta}_i &= [\sqrt{\rho_{oi}}\,\theta_i^T, \rho_i]^T \\
\bar{\phi}_i &= \bar{\theta}_i - \bar{\theta}_i^* = [\sqrt{\rho_{oi}}\,\phi_i^T, \psi_i]^T \\
\bar{\zeta}_i &= [\sqrt{\rho_{oi}}\,\zeta_i^T, \xi_i]^T
\end{aligned}$$

then Eq. (5.3) becomes

$$\epsilon_{1i} = \bar{\phi}_i^T\bar{\zeta}_i + \rho_{oi}W_{\Delta_i}(s)y \tag{5.4}$$

Now consider the cost function $J_i(\bar{\phi}_i) = \frac{\epsilon_{1i}^2}{2m_i} + \sigma_i\bar{\theta}_i^T\bar{\phi}_i$ where m_i is a normalizing signal [6] to be designed in a way so that $\frac{\|\bar{\zeta}_i\|^2}{m_i}$ and $\frac{\|W_{\Delta_i}(s)[y]\|^2}{m_i}$ are bounded. This will guarantee that the cost function J_i cannot be driven to infinity by $\bar{\zeta}_i$ or $W_{\Delta_i}(s)[y]$.

Now from the cost function $J_i(\bar{\phi}_i)$, using the steepest descent method, we obtain

$$\dot{\bar{\phi}}_i = -\frac{\gamma_i\epsilon_{1i}\bar{\zeta}_i}{m_i} - \gamma_i\sigma_i\bar{\theta}_i , \quad \gamma_i > 0$$

or

$$\dot{\bar{\theta}}_i = -\frac{\gamma_i\epsilon_{1i}\bar{\zeta}_i}{m_i} - \gamma_i\sigma_i\bar{\theta}_i , \quad \gamma_i > 0 \tag{5.5}$$

as the adaptive law for estimating $\bar{\theta}_i$.

In Eq. (5.5) above, a suitable choice for the normalizing signal m_i is

$$\dot{m}_i = -\delta_{oi}m_i + u_i^2 + \frac{1}{N} \sum_{j=1}^{N} y_j^2 + q_{m3i}^2 , \qquad (5.6)$$

$$q_{m3i} > 0 \ , \ m_i(0) \geq \frac{q_{m3i}^2}{\delta_{oi}}$$

where $\delta_{oi} \leq \delta_{oi}^*$ and δ_{oi}^* is chosen such that $W_{mi}(s), \hat{G}_i(s), \Delta_{ij}(s), j = 1, 2, ., N$ are all analytic in $\Sigma \frac{\delta_{oi}^*}{2}$. Such a choice is possible since δ_{Δ_i} is assumed to be known (see Section 3) and $W_{mi}(s), \hat{G}_i(s)$ are design transfer matrices.

The term $-\gamma_i \sigma_i \bar{\theta}_i$ in Eq. (5.5) arises from the use of the so-called switching $-\sigma$ modification discussed and analyzed by Ioannou et al. [3]. By choosing

$$\sigma_i = \begin{cases} 0 & \text{if } ||\bar{\theta}_i|| < \bar{M}_{oi} \\ \sigma_{oi} & \text{if } ||\bar{\theta}_i|| \geq \bar{M}_{oi} \end{cases} \qquad (5.7)$$

where $\sigma_{oi} > 0$ is a design parameter and $\bar{M}_{oi} \geq 2||\bar{\theta}_i^*||$, it can be shown following [3,6] that this modification guarantees zero residual tracking errors when the unmodelled interconnections are removed.

Since ρ_{oi} is unknown, the signal $\bar{\zeta}_i$ is not available from measurements and so, Eq. (5.5), in its present form, cannot be implemented. However, we note that Eq. (5.5) is equivalent to

$$\left. \begin{array}{l} \dot{\theta}_i = -\frac{\gamma_i \epsilon_{1i} \zeta_i}{m_i} - \gamma_i \sigma_i \theta_i \\ \dot{\rho}_i = -\frac{\gamma_i \epsilon_{1i} \xi_i}{m_i} - \gamma_i \sigma_i \rho_i \end{array} \right\} \qquad (5.8)$$

and since the above equation does not involve ρ_{oi} , it can be easily implemented. However, Eq. (5.8) does not yield an estimate for $\bar{\theta}_i$ and since ρ_{oi} is unknown, the switching in Eq. (5.7) cannot be carried out using the estimates obtained from Eq. (5.8). Instead, it can be shown that if

$$M_{oi} \geq 2 \max \left(\sqrt{c_{oi}^*} , \frac{1}{\sqrt{c_{oi}^*}} \right) \left\| \begin{matrix} \theta_i^* \\ \rho_{oi} \end{matrix} \right\| \text{ and if}$$

$$\sigma_i = \begin{cases} 0 & \text{if } \left\| \begin{bmatrix} \theta_i \\ \rho_i \end{bmatrix} \right\| < M_{oi} \\[3ex] \sigma_{oi} & \text{if } \left\| \begin{bmatrix} \theta_i \\ \rho_i \end{bmatrix} \right\| \geq M_{oi} \end{cases} \tag{5.9}$$

where θ_i, ρ_i are as obtained from Eq. (5.8), then σ_i in Eq. (5.9) is greater than zero only when $\|\bar{\theta}_i\| \geq 2\|\bar{\theta}_i^*\|$ so that the essential requirement for guaranteeing zero residual tracking errors in the absence of unmodelled interconnections is still met. Thus Eqs. (5.6), (5.8), (5.9) together give us the robust adaptive law for estimating $\theta_i(t), \rho_i(t)$.

The estimate of $\theta_i(t)$ obtained from Eq. (5.8) is now used in Eq. (4.3) to generate the control $u_i(t)$. We have thus obtained a model reference adaptive controller for each isolated subsystem. By applying N such controllers to the system described by Eq. (3.1), we obtain a decentralized MRAC scheme. The stability properties of this scheme will be analyzed in the next section.

Remark 5.1. As already mentioned, the choice of the normalizing signal $m_i(t)$ must be such that $\frac{\|\zeta_i\|^2}{m_i}$ and $\frac{\|W_{\Delta_i}(s)[y]\|^2}{m_i}$ are bounded. That the $m_i(t)$ given by Eq. (5.6) satisfies these properties will become clear in the course of the stability analysis in the next section.

Remark 5.2. The adaptive law given by Eq. (5.8) is a slight variation of the one used in [5]. The only difference is that here the local normalizing signal m_i represents an exponentially weighted L^2 norm of not only the local input signal but also that of the root mean square value of the subsystem outputs. This partial exchange of information between the subsystems guarantees that $\frac{\|W_{\Delta_i}(s)[y]\|^2}{m_i}$ is bounded so that the unmodelled interconnections effectively enter the local adaptive law as a bounded disturbance.

6 Stability Analysis

In this section, we analyze the stability properties of the decentralized MRAC scheme proposed in the last section. In our analysis, we will be using the decomposition aggregation method of analysis for large scale systems [10]. To do so, the properties of each local subsystem are first analyzed one at a time and then aggregated together. Accordingly, we first begin with the analysis of the individual subsystem properties.

6.1 Properties of the ith Subsystem

The analysis of the properties of the ith subsystem can again be broken up into two parts:

1. Properties resulting only from the adaptive law of Eqs. (5.6), (5.8), (5.9) and which are, therefore, independent of the control structure; and

2. Properties resulting from the choice of the specific control structure.

We now consider them one by one.

6.1.1 Properties resulting from the adaptive law

The properties resulting only from the adaptive law are conveniently stated in terms of the following two lemmas.

Lemma 6.1. The parameter error $\bar{\phi}_i(t) \triangleq \bar{\theta}_i - \bar{\theta}_i^* = [\sqrt{\rho_{oi}} \ \phi_i^T \ \psi_i]^T$ is uniformly bounded and $\exists \ t_{oi} > 0$ such that

$$\|\bar{\phi}_i(t)\|^2 \leq c_{\bar{\phi}_i} + \epsilon_t \quad \forall \ t \geq t_{oi}$$

where

$$c_{\bar{\phi}_i} \triangleq \max\{\max\left(\frac{1}{c_{oi}^*},1\right)4M_{oi}^2 , \frac{(1+\lambda_1)c_{\Delta_i}M_{oi}}{2(c_{oi}^*)^2\sigma_{oi}(2M_{oi}-\|\Theta_i^*\|)}\}$$

$$\Theta_i^* \triangleq [\theta_i^{*T},\rho_{oi}]^T$$

$$c_{\Delta_i} \triangleq N \inf_{p \geq \delta_{oi}} (\||[P_{W_{\Delta_i}}](s - \frac{\delta_{oi}}{2})\|_\infty^2)$$

$$W_{\Delta_i}(s) = W_{m_i}(s)[\frac{\Gamma_i(s) - g_{1i}(s,\theta_{1i}^*)}{\Gamma_i(s)k_{P_i}Z_i(s)}]R_i(s)\Delta_i(s)$$

$$\Delta_i(s) = [\Delta_{i1}(s),\Delta_{i2}(s),\Delta_{i3}(s),..,\Delta_{iN}(s)]$$

$$\lambda_1 > 0 \text{ is arbitrary}$$

and ϵ_t is an exponentially decaying term which can be made arbitrarily small by choosing t_{oi} to be large enough.

Proof: From Eq. (5.4), we have

$$\epsilon_{1i} = \bar{\phi}_i^T \bar{\zeta}_i + \rho_{oi}W_{\Delta_i}(s)[y]$$

Now consider the positive definite function

$$V_i = \frac{\bar{\phi}_i^T \bar{\phi}_i}{2\gamma_i}$$

Then

$$\dot{V}_{i(5.8)} = -\frac{(\bar{\phi}_i^T \bar{\zeta}_i)^2}{m_i} - \rho_{oi}\frac{\bar{\phi}_i^T \bar{\zeta}_i}{m_i} W_{\Delta_i}(s)[y] - \sigma_i\bar{\phi}_i^T \bar{\theta}_i \qquad (6.1)$$

By a simple completion of squares, we obtain

$$\dot{V}_i \leq \frac{1}{4(c_{oi}^*)^2} \frac{\|W_{\Delta_i}(s)[y]\|^2}{m_i} - \sigma_i\bar{\phi}_i^T \bar{\theta}_i \qquad (6.2)$$

Since $W_{\Delta_i}(s)$ is strictly proper, it follows from Lemma 2.2 that

$$||W_{\Delta_i}(s)[y]||^2 \leq (1+\lambda_1)c_{\Delta_i}m_i + (1+\frac{1}{\lambda_1})\epsilon_t^2$$

where

$$c_{\Delta_i} \triangleq N \inf_{p \geq \delta_{oi}} (||[P_{W_{\Delta_i}}](s - \frac{\delta_{oi}}{2})||_\infty^2)$$

and $\lambda_1 > 0$ is arbitrary.

Furthermore, for $V_i \geq \max\,[\,\frac{1}{c_{oi}^*},1\,]\,\frac{2M_{oi}^2}{\gamma_i}$ and with Θ_i defined by

$$\Theta_i \triangleq [\theta_i^T, \rho_i]^T$$

we have

$$||\Theta_i|| \geq M_{oi} \text{ so that } \sigma_i = \sigma_{oi}.$$

Thus for $V_i \geq \max[\,\frac{1}{c_{oi}^*},\,1\,]\,\frac{2M_{oi}^2}{\gamma_i}$, we have

$$
\begin{aligned}
\sigma_i \bar{\phi}_i^T \bar{\theta}_i &= \sigma_i \bar{\phi}_i^T (\bar{\phi}_i + \bar{\theta}_i^*) \\
&\geq \sigma_i ||\bar{\phi}_i||^2 - \sigma_i ||\bar{\phi}_i||\,||\bar{\theta}_i^*|| \\
&= \sigma_{oi} ||\bar{\phi}_i||^2 - \sigma_{oi}\,||\bar{\phi}_i||^2\,\frac{||\bar{\theta}_i^*||}{||\bar{\phi}_i||} \\
&= \sigma_{oi} ||\bar{\phi}_i||^2\,[1 - \frac{||\bar{\theta}_i^*||}{||\bar{\phi}_i||}]
\end{aligned}
\tag{6.3}
$$

Now we note that

$$
\begin{aligned}
||\bar{\theta}_i^*|| &= ||[\sqrt{\rho_{oi}}\,\theta_i^{*T}, \rho_{oi}]|| \\
&\leq \max[1, \sqrt{\rho_{oi}}]||[\theta_i^{*T}, \rho_{oi}]^T|| \\
&= \max[1, \frac{1}{\sqrt{c_{oi}^*}}]||\Theta_i^*||
\end{aligned}
$$

$$
\begin{aligned}
\text{Also } V_i &\geq \max[\,\frac{1}{c_{oi}^*},\,1\,]\,\frac{2M_{oi}^2}{\gamma_i} \\
\implies ||\bar{\phi}_i|| &\geq 2M_{oi}\max[1, \frac{1}{\sqrt{c_{oi}^*}}\,]
\end{aligned}
$$

Substituting these expressions in Eq. (6.3), we obtain

$$\sigma_i \bar{\phi}_i^T \bar{\theta}_i \geq \sigma_{oi} ||\bar{\phi}_i||^2 [1 - \frac{||\Theta_i^*||}{2M_{oi}}]$$

$$= \sigma_{1i} V_i \text{ where } \sigma_{1i} \triangleq \frac{\sigma_{oi}[2M_{oi} - ||\Theta_i^*||]\gamma_i}{M_{oi}}$$

Hence, for $V_i \geq max[\frac{1}{c_{oi}^*}, 1] \frac{2M_{oi}^2}{\gamma_i}$,

$$\dot{V}_i \leq -\sigma_{1i} V_i + \frac{1}{4(c_{oi}^*)^2} [(1 + \lambda_1)c_{\Delta_i} + (1 + \frac{1}{\lambda_1})\epsilon_t^2 \frac{\delta_{oi}}{q_{m3i}^2}] \qquad (6.4)$$

Hence V_i, and therefore, $\bar{\phi}_i, \phi_i$ are uniformly bounded. Since this is true $\forall\ i = 1, 2, 3, \ldots, N$, it follows that all the closed loop signals are in L_∞^e. We now consider a time $t_{1i} \geq 0$ such that $V_i(t_{1i}) \geq max[\frac{1}{c_{oi}^*}, 1] \frac{2M_{oi}^2}{\gamma_i}$. Then integrating Eq. (6.4) and using the definition of $V_i(t)$, the desired result follows.

Lemma 6.2: For $t > s \geq t_{oi}$, we have

$$(i)\quad \int_s^t \frac{(\bar{\phi}_i^T \bar{\zeta}_i)^2}{m_i} d\tau \leq (1 + \lambda_1) \frac{1}{(c_{oi}^*)^2} c_{\Delta_i}(t - s) + c_{i2}$$

$$(ii)\quad \int_s^t ||\dot{\bar{\phi}}_i||^2 d\tau \leq (1 + \lambda_1) \frac{\gamma_i^2 c_{\Delta_i}}{(c_{oi}^*)^2} [2\sqrt{c_{\bar{z}_i}} + \sqrt{\frac{\sigma_{oi}}{2}}]^2 (t - s) + c_{i3}$$

where λ_1 and c_{Δ_i} are as defined in Lemma 6.1.

$c_{i2}, c_{i3} > 0$ are finite constants

$$c_{\bar{z}_i} \triangleq [\frac{1}{c_{oi}^*} \{(1 + \lambda_5)c_{zi} + \frac{\delta_{oi}}{q_{m3i}^2} c_{y_{mi}}\}] +$$

$$[\sqrt{[\frac{M_{oi}}{2} + \sqrt{c_{\phi_i}}]^2[(1 + \lambda_5)c_{zi} + \frac{\delta_{oi}}{q_{m3i}^2} c_{y_{mi}}]} + \sqrt{(1 + \lambda_6)c_{W_{mi}}}]^2$$

$$c_{zi} \triangleq \inf_{p \geq \delta_{oi}} (||[P_{W_{mi}}\bar{G}_i](s - \frac{\delta_{oi}}{2}) \begin{bmatrix} 1 & 0 \\ 0 & \sqrt{N} \end{bmatrix} ||_\infty^2)$$

$$c_{W_{mi}} \triangleq \inf_{p \geq \delta_{oi}} (||[P_{W_{mi}}](s - \frac{\delta_{oi}}{2})||_\infty^2)$$

c_{ymi} is the upper bound for $y_{mi}^2(t)$.

$c_{\phi i} \triangleq c_{oi}^* c_{\bar{\phi}_i}, c_{\bar{\phi}_i}$ being as defined in Lemma 6.1

and $\lambda_5, \lambda_6 > 0$ are arbitrary.

Proof: Now integrating Eq. (6.2) from s to t and using the uniform boundedness of V_i we obtain

$$\int_s^t \sigma_i \bar{\phi}_i^T \bar{\theta}_i d\tau \leq \frac{1}{4} \frac{(1 + \lambda_1)}{(c_{oi}^*)^2} c_{\Delta_i}(t - s) + c_{i1} \qquad (6.5)$$

where $c_{i1} > 0$ is a finite constant.

Similarly, starting with Eq. (6.1), after a simple completion of squares and using the uniform boundedness of V_i it follows that

$$\int_s^t \frac{(\bar{\phi}_i^T \bar{\zeta}_i)^2}{m_i} d\tau \leq (1 + \lambda_1) \frac{1}{(c_{oi}^*)^2} c_{\Delta_i}(t - s) + c_{i2} \qquad (6.6)$$

where $c_{i2} > 0$ is a finite constant.

This completes the proof of (i).

Now from the adaptive law of Eq. (5.8), we have

$$\dot{\bar{\phi}}_i = -\frac{\gamma_i \epsilon_{1i} \bar{\zeta}_i}{m_i} - \gamma_i \sigma_i \bar{\theta}_i$$

Substituting for ϵ_{1i} from Eq. (5.4) we obtain

$$\dot{\bar{\phi}}_i = -\gamma_i \sigma_i \bar{\theta}_i - \frac{\gamma_i (\bar{\phi}_i^T \bar{\zeta}_i) \bar{\zeta}_i}{m_i} - \frac{1}{c_{oi}^*} \frac{\gamma_i \bar{\zeta}_i}{m_i} W_{\Delta_i}(s)[y]$$

Defining $p_1 = (1 + \lambda_2), p_2 = (1 + \frac{1}{\lambda_2})(1 + \lambda_3), p_3 = (1 + \frac{1}{\lambda_2})(1 + \frac{1}{\lambda_3})$ where $\lambda_2, \lambda_3 > 0$ are arbitrary and using Eq. (2.1) twice we obtain

$$\| \dot{\bar{\phi}}_i \|^2 \leq p_1 \gamma_i^2 \sigma_i^2 \|\bar{\theta}_i\|^2 + p_2 \gamma_i^2 \frac{(\bar{\phi}_i^T \bar{\zeta}_i)^2}{m_i} \frac{\|\bar{\zeta}_i\|^2}{m_i}$$
$$+ p_3 \gamma_i^2 \frac{1}{c_{oi}^{*2}} \frac{\|\bar{\zeta}_i\|^2}{m_i} \frac{\|W_{\Delta_i}(s)[y]\|^2}{m_i} \qquad (6.7)$$

Now, from the adaptive law, we obtain

$$
\begin{aligned}
\sigma_i \bar{\phi}_i^{\ T} \bar{\theta}_i &= \sigma_i \bar{\theta}_i^{\ T} (\bar{\theta}_i - \bar{\theta}_i^{\ *}) \\
&\geq \sigma_i \|\bar{\theta}_i\| (\|\bar{\theta}_i\| - \|\bar{\theta}_i^{\ *}\|) \\
&\geq \sigma_i \|\bar{\theta}_i\| (2\|\bar{\theta}_i^{\ *}\| - \|\bar{\theta}_i^{\ *}\|)
\end{aligned}
$$

$$
[\text{ since } \sigma_i > 0 \implies \|\bar{\theta}_i\| \geq 2\|\bar{\theta}_i^{\ *}\| \ ]
$$

Thus

$$
\sigma_i \|\bar{\theta}_i\| \leq \frac{\sigma_i \bar{\phi}_i^{\ T} \bar{\theta}_i}{\|\bar{\theta}_i^{\ *}\|}
$$

and

$$
\begin{aligned}
\sigma_i^2 \|\bar{\theta}_i\|^2 &= \sigma_i^2 \bar{\theta}_i^{\ T} \bar{\theta}_i \\
&= \sigma_i^2 \bar{\theta}_i^{\ T} [\bar{\phi}_i + \bar{\theta}_i^{\ *}] \\
&\leq \sigma_i^2 \bar{\theta}_i^{\ T} \bar{\phi}_i + \sigma_i^2 \|\bar{\theta}_i\| \, \|\bar{\theta}_i^{\ *}\| \\
&\leq 2\sigma_{oi} \sigma_i \bar{\theta}_i^{\ T} \bar{\phi}_i
\end{aligned}
\tag{6.8}
$$

Now

$$
\zeta_i = W_{m_i}(s) \begin{bmatrix} \hat{G}_i(s) \begin{bmatrix} u_i \\ y_i \end{bmatrix} \\ r_i \end{bmatrix}
$$

Hence, from Lemma 2.2, it follows that

$$
\|\zeta_i\|^2 \leq (1 + \lambda_5) c_{zi} m_i + c_{ymi} + (1 + \frac{1}{\lambda_5}) \epsilon_t^2
\tag{6.9}
$$

where $\lambda_5 > 0$ is arbitrary and c_{zi}, c_{ymi} are as defined in the statement of the Lemma.

From the definition of $\bar{\zeta}_i$, it follows that $\forall \, t \geq t_{oi}$,

$$
\|\bar{\zeta}_i\|^2 = \frac{1}{c_{oi}^*} \|\zeta_i\|^2 + \|\xi_i\|^2
$$

$$= \frac{1}{c_{oi}^*}||\zeta_i||^2 + ||\theta_i^T\zeta_i - W_{mi}(s)u_i||^2$$

$$\leq \frac{1}{c_{oi}^*}||\zeta_i||^2 + (1+\lambda_4)||\theta_i||^2\,||\zeta_i||^2 + (1+\frac{1}{\lambda_4})||W_{mi}(s)[u_i]||^2$$

(using Eq. (2.1))

$$\leq \frac{1}{c_{oi}^*}[(1+\lambda_5)c_{zi}m_i + c_{ymi} + (1+\frac{1}{\lambda_5})\epsilon_t^2]$$

$$+ (1+\lambda_4)[\frac{M_{oi}}{2} + \sqrt{c_{\phi_i}}]^2[(1+\lambda_5)c_{zi}m_i + c_{ymi} + (1+\frac{1}{\lambda_5})\epsilon_t^2]$$

$$+ (1+\frac{1}{\lambda_4})[(1+\lambda_6)c_{W_{mi}}m_i + (1+\frac{1}{\lambda_6})\epsilon_t^2]$$

where we used Lemma 2.2.

Thus

$$\frac{||\bar\zeta_i||^2}{m_i} \leq \frac{1}{c_{oi}^*}[(1+\lambda_5)c_{zi} + c_{ymi}\frac{\delta_{oi}}{q_{m3i}^2} + (1+\frac{1}{\lambda_5})\frac{\epsilon_t^2}{m_i}]$$

$$+(1+\lambda_4)[\frac{M_{oi}}{2} + \sqrt{c_{\phi_i}}]^2[(1+\lambda_5)c_{zi} + c_{ymi}\frac{\delta_{oi}}{q_{m3i}^2} + (1+\frac{1}{\lambda_5})\frac{\epsilon_t^2}{m_i}]$$

$$+(1+\frac{1}{\lambda_4})(1+\lambda_6)c_{W_{mi}} + (1+\frac{1}{\lambda_4})(1+\frac{1}{\lambda_6})\frac{\epsilon_t^2}{m_i}$$

where we used the fact that $m_i(t) \geq m_i(0) \geq \frac{q_{m3i}^2}{\delta_{oi}}$.

Using Eq. (2.2), it now follows that

$$\frac{||\bar\zeta_i||^2}{m_i} \leq c_{\bar z_i} + \epsilon_t \qquad (6.10)$$

where ϵ_t is an exponentially decaying term. The expression for c_{ϕ_i} follows from Lemma 6.1 and the definition of $\bar\phi_i$.

Now using Eqs. (6.5), (6.6), (6.8), (6.10) and the fact that

$$\frac{||W_{\Delta_i}(s)[y]||^2}{m_i} \leq (1+\lambda_1)c_{\Delta_i} + (1+\frac{1}{\lambda_1})\epsilon_t^2\frac{\delta_{oi}}{q_{m3i}^2}$$

in Eq. (6.7), we obtain

$$\int_s^t ||\dot{\bar\phi}_i||^2 d\tau \leq 2p_1\gamma_i^2\sigma_{oi}[\frac{(1+\lambda_1)\,c_{\Delta_i}(t-s)}{4(c_{oi}^*)^2}]$$

$$+p_2\gamma_i^2 c_{\bar z_i}(1+\lambda_1)\frac{1}{(c_{oi}^*)^2}c_{\Delta_i}(t-s)$$

$$+p_3\gamma_i^2 c_{\bar{z}_i}(1+\lambda_1)\,\frac{1}{(c_{oi}^*)^2}\,c_{\Delta_i}(t-s)+c_{i3}$$

where $c_{i3}>0$ is a constant.

Now noting that
$$
\begin{aligned}
p_2+p_3 &= (1+\frac{1}{\lambda_2})(1+\lambda_3)+(1+\frac{1}{\lambda_2})(1+\frac{1}{\lambda_3})\\
&= (1+\frac{1}{\lambda_2})[2+\lambda_3+\frac{1}{\lambda_3}]\\
&\geq 4(1+\frac{1}{\lambda_2})
\end{aligned}
$$

we obtain

$$\int_s^t \|\dot{\phi}_i\|^2 d\tau \leq (1+\lambda_2)2\gamma_i^2\sigma_{oi}(1+\lambda_1)\,\frac{c_{\Delta_i}}{4}\,\frac{1}{(c_{oi}^*)^2}(t-s)$$

$$+4(1+\frac{1}{\lambda_2})\gamma_i^2 c_{\bar{z}_i}\frac{1}{(c_{oi}^*)^2}\,(1+\lambda_1)c_{\Delta_i}(t-s)+c_{i3}$$

Now using Eq.(2.2) to eliminate λ_2 from the above equation, the desired expression for the average of $\|\dot{\phi}_i\|^2$ follows.

Before proceeding further, we make the following definitions.

$$
\begin{aligned}
\text{Define } \Phi_i &\triangleq [(\rho_{oi}+\psi_i)\phi_i^T,-\psi_i]^T\\
Z_i &\triangleq [\zeta_i^T,W_{mi}(s)[\phi_i^T w_i]]^T\\
\Omega_i &\triangleq W_{mi}^{-1}(s)Z_i
\end{aligned}
$$

Thus

$$\|\dot{\Phi}_i\|^2 \leq \|\dot{\psi}_i\phi_i^T+(\rho_{oi}+\psi_i)\dot{\phi}_i^{\,T}\|^2+|\dot{\psi}_i|^2$$

Using Eq. (2.1) we obtain that for $t\geq t_{oi}$

$$
\begin{aligned}
\|\dot{\Phi}_i\|^2 &\leq (1+\epsilon_2)c_{oi}^* c_{\bar{\phi}_i}|\dot{\psi}_i|^2+(1+\frac{1}{\epsilon_2})[\rho_{oi}+\sqrt{c_{\bar{\phi}_i}}]^2\|\dot{\phi}_i\|^2+|\dot{\psi}_i|^2\\
&\leq \max[(1+\frac{1}{\epsilon_2})[\rho_{oi}+\sqrt{c_{\bar{\phi}_i}}]^2\frac{1}{\rho_{oi}},1+(1+\epsilon_2)c_{oi}^* c_{\bar{\phi}_i}]\\
&\quad \{\rho_{oi}\|\dot{\phi}_i\|^2+\psi_i^2\}\\
&\leq C_{\dot{\Phi}_i}\|\dot{\phi}_i\|^2
\end{aligned}
$$

where

$$C_{\dot{\Phi}_i} = \inf_{\epsilon_2 > 0} \max\{(1 + \frac{1}{\epsilon_2})[\rho_{oi} + \sqrt{c_{\bar{\phi}_i}}]^2 \frac{1}{\rho_{oi}}, 1 + (1 + \epsilon_2)c_{oi}^* c_{\bar{\phi}_i}\}$$

In view of the above inequality and also the fact that $\Phi_i^T Z_i = \bar{\phi}_i^T \bar{\zeta}_i$, Lemma 6.2 can be restated as follows:

Lemma 6.2': For $t > s \geq t_{oi}$, we have

$$(i) \quad \int_s^t \frac{(\Phi_i^T Z_i)^2}{m_i} \, d\tau \leq (1 + \lambda_1) \frac{1}{(c_{oi}^*)^2} c_{\Delta_i}(t - s) + c_{i2}$$

$$(ii) \quad \int_s^t \|\dot{\Phi}_i\|^2 d\tau \leq C_{\dot{\Phi}_i}(1 + \lambda_1) \frac{\gamma_i^2 c_{\Delta_i}}{(c_{oi}^*)^2} [2\sqrt{c_{\bar{z}_i}} + \sqrt{\frac{\sigma_{oi}}{2}}]^2 (t - s) + c_{i3}$$

where

$$C_{\dot{\Phi}_i} = \inf_{\epsilon_2 > 0} \max\{(1 + \frac{1}{\epsilon_2})[\rho_{oi} + \sqrt{c_{\bar{\phi}_i}}]^2 \frac{1}{\rho_{oi}}, 1 + (1 + \epsilon_2)c_{oi}^* c_{\bar{\phi}_i}\}$$

and all other symbols are as defined in Lemma 6.2.

To avoid the proliferation of symbols, we have retained the same constants c_{i2}, c_{i3} as before.

We now proceed to examine the properties that follow from the choice of the specific control structure, that is, the model reference control structure in this case.

6.1.2 Properties Resulting from the Specific Control Structure

These properties are best described in terms of the following Lemmas.

Lemma 6.3: The ith subsystem described by Eq. (3.1) and controlled by Eqs. (4.3), (5.8) satisfies the equation

$$\begin{bmatrix} u_i \\ y_i \end{bmatrix} = H_i(s) \begin{bmatrix} \Phi_i^T \Omega_i + r_i \\ \Delta_i(s)y \end{bmatrix}$$

where

$$\Delta_i(s) \triangleq [\Delta_{i1}(s), \Delta_{i2}(s), \ldots, \Delta_{iN}(s)]$$

$$y \triangleq [y_1, y_2, \ldots, y_N]^T$$

and

$$H_i(s) \triangleq \left[\begin{array}{cc} \dfrac{R_i(s)}{k_{Pi} Z_i(s)} W_{mi}(s) & \dfrac{g_{2i}(s, \theta_{2i}^*)}{k_{mi}} \dfrac{R_i(s)}{Z_i(s)} \dfrac{W_{mi}(s)}{\Gamma_i(s)} \\ W_{mi}(s) & \dfrac{R_i(s)(\Gamma_i(s) - g_{1i}(s, \theta_{1i}^*))}{k_{mi} Z_i(s) \Gamma_i(s)} W_{mi}(s) \end{array} \right]$$

with $g_{1i}(s, \theta_{1i}^*), g_{2i}(s, \theta_{2i}^*)$ being as defined in Lemma 4.1.

Proof: The proof can be obtained by starting with Eqs. (3.1), (4.3), using the fact that $\theta_i = \phi_i + \theta_i^*$ and Eq. (4.5).

Now suppose that $H_i(s)$, as defined in Lemma 6.3, is analytic in $\Sigma \frac{\delta_{oi}'}{2}$. We define $\delta_0' = \min_i \delta_{oi}'$ and $\delta_0 = \min_i \delta_{oi}$. Then for any $\delta \in (0, \min[\delta_0, \delta_0'])$, define the fictitious normalizing signal $m_{fi}(t)$ by

$$m_{fi}(t) = e^{-\delta t} m_i(0) + \int_0^t e^{-\delta(t-\tau)} [u_i^2 + \frac{1}{N} \sum_{j=1}^{N} y_j^2 + q_{m3i}^2] d\tau \qquad (6.11)$$

We now proceed to calculate the exponentially weighted average of $(\Phi_i^T \Omega_i)^2$ after the decay of the initial transients. We first recall from the proof of Lemma 6.1 that the inequality $||\bar{\phi}_i(t)||^2 \le c_{\bar{\phi}_i} + \epsilon_t$ holds only for $t \ge t_{oi}$ where $t_{oi} > 0$ is finite.

Define $t_o := \max_i t_{oi}$ and consider

$$\int_{t_o}^t e^{-\delta(t-\tau)} (\Phi_i^T \Omega_i)^2 d\tau \qquad (6.12)$$

where $t \ge t_o$.

For the purpose of proving stability, we need to show that the time average of the integral in Eq. (6.12) can be made arbitrarily small provided the interconnection strengths are weak enough. This issue will be formally addressed in Lemma 6.4. For the present, following [5], we give an intuitive discussion of the idea involved in the proof.

Consider a fictitious filter [5] given by

$$\Lambda_i(s) = [\,\frac{a_i}{s + a_i}\,]^{n_i^*} \quad , \quad a_i > \frac{\delta}{2}.$$

Define $\Lambda_{1i}(s) = \frac{1-\Lambda_i(s)}{s}$. Clearly, for large $a_i, \|\Lambda_{1i}(s)\|_\infty = O(\,a_i^{-n_i^*})$. Now using the identity

$$1 = s\Lambda_{1i}(s) + \Lambda_i(s)$$

we obtain

$$\Phi_i^T \Omega_i = \Lambda_{1i}(s)[\dot{\Phi}_i^T \Omega_i + \Phi_i^T \left\{ \begin{array}{c} s\hat{G}_i(s) \begin{bmatrix} u_i \\ y_i \end{bmatrix} \\ \dot{r}_i \\ \dot{\phi}_i^T \begin{bmatrix} w_{1i} \\ w_{2i} \\ r_i \end{bmatrix} + \phi_i^T \left\{ \begin{array}{c} s\hat{G}_i(s) \begin{bmatrix} u_i \\ y_i \end{bmatrix} \\ \dot{r}_i \end{array} \right\} \end{array} \right\}]$$
$$+ \Lambda_i(s)[\Phi_i^T \Omega_i] \tag{6.13}$$

Also, from the operator identity $s\phi = \phi s + \dot{\phi}$, we obtain

$$W_{mi}(s)[\Phi_i^T \Omega_i] = \Phi_i^T W_{mi}(s)\Omega_i + W_{M_{1i}}(s)\{(W_{M_{2i}}(s)[\Omega_i^T])\dot{\Phi}_i\} \tag{6.14}$$

where $W_{M_{1i}}(s), W_{M_{2i}}(s)$ are strictly proper stable transfer matrices which depend only on $W_{mi}(s)$. Using Eq. (6.14) in Eq. (6.13), we obtain

$$\Phi_i^T \Omega_i = \Lambda_{1i}(s)[\dot{\Phi}_i^{\,T} \Omega_i + \Phi_i^T \left\{ \begin{array}{c} s\hat{G}_i(s) \begin{bmatrix} u_i \\ y_i \end{bmatrix} \\ \dot{r}_i \\ \dot{\phi}_i^{\,T} \begin{bmatrix} w_{1i} \\ w_{2i} \\ r_i \end{bmatrix} + \phi_i^T \left\{ \begin{array}{c} s\hat{G}_i(s) \begin{bmatrix} u_i \\ y_i \end{bmatrix} \\ \dot{r}_i \end{array} \right\} \end{array} \right\}]$$

$$+ \Lambda_i(s)W_{mi}^{-1}(s)[\Phi_i^T Z_i + W_{M_{1i}}(s)\{(W_{M_{2i}}(s)[\Omega_i^T])\dot{\Phi}_i\}] \qquad (6.15)$$

Since $||\Lambda_{1i}(s)||_\infty = O(a_i^{-n_i^*})$, it follows that the contribution of the first term to the integral in Eq. (6.12) can be made arbitrarily small by choosing a_i to be large enough. This will increase the contribution of the second term but, in view of Lemma 6.2', the former is proportional to the size of the unmodelled interconnections and, therefore, can be made smaller by allowing weaker interconnections. We now formalize the above intuitive discussion in the form of a Lemma.

Lemma 6.4: For $t \geq t_o$

$$\int_{t_o}^t e^{-\delta(t-\tau)}[\Phi_i^T \Omega_i]^2 d\tau \leq p_1(q)K_{1i}m_{fi}+p_2(q)K_{2i}\int_{t_o}^t e^{-\delta(t-\tau)}\frac{(\Phi_i^T Z_i)^2}{m_i} m_{fi}d\tau$$

$$+p_3(q)[\sqrt{1+\lambda_8}\sqrt{K_{3i}}+\sqrt{1+\lambda_9}\sqrt{K_{4i}}]^2 \int_{t_o}^t e^{-\delta(t-\tau)}||\dot{\Phi}_i||^2 m_{fi}d\tau + c_i$$

where $K_{1i} = ||\Lambda_{1i}(s-\frac{\delta}{2})||_\infty^2 C_{\Phi_i}[||(s-\frac{\delta}{2})\hat{G}_i(s-\frac{\delta}{2})\begin{bmatrix} 1 & 0 \\ 0 & \sqrt{N} \end{bmatrix}||_\infty^2$

$$+\{\sqrt{c_{\dot{\phi}_i}}\,||\hat{G}_i(s-\frac{\delta}{2})\begin{bmatrix} 1 & 0 \\ 0 & \sqrt{N} \end{bmatrix}||_\infty$$

$$+\sqrt{c_{\dot{\phi}_i}}\,||(s-\frac{\delta}{2})\hat{G}_i(s-\frac{\delta}{2})\begin{bmatrix} 1 & 0 \\ 0 & \sqrt{N} \end{bmatrix}||_\infty\}^2]$$

$$K_{2i} = ||\Lambda_i(s-\frac{\delta}{2})W_{mi}^{-1}(s-\frac{\delta}{2})||_\infty^2$$

$$K_{3i} = ||\Lambda_{1i}(s-\frac{\delta}{2})||_\infty^2 C_{\Omega_i}$$

$$K_{4i} = C_{M_{2i}}||\Lambda_i W_{mi}^{-1}W_{M_{1i}}(s-\frac{\delta}{2})||_\infty^2$$

$$C_{\Omega_i} = (1+c_{\dot{\phi}_i})\inf_{p\geq\delta}\{||[P_{\hat{G}_i}](s-\frac{\delta}{2})\begin{bmatrix} 1 & 0 \\ 0 & \sqrt{N} \end{bmatrix}||_\infty^2\}$$

$$C_{M2i} = \sum_{j=1}^{2} \inf_{p \geq \delta} ||[P_{W_{M_{2i}}} \hat{G}_{ij}](s - \frac{\delta}{2}) \begin{bmatrix} 1 & 0 \\ 0 & \sqrt{N} \end{bmatrix} ||_{\infty}^2$$

$$+ \inf_{p \geq \delta} [||[P_{W_{M_{2i}}}](s - \frac{\delta}{2})||_{\infty}^2] c_{\phi_i} ||\hat{G}_i(s - \frac{\delta}{2}) \begin{bmatrix} 1 & 0 \\ 0 & \sqrt{N} \end{bmatrix} ||_{\infty}^2$$

$$\hat{G}_{ij}(s) = \text{the jth row of } \hat{G}_i(s)$$

$$c_{\dot{\phi}_i} = c_{oi}^* \gamma_i^2 [\sigma_{oi} \{ \max(1, \frac{1}{\sqrt{c_{oi}^*}}) \frac{M_{oi}}{2} + \sqrt{c_{\bar{\phi}_i}} \} + \sqrt{c_{\bar{\phi}_i}} \, c_{\bar{z}_i}$$

$$+ \frac{1}{c_{oi}^*} \sqrt{c_{\bar{z}_i}(1 + \lambda_1) c_{\Delta_i}}]^2$$

$$c_{\phi_i} = c_{oi}^* c_{\bar{\phi}_i}$$

$$C_{\Phi_i} = c_{\bar{\phi}_i} [1 + \{\frac{1}{\sqrt{c_{oi}^*}} + \sqrt{c_{\phi_i}}\}^2]$$

$c_{\bar{\phi}_i}$, c_{Δ_i}, λ_1 and $c_{\bar{z}_i}$ are as defined in Lemmas 6.1, 6.2,

$$p_1(q) = (1 + q_1), \quad p_2(q) = (1 + \frac{1}{q_1})(1 + q_2),$$

$$p_3(q) = (1 + \frac{1}{q_1})(1 + \frac{1}{q_2})$$

$q_1, q_2, q_3, \lambda_8, \lambda_9 > 0$ are arbitrary

and $c_i > 0$ is a constant.

Proof: Using the operator identity

$$1 = s \Lambda_{1i}(s) + \Lambda_i(s)$$

we obtain

$$\Phi_i^T \Omega_i = \Lambda_{1i}(s) \left\{ \dot{\Phi}_i^T \Omega_i + \Phi_i^T \left[s\hat{G}_i(s) \begin{bmatrix} u_i \\ y_i \end{bmatrix} \right. \right.$$
$$\left. \left. \dot{\phi}_i^T \begin{bmatrix} w_{1i} \\ w_{2i} \\ r_i \end{bmatrix} + \phi_i^T \left\{ s\hat{G}_i(s) \begin{bmatrix} u_i \\ y_i \end{bmatrix} \right\} \right] \right\}$$

$$+\Lambda_i(s)[\Phi_i^T\Omega_i]$$

Also

$$W_{mi}(s)[\Phi_i^T\Omega_i] = \Phi_i^T W_{mi}(s)[\Omega_i] + W_{M_{1i}}(s)\{(W_{M_{2i}}(s)[\Omega_i^T])\dot{\Phi}_i\}$$

for some stable strictly proper transfer matrices $W_{M_{1i}}(s)$, $W_{M_{2i}}(s)$.
Thus

$$\Phi_i^T\Omega_i = \Lambda_{1i}(s)\left\{ \dot{\Phi}_i^T\Omega_i + \Phi_i^T\left[\begin{array}{c} s\hat{G}_i(s)\begin{bmatrix} u_i \\ y_i \end{bmatrix} \\ \dot{r}_i \\ \phi_i^T\begin{bmatrix} w_{1i} \\ w_{2i} \\ r_i \end{bmatrix} + \phi_i^T\left\{ s\hat{G}_i(s)\begin{bmatrix} u_i \\ y_i \end{bmatrix} \\ \dot{r}_i \right\} \end{array} \right] \right\}$$

$$+\Lambda_i(s)W_{m_i}^{-1}(s)[\Phi_i^T Z_i + W_{M_{1i}}(s)\{(W_{M_{2i}}(s)[\Omega_i^T])\dot{\Phi}_i\}]$$

With $p_1(q), p_2(q)$ and $p_3(q)$ as already defined and using Eq. (2.1) twice,
we obtain

$$(\Phi_i^T\Omega_i)^2$$

$$\leq p_1(q)\|\Lambda_{1i}(s)\left\{ \Phi_i^T\left[\begin{array}{c} s\hat{G}_i(s)\begin{bmatrix} u_i \\ y_i \end{bmatrix} \\ \dot{r}_i \\ \dot{\phi}_i^T\begin{bmatrix} w_{1i} \\ w_{2i} \\ r_i \end{bmatrix} + \phi_i^T\left\{ s\hat{G}_i(s)\begin{bmatrix} u_i \\ y_i \end{bmatrix} \\ \dot{r}_i \right\} \end{array} \right] \right\}\|^2$$

$$+p_2(q)\ \| \Lambda_i(s)W_{m_i}^{-1}(s)[\Phi_i^T Z_i] \|^2$$

$$+p_3(q)\|\Lambda_{1i}(s)[\dot{\Phi}_i^T\Omega_i] + \Lambda_i(s)W_{m_i}^{-1}(s)W_{M_{1i}}(s)\{(W_{M_{2i}}(s)[\Omega_i^T])\dot{\Phi}_i\}\|^2 \quad (6.16)$$

Now, from Lemma 6.1, it follows that $\forall\, t \geq t_{oi}$

$$\|\phi_i\|^2 \leq c_{\phi_i} + \epsilon_t \tag{6.17}$$

$$\|\Phi_i\|^2 \leq C_{\Phi_i} + \epsilon_t \tag{6.18}$$

where ϵ_t is an exponentially decaying term and c_{ϕ_i} and C_{Φ_i} are as defined in the statement of the lemma.

We now obtain an upper bound on $\|\dot{\phi}_i\|^2$ after the decay of the initial transients.

From Eqs. (5.4) - (5.5), using twice Eq. (2.1), we obtain

$$
\begin{aligned}
\|\dot{\bar{\phi}}_i\|^2 \;\leq\;& r_1 \gamma_i^2 \sigma_i^2 \|\bar{\theta}_i^* + \bar{\phi}_i\|^2 + r_2 \gamma_i^2\,\|\bar{\phi}_i\|^2\,\frac{\|\bar{\zeta}_i\|^2}{m_i}\,\frac{\|\bar{\zeta}_i\|^2}{m_i} \\
&+ r_3 \gamma_i^2\,\frac{1}{(c_{oi}^*)^2}\,\frac{\|\bar{\zeta}_i\|^2}{m_i}\,\frac{\|W_{\Delta_i}(s)[y]\|^2}{m_i}
\end{aligned}
$$

where

$$r_1 \overset{\Delta}{=} 1 + s_1\;,\quad r_2 \overset{\Delta}{=} (1 + \frac{1}{s_1})(1 + s_2)\;,\quad r_3 \overset{\Delta}{=} (1 + \frac{1}{s_1})(1 + \frac{1}{s_2})$$

and $\quad s_1, s_2, s_3 > 0$ are arbitrary.

Thus $\forall\, t \geq t_{oi}$,

$$
\begin{aligned}
\|\dot{\bar{\phi}}_i(t)\|^2 \;\leq\;& \gamma_i^2 [\sigma_{oi}\{\, \max(1, \frac{1}{\sqrt{c_{oi}^*}})\,\frac{M_{oi}}{2} + \sqrt{c_{\bar{\phi}_i}}\} \\
&+ \sqrt{c_{\bar{\phi}_i}}\; c_{z_i} + \frac{1}{c_{oi}^*}\,\sqrt{c_{z_i}(1 + \lambda_1)c_{\Delta_i}}\,]^2
\end{aligned}
$$

where $c_{\Delta_i}, c_{\bar{\phi}_i}$ are as defined in Lemma 6.1 and c_{z_i} is as defined in Lemma 6.2. In arriving at the above inequality, we made use of Eq. (2.3) to eliminate r_1, r_2 and r_3 .

Since $\dot{\bar{\phi}}_i = [\sqrt{\rho_{oi}}\;\dot{\phi}_i^{\,T}, \dot{\psi}_i]^T$ it follows that

$$
\begin{aligned}
\|\dot{\bar{\phi}}_i\|^2 \;&=\; \rho_{oi}\|\dot{\phi}_i\|^2 + |\dot{\psi}_i|^2 \\
\Rightarrow \|\dot{\phi}_i\|^2 \;&\leq\; \frac{1}{\rho_{oi}}\,\|\dot{\bar{\phi}}_i\|^2
\end{aligned}
$$

Thus $\forall\, t \geq t_{oi}$

$$\|\dot{\phi}_i\|^2 \leq c_{oi}^* \gamma_i^2 [\sigma_{oi}\{\ \max(1, \sqrt{\frac{1}{c_{oi}^*}})\, \frac{M_{oi}}{2} + \sqrt{c_{\dot{\phi}_i}}\}$$
$$+\sqrt{c_{\dot{\phi}_i}}\, c_{\bar{z}_i} + \frac{1}{c_{oi}^*}\, \sqrt{c_{\bar{z}_i}(1 + \lambda_1)c_{\Delta_i}}]^2$$

With $c_{\dot{\phi}_i}$ as defined in the statement of the Lemma, we have $\forall\, t \geq t_{oi}$

$$\|\dot{\phi}_i\|^2 \leq c_{\dot{\phi}_i} \qquad (6.19)$$

We now start evaluating $\int_{t_0}^t e^{-\delta(t-\tau)}(\Phi_i^T \Omega_i)^2 d\tau$, term by term, using Eq. (6.16).

For the first term, using Lemma 2.1 and Eq. (2.1), we obtain

$$p_1(q)\int_{t_0}^t \left\| \Lambda_{1i}(s) \left\{ \Phi_i^T \left[\begin{array}{c} s\hat{G}_i(s)\begin{bmatrix} u_i \\ y_i \end{bmatrix} \\ \dot{r}_i \\ \dot{\phi}_i^T \begin{bmatrix} w_{1i} \\ w_{2i} \\ r_i \end{bmatrix} + \phi_i^T \left\{ s\hat{G}_i(s)\begin{bmatrix} u_i \\ y_i \end{bmatrix} \\ \dot{r}_i \right\} \end{array} \right] \right\} \right\|^2 e^{-\delta(t-\tau)}d\tau$$

$$\leq\ p_1(q)\|\Lambda_{1i}(s - \frac{\delta}{2})\|_\infty^2 C_{\Phi_i}[\int_{t_0}^t \|s\hat{G}_i(s)\begin{bmatrix} u_i \\ y_i \end{bmatrix}\|^2 e^{-\delta(t-\tau)}d\tau$$

$$+\int_{t_0}^t e^{-\delta(t-\tau)}\dot{r}_i^{\,2} d\tau$$

$$+(1 + \lambda_7)\int_{t_0}^t \|\dot{\phi}_i\|^2 \{\|\hat{G}_i(s)\begin{bmatrix} u_i \\ y_i \end{bmatrix}\|^2 + r_i^2\}e^{-\delta(t-\tau)}d\tau$$

$$+(1 + \frac{1}{\lambda_7})c_{\phi_i}\{\int_{t_0}^t \|s\hat{G}_i(s)\begin{bmatrix} u_i \\ y_i \end{bmatrix}\|^2 e^{-\delta(t-\tau)}d\tau$$

$$+\int_{t_0}^t e^{-\delta(t-\tau)}\dot{r}_i^{\,2} d\tau\}] + c_{1i}', \ c_{1i}' > 0$$

$$\leq\ p_1(q)K_{1i}m_{fi} + c_{1i}\,, \ c_{1i} > 0 \qquad (6.20)$$

where we made use of Lemma 2.1 again and Eqs. (6.17) - (6.19). The elimination of λ_7 from the expression for K_{1i} was achieved using Eq. (2.2).

For the second term, using Lemma 2.1, and the fact that $m_{fi}(t) \geq m_i(t)$, we obtain

$$p_2(q) \int_{t_0}^{t} e^{-\delta(t-\tau)} \|\Lambda_i(s) W_{mi}^{-1}(s) [\Phi_i^T Z_i]\|^2 d\tau$$

$$\leq p_2(q) K_{2i} \int_{t_0}^{t} e^{-\delta(t-\tau)} \frac{(\Phi_i^T Z_i)^2}{m_i} m_{fi} d\tau + c_{2i} \, , \, c_{2i} > 0 \quad (6.21)$$

We now proceed to evaluate the third term in Eq. (6.16).

Now from the definition of Ω_i , we obtain

$$\|\Omega_i\|^2 \leq [1 + \|\phi_i\|^2] \|\hat{G}_i(s) \begin{bmatrix} u_i \\ y_i \end{bmatrix} \|^2 + [1 + \|\phi_i\|^2] r_i^2$$

Then from Lemma 2.2, it follows that

$$\|\Omega_i\|^2 \leq (1 + \lambda_8) C_{\Omega_i} m_{fi} + (1 + c_{\phi_i}) c_{r_i} + (1 + \frac{1}{\lambda_8}) \epsilon_t^2 \quad (6.22)$$

where $\lambda_8 > 0$ is arbitrary and C_{Ω_i} is as defined in the statement of the Lemma.

Now we first note that

$$\|W_{M_{2i}}(s) [\Omega_i^T]\|^2 \leq \|W_{M_{2i}}(s) w_{1i}\|^2 + \|W_{M_{2i}}(s) w_{2i}\|^2$$

$$+ \|W_{M_{2i}}(s) r_i\|^2 + \|W_{M_{2i}}(s) \begin{bmatrix} \phi_i^T \begin{pmatrix} w_{1i} \\ w_{2i} \\ r_i \end{pmatrix} \end{bmatrix} \|^2$$

$$= \|W_{M_{2i}}(s) \hat{G}_{i1}(s) \begin{bmatrix} u_i \\ y_i \end{bmatrix} \|^2$$

$$+ \|W_{M_{2i}}(s) \hat{G}_{i2}(s) \begin{bmatrix} u_i \\ y_i \end{bmatrix} \|^2 + \|W_{M_{2i}}(s) r_i\|^2$$

$$+\|W_{M_{2i}}(s)\left[\phi_i^T\left\{\hat{G}_i(s)\begin{bmatrix}u_i\\y_i\end{bmatrix}\right\}\right]\|^2$$

where $\hat{G}_{ij}(s), j = 1, 2$ denote the jth row of $\hat{G}_i(s)$.

Then using Lemmas 2.1 and 2.2, it follows that

$$\|W_{M_{2i}}(s)[\Omega_i^T]\|^2 \le (1 + \lambda_9)C_{M_{2i}}m_{fi} + (1 + \frac{1}{\lambda_9})\epsilon_t^2 + c_{W_{M_{2i}}} \qquad (6.23)$$

where $\lambda_9 > 0$ is arbitrary, $c_{W_{M_{2i}}}$ is a positive constant and $C_{M_{2i}}$ is as defined in the statement of the Lemma.

Using Eqs. (6.22), (6.23), (2.3) we obtain

$$p_3(q)\int_{t_0}^t e^{-\delta(t-\tau)}$$

$$\|\Lambda_{1i}(s)[\dot{\Phi}_i^T\Omega_i] + \Lambda_i(s)W_{mi}^{-1}(s)W_{M_{1i}}(s)\{(W_{M_{2i}}(s)[\Omega_i^T]\dot{\Phi}_i\}\|^2d\tau$$

$$\le p_3(q)[\sqrt{1+\lambda_8}\sqrt{K_{3i}} + \sqrt{1+\lambda_9}\sqrt{K_{4i}}]^2$$

$$\int_{t_0}^t e^{-\delta(t-\tau)}\|\dot{\Phi}\|^2m_{fi}d\tau$$

$$+ c_{3i} \qquad (6.24)$$

Combining Eqs. (6.20), (6.21), (6.24) we obtain

$$\int_{t_0}^t e^{-\delta(t-\tau)}[\Phi_i^T\Omega_i]^2d\tau$$

$$\le p_1(q)K_{1i}m_{fi} + p_2(q)K_{2i}\int_{t_0}^t e^{-\delta(t-\tau)}\frac{(\Phi_i^TZ_i)^2}{m_i}m_{fi}d\tau$$

$$+p_3(q)[\sqrt{(1+\lambda_8)}\sqrt{K_{3i}} + \sqrt{1+\lambda_9}\sqrt{K_{4i}}]^2$$

$$\int_{t_0}^t e^{-\delta(t-\tau)}\|\dot{\Phi}_i\|^2m_{fi}d\tau + c_i \qquad (6.25)$$

where $c_i \stackrel{\Delta}{=} c_{1i} + c_{2i} + c_{3i}$.

This completes our analysis of the properties of the individual local subsystems. We will now aggregate these properties using a "fictitious vector normalizing signal" to be defined shortly.

6.2   Aggregation of the Subsystem Properties

In the analysis of large scale systems, the aggregation of the individual subsystem properties is usually carried out using a vector Lyapunov function [10]. In this paper, the aggregation of the individual subsystem properties will be carried out using a fictitious signal vector which we call the "fictitious vector normalizing signal" and which is defined as follows.

Let $m_{fi}(t)$ be as described by Eq. (6.11) and consider the time varying vector

$$m_f(t) \triangleq [m_{f_2}(t), m_{f_2}(t), \ldots, m_{f_N}(t)]^T \tag{6.26}$$

Then $m_f(t)$ is called the "fictitious vector normalizing signal" and it plays a crucial role in establishing the stability of the decentralized MRAC scheme. We now state the main result of this paper.

Theorem 6.1: Let $K_{1i}, K_{21}, K_{3i}, K_{4i}, c_{\Delta_i}, c_{z_i}, H_i(s), \Delta_i(s)$ be as already defined. Let $K_j = \max_{i \in [1,N]} K_{ji}, j = 1,2,3,4$. Then a sufficient condition for the closed loop stability of Eqs. (3.1), (4.3), (5.6), (5.8) and Eq. (5.9) is given by

$$
\begin{aligned}
\sup_{\substack{\delta \in (0, \min[\delta_0, \delta_0']), \\ a_i > \frac{\delta}{2}}} \Big\{ &\frac{1}{\max_i \|H_i(s - \frac{\delta}{2})\|_\infty^2} - \max_i \|\Delta_i(s - \frac{\delta}{2})\|_\infty^2 N \\
&- \frac{1}{\delta}[\sqrt{\delta K_1'} + \sqrt{[\sum_{i=1}^{N} \frac{c_{\Delta_i}}{(c_{oi}^*)^2}]} K_2 + \\
(\sqrt{K_3'} + \sqrt{K_4'})&\sqrt{\sum_{i=1}^{N} \{ C_{\dot{\Phi}_i} \frac{\gamma_i^2 c_{\Delta_i}}{(c_{oi}^*)^2} \} \{ 2\sqrt{c_{z_i}'} + \sqrt{\frac{\sigma_{oi}}{2}} \}^2]^2 \Big\} \\
&> 0
\end{aligned}
$$

where $K_1' = K_1|_{\lambda_1 = \lambda_5 = \lambda_6 = 0}$

$$K_3' = K_3|_{\lambda_1=0}$$

$$K_4' = K_4|_{\lambda_1=0}$$

$$C_{\dot{\Phi}_i}' = C_{\dot{\Phi}_i}|_{\lambda_1=0}$$

$$c_{z_i}' = c_{z_i}|_{\lambda_5=\lambda_6=\lambda_1=0}$$

Proof: From Eq. (6.26), we obtain

$$m_f(t) = e^{-\delta t}m(0) + \int_0^t e^{-\delta(t-\tau)} \begin{bmatrix} u_1^2 + \frac{1}{N}\sum_{j=1}^N y_j^2 + q_{m_{31}}^2 \\ \vdots \\ u_N^2 + \frac{1}{N}\sum_{j=1}^N y_j^2 + q_{m_{3N}}^2 \end{bmatrix} d\tau$$

Then

$$\|m_f(t)\|_1 \leq e^{-\delta t}\|m(0)\|_1 + \int_0^t e^{-\delta(t-\tau)} \sum_{i=1}^N [u_i^2 + y_i^2]d\tau + C$$

for some $C > 0$

$$\leq C_1$$
$$+ \sum_{i=1}^N \|H_i(s - \frac{\delta}{2})\|_\infty^2 \int_0^t e^{-\delta(t-\tau)}[\|\Phi_i^T \Omega_i + r_i\|^2 + \|\Delta_i(s)[y]\|^2]d\tau$$

where we made use of Lemmas 2.1 and 6.3.

Now, using Eq. (2.1), we obtain

$$\|\Phi_i^T \Omega_i + r_i\|^2 \leq (1+\epsilon)\|\Phi_i^T \Omega_i\|^2 + (1 + \frac{1}{\epsilon})r_i^2$$

where $\epsilon > 0$ is arbitrary

Thus

$$\|m_f(t)\|_1 \leq C_2 + \max_i[\|H_i(s - \frac{\delta}{2})\|_\infty^2]\max_i[\|\Delta_i(s - \frac{\delta}{2})\|_\infty^2]N\|m_f(t)\|_1$$
$$+ \max_i[\|H_i(s - \frac{\delta}{2})\|_\infty^2](1+\epsilon)\sum_{i=1}^N \int_{t_0}^t e^{-\delta(t-\tau)}(\Phi_i^T \Omega_i)^2 d\tau$$

where $C_2 > 0$ depends on $\frac{1}{\epsilon}$ and the bound for $r_i(t)$.

Using Lemma 6.4, we obtain

$$
\begin{aligned}
\|m_f(t)\|_1 \;\leq\; & C_f \\
+ \; & \max_i\{\|H_i(s - \tfrac{\delta}{2})\|_\infty^2\} \cdot \max_i\{\|\Delta_i(s - \tfrac{\delta}{2})\|_\infty^2\} N \|m_f(t)\|_1 \\
& + \max_i \|H_i(s - \tfrac{\delta}{2})\|_\infty^2 (1 + \epsilon)\{ p_1(q) \sum_{i=1}^{N} K_{1i} m_{fi} \\
& + p_2(q) \int_{t_0}^{t} e^{-\delta(t-\tau)} \sum_{i=1}^{N} K_{2i}\, \frac{(\Phi_i^T Z_i)^2}{m_i}\, m_{fi}\, d\tau \\
& + p_3(q) \int_{t_0}^{t} e^{-\delta(t-\tau)} \sum_{i=1}^{N} \{[\sqrt{1+\lambda_8}\sqrt{K_{3i}} + \sqrt{1+\lambda_9}\sqrt{K_{4i}}]^2 \\
& \|\dot{\Phi}_i\|^2 m_{fi}\} d\tau \} \\
\leq\; & C_f \\
& + \max_i\{\|H_i(s - \tfrac{\delta}{2})\|_\infty^2\} \max_i\{\|\Delta_i(s - \tfrac{\delta}{2})\|_\infty^2\} N \|m_f(t)\|_1 \\
& + (1 + \epsilon) \max_i\{\|H_i(s - \tfrac{\delta}{2})\|_\infty^2\} \\
\{ \; p_1(q) \;\; & K_1 \|m_f(t)\|_1 + p_2(q) K_2 \int_{t_0}^{t} e^{-\delta(t-\tau)} \sum_{i=1}^{N} \frac{(\Phi_i^T Z_i)^2}{m_i}\, \|m_f\|_1 d\tau \\
+ \; p_3(q) \;\; & [\sqrt{1+\lambda_8}\sqrt{K_3} + \sqrt{1+\lambda_9}\sqrt{K_4}]^2 \\
& \int_{t_0}^{t} e^{-\delta(t-\tau)} \sum_{i=1}^{N} \|\dot{\Phi}_i\|^2 \|m_f\|_1 d\tau \}
\end{aligned}
$$

where $C_f > 0$ is a constant.

Thus

$$
\begin{aligned}
\|m_f(t)\|_1 \;\leq\; & C + \frac{(1+\epsilon)}{A} \max_i \|H_i(s - \tfrac{\delta}{2})\|_\infty^2 \\
& [p_2(q) K_2 \int_{t_0}^{t} e^{-\delta(t-\tau)} \sum_{i=1}^{N} \frac{(\Phi_i^T Z_i)^2}{m_i}\, \|m_f\|_1 d\tau \\
& + p_3(q)\{\sqrt{1+\lambda_8}\sqrt{K_3} + \sqrt{1+\lambda_9}\sqrt{K_4}\}^2 \\
& \int_{t_0}^{t} e^{-\delta(t-\tau)} \sum_{i=1}^{N} \|\dot{\Phi}_i\|^2 \|m_f\|_1 d\tau]
\end{aligned}
$$

where

$$A \triangleq [1 - \max_i\{\|H_i(s - \frac{\delta}{2})\|_\infty^2\} \cdot \max_i\{\|\Delta_i(s - \frac{\delta}{2})\|_\infty^2\}N$$

$$-(1 + \epsilon)\max_i\{\|H_i(s - \frac{\delta}{2})\|_\infty^2\}p_1(q)K_1]$$

Since $\frac{(\Phi_i^T Z_i)^2}{m_i}$ and $\|\dot{\Phi}_i\|^2$ are uniformly bounded, using Lemma 6.2', the Bellman-Gronwall Lemma [11] and eliminating $p_1(q), p_2(q), p_3(q)$ using Eq. (2.3), the desired stability condition follows.

Remark 6.1. In arriving at the above stability condition, we have set $\epsilon = \lambda_1 = \lambda_5 = \lambda_6 = \lambda_8 = \lambda_9 = 0$. This is allowed since we have a strict inequality so that if the latter holds for zero values of $\epsilon, \lambda_1, \lambda_5, \lambda_6, \lambda_8$ and λ_9, then it must also hold for sufficiently small positive values of these parameters.

Remark 6.2. In the above stability condition, δ and a_i are free parameters and so, for a qualitative result, it is sufficient that the stability condition hold for some combination of these values. However, in order to maximize the size of the allowable unmodelled interconnections, the optimization with respect to δ and $a_i, i = 1, 2, 3, \ldots, N$ would have to be carried out.

Remark 6.3. In the absence of interconnections, our stability condition can always be satisfied by choosing $a \triangleq \min_i a_i$ to be large enough. This will result in a small enough K_1 so that the stability condition holds. This observation is in agreement with the well-known result that in the absence of dynamic uncertainty, an adaptive controller should be able to tolerate any amount of parametric uncertainty.

Remark 6.4. By setting $\gamma_i = 0$ in the stability condition, we see that provided each of the isolated subsystems have minimum phase transfer functions, the stability condition can always be satisfied for some non zero, stable interconnections such that $\max_i \|\Delta_i(s - \frac{\delta}{2})\|^2_\infty$ is small enough. Since we have a strict inequality, the stability condition would also hold for some non-zero γ_i's provided the latter are small enough. Thus, using the new methodology, we have been able to arrive at the same kind of qualitative results as in [1].

7 Conclusions

In this paper, we have proposed and analyzed a decentralized model reference adaptive control scheme for an interconnected system. Our algorithm does not impose any restrictions on the relative degree of the transfer functions of the isolated subsystems. The analysis imposes an H_∞ norm bound on the size of the allowable unmodelled interconnections. If these bounds are satisfied then closed loop stability is guaranteed. The stability condition derived by us here is only a sufficient one and, therefore, it may be somewhat conservative.

By following a method of analysis somewhat different from the standard ones used in Adaptive Control, we have been able to derive a stability condition which gives us a quantitative feel about how to vary the free parameters of the adaptive controller in order to allow for larger interconnection strengths. The precise methodology to be used is currently under investigation. We do, however, believe that the result in this paper narrows the gap between the analysis of adaptive decentralized schemes in [1], [2] and non-adaptive decentralized schemes in [4].

The adaptive algorithm proposed in this paper does require the exchange of output signals between subsystems. While this may be difficult to achieve in cases where, for example, the individual subsystems are separated by large distances, it is clear that the mere exchange of output signals between the subsystems does not really increase the computational complexity of each local controller. Consequently, our scheme would certainly be preferred over a multi-input multi-output adaptive controller design for a predominantly diagonal multivariable system since, in such a situation, the exchange of output signals, as required by our scheme, can be easily carried out.

8   References

1. P. A. Ioannou, "Decentralized Adaptive Control of Interconnected Systems," IEEE Trans. on Automat. Contr., Vol. AC-31, 291-298 (April 1986).

2. D. T. Gavel and D. D. Siljak, "Decentralized Adaptive Control: Structural Conditions for Stability," IEEE Trans. on Automat. Contr., Vol. AC-34, 413-425 (April 1989).

3. P. A. Ioannou and K. S. Tsakalis, "A Robust Direct Adaptive Controller," IEEE Trans. on Automat. Contr., Vol. AC-31, 1033-1043 (November 1986).

4. P. Grosdidier and M. Morari, "Interaction Measures for Systems under Decentralized Control," Automatica, Vol. 22, No. 3, 309-319 (1986).

5. K. S. Tsakalis, "Robustness of Model Reference Adaptive Controllers: Input-Output Properties," Department of Electrical and Computer Engineering, Arizona State University, Report No. 89-03-01.

6. P. Ioannou and J. Sun, "Theory and Design of Robust Direct and Indirect Adaptive Control Schemes," International Journal of Control, Vol. 47, No. 3, 775-813 (1988).

7. K. S. Narendra and A. M. Annaswamy, "Stable Adaptive Systems," Prentice Hall, 1989.

8. S. Sastry and M. Bodson, "Adaptive Control: Stability, Convergence and Robustness," Prentice Hall, 1989.

9. M. G. Safonov, A. J. Laub, and G. L. Hartmann, "Feedback Properties of Multivariable Systems: The Role and Use of the Return Difference Matrix," IEEE Trans. on Automat. Contr., Vol. AC-26, No. 1, 47-65 (February 1981)

10. D. D. Siljak, "Large Scale Dynamical Systems: Stability and Structure," Amsterdam, The Netherlands: North Holland, 1978.

11. C. A. Desoer and M. Vidyasagar, "Feedback Systems: Input-Output Properties," Academic Press, New York, 1975.

Appendix A

A.1 Proof of Lemma 2.1: [5]

Now $y = H(s)[u]$.

Since $H(s)$ is casual, then for $\tau \in [0, t]$, we have

$$y(\tau) = H(s)[u_t]$$

where u_t is the truncation of u to the interval $[0, t]$.

Now

$$\int_0^t e^{-\delta(t-\tau)}\|y(\tau)\|^2 d\tau = \int_0^t e^{-\delta(t-\tau)}\|H(s)[u_t]\|^2 d\tau$$

$$= e^{-\delta t}\int_0^t \|e^{\delta\frac{\tau}{2}}H(s)[u_t]\|^2 d\tau$$

$$= e^{-\delta t}\int_0^t \|H(s-\frac{\delta}{2})[u_t e^{\delta\frac{\tau}{2}}]\|^2 d\tau$$

$$(\text{ Since } e^{at}H(s)[u(t)] = H(s-a)[e^{at}u(t)])$$

$$\le e^{-\delta t}\int_0^\infty \|H(s-\frac{\delta}{2})[u_t e^{\delta\frac{\tau}{2}}]\|^2 d\tau$$

$$= e^{-\delta t}\frac{1}{2\pi}\int_{-\infty}^\infty \|H(j\omega-\frac{\delta}{2})\|^2\|\mathcal{F}[u_t e^{\delta\frac{\tau}{2}}]\|^2 d\omega$$

Here $\mathcal{F}(.)$ denotes the Fourier Transform and we have just used the Parseval's Identity.

$$\le e^{-\delta t}\|H(s-\frac{\delta}{2})\|_\infty^2 \int_{-\infty}^\infty \|\mathcal{F}[u_t e^{\delta\frac{\tau}{2}}]\|^2 \frac{d\omega}{2\pi}$$

$$= e^{-\delta t}\|H(s-\frac{\delta}{2})\|_\infty^2 \int_0^\infty \|u_t e^{\delta\frac{\tau}{2}}\|^2 d\tau$$

$$(\text{Using Parseval's Identity once again.})$$

$$= e^{-\delta t}\|H(s-\frac{\delta}{2})\|_\infty^2 \int_0^t \|u e^{\delta\frac{\tau}{2}}\|^2 d\tau$$

$$[\text{ Since } u_t(\tau) = 0 \text{ for } \tau > t]$$

$$= \|H(s-\frac{\delta}{2})\|_\infty^2 \int_0^t e^{-\delta(t-\tau)}\|u\|^2 d\tau$$

Thus, $\int_0^t e^{-\delta(t-\tau)}\|y(\tau)\|^2 d\tau \le \|H(s-\frac{\delta}{2})\|_\infty^2 \int_0^t e^{-\delta(t-\tau)}\|u(\tau)\|^2 d\tau.$

A.2. Proof of Corollary 2.1:

Corollary 2.1 follows as an immediate consequence of Lemma 2.1 and so no additional proof is necessary.

A.3. Proof of Lemma 2.2:

$$\text{Let}[P_H](s) = \frac{(s+p)H(s)}{\sqrt{p}} \text{ where } p \ge \delta$$

Then $\|y\|^2 = \|H(s)[u]\|^2$

$$= p\|\frac{1}{(s+p)}[P_H](s)[u]\|^2$$

$$= p\|\int_0^t e^{-p(t-\tau)}([P_H](s)[u])(\tau)d\tau + \frac{1}{\sqrt{p}}\epsilon_t\|^2$$

where $\frac{1}{\sqrt{p}}\epsilon_t$ is an exponentially decaying term that accounts for the initial conditions.

Thus using Eq. (2.1), we obtain

$$\|y\|^2 \leq (1+\lambda)p\|\int_0^t e^{-p(t-\tau)}([P_H](s)[u])(\tau)d\tau\|^2 + (1+\frac{1}{\lambda})\epsilon_t^2$$

$$\leq (1+\lambda)p\|\int_0^t e^{-\frac{p}{2}(t-\tau)}e^{-\frac{p}{2}(t-\tau)}([P_H](s)[u])(\tau)d\tau\|^2 + (1+\frac{1}{\lambda})\epsilon_t^2$$

Now from the Holder Inequality, it follows that

$$\|y\|^2 \leq (1+\lambda)\int_0^t e^{-\delta(t-\tau)}\|([P_H](s)[u])(\tau)\|^2 d\tau + (1+\frac{1}{\lambda})\epsilon_t^2$$

Since $p \geq \delta$ is arbitrary, we obtain

$$\|y\|^2 \leq (1+\lambda)\inf_{p\geq\delta}\|[P_H](s-\frac{\delta}{2})\|_\infty^2\int_0^t e^{-\delta(t-\tau)}\|u\|^2 d\tau + (1+\frac{1}{\lambda})\epsilon_t^2$$

where we used Lemma 2.1.

This completes the proof of Lemma 2.2.

Robust Recursive Estimation of States and Parameters of Bilinear Systems

Heping Dai
Naresh K. Sinha

Department of Electrical and Computer Engineering
McMaster University
Hamilton, Ontario
Canada L8S 4L7

Many methods for state and parameter estimation are well developed and widely used in many different areas. These are based only upon the observations and prior assumptions made about the underlying scenarios. Usually, these assumptions are convenient mathematical rationalizations based on some fuzzy knowledge or belief.

In practice, these assumptions do not hold in a variety of physical situations. Often there is a likelihood of large errors, called outliers, in the collected data from physical processes, due to various reasons, such as the failure of transducers, analog to digital conversion errors, large disturbances or even due to problems in data transmission. Such errors are quite large in magnitude and usually it is very difficult to pick them out before processing the data. During the last couple of decades, people have become increasingly aware that the most frequently used estimation procedures are extremely sensitive to minor deviations from their assumptions. In particular, the deviations from assumptions of an underlying normal distribution may cause a catastrophic failure of those procedures which are optimised under the normal distribution assumption. Clearly, from the practical point of view, it is very important to develop certain algorithms in such a way that they can deal with these outliers successfully without producing bias in estimates, while preserving good convergence properties.

This chapter, together with the following chapter, is an attempt to solve the robust estimation problem of dynamic systems. Following Huber's minimax principle [1], we present six robust recursive methods for state and parameter estimation of bilinear systems. The chapter is composed of nine sections. In the first section, we shall start with a brief introduction to robust estimation technique developed in robust statistics and then we shall introduce a definition of robust identification. A survey of previous work on robust identification of systems will be given at the end of the section. The problem of robust identification of bilinear systems will be formulated in the second

CONTROL AND DYNAMIC SYSTEMS, VOL. 53

173

section. From sections III to VIII, six robust recursive methods for state and parameter estimation of bilinear systems will be described in detail. These are the robust recursive least squares method with modified weights, the robust recursive instrumental variable method with modified weights, the robust recursive output error method with modified weights, the robust bootstrap method, the robust extended least squares method with modified weights, and the robust recursive prediction error method with modified weights. Conclusions with some interesting remarks will be summarized in the last section.

I. INTRODUCTION TO ROBUST IDENTIFICATION

Basically, algorithms for robust identification of systems are extensions of estimation techniques in robust statistics to the state and parameter estimation problem. In order to have a deep and better understanding of the robust identification problem, we shall first introduce a brief description of robust estimation techniques. This will be followed by a general formulation for robust identification. A short survey of previous work in this area will be included.

I.A.   Robust Estimation

Since 1960, statisticians have been seeking various special techniques (i.e. robust techniques) which will make the generated estimates insensitive to deviations from the assumptions of estimation methods. After the important contribution of Huber [2], many results have appeared in the theory of the robustness of statistical estimates and inferences. In robust statistics, the word robust may be loaded with many connotations. Here, in a relatively narrow sense, robustness signifies insensitivity to small deviations from the assumptions of noise distributions. This kind of robustness has been called "distributional robustness" in the literature.

Based on robust statistics, a good robust procedure should possess the following desirable characteristics:

(1)   It should have a reasonably good (optimal or nearly optimal) efficiency for the assumed model.

(2)   It should be robust in the sense that small deviations from the model assumptions should impair the performance only slightly. This implies that the approximated model, described in the sense of asymptotic variance of estimates, should be close to the nominal model.

(3)   Somewhat large deviations from the model should not cause a catastrophe.

The primary goal of these requirements is to safeguard against the occurrence of gross errors in a small fraction of the observations. According to Huber [2], an optimal robust procedure is defined in such a way that it minimizes the maximum degradations, and therefore, it is a minimax procedure of some kind.

For practical implementation, certain asymptotic performance criteria (like asymptotic variance) should be employed to process data from real-life situations. One of the key conditions is convergence, otherwise the robustness of the procedure for any finite number of observations N cannot be guaranteed no matter how large N is. Some asymptotic theories have been developed and work well only if there is a high degree of symmetry of criteria (left-right symmetry, translation invariance, etc.). It has been shown that these estimates, called asymptotic minimax estimates, coincide with certain finite-sample minimax estimates although they are derived under quite different assumptions [1].

Various statistical methods are currently used for solving robust estimation problems. For example, there are three basic types of estimates which are called M-, L-, and R-estimates, respectively. They correspond to maximum likelihood type estimates, linear combinations of order statistics, and estimates derived from rank tests.

Among these methods, M-estimates are the most flexible. They can be easily generalized to multiparameter problems even though they are not automatically scale invariant and have to be supplemented by an auxiliary estimate of scale. In fact, many applications of M-estimates have been carried out in a variety of estimation problems. The basic idea is to find out any estimate in such a way that a summation of arbitrary functions $\rho(\theta)$'s is minimized. Usually, the selection of $\rho(\theta)$ is not unique and may lead to different estimates. In particular, the choice

$$\rho(\theta) = -\log f(\theta) \qquad (1)$$

gives the ordinary maximum likelihood estimates.

I.B.   Robust Identification of Systems

As mentioned earlier, robust system identification is simply the application of robust estimation techniques to the problem of state and parameter estimation. The corresponding algorithms for system identification are called "robust identification algorithms." In this chapter and the following chapter, the development of robust identification methods is based on the minimization of a sum of less rapidly increasing functions of the form Eq.(1). A detailed discussion about the criterion will be given in the next chapter.

To describe the formulation of robust identification, we shall first introduce several notations. Let $\Im$ be a set of estimates, $\wp$ be a class of distributions i.e. probability density functions, and $V(T,F)$ be an asymptotic variance of $T \in \Im$ when the distribution $F \in \wp$. Consider the cost function $V(T,F)$ which is the payoff where $T \in \Im$ and $F \in \wp$ are chosen. According to Huber's minimax principle, this game has a saddle point pair (T_0, F_0) if T_0 and F_0 satisfy

$$\underset{T \in \Im}{Min} \ \underset{F \in \wp}{Max} \ V(T,F) = V(T_0, F_0) = \underset{F \in \wp}{Max} \ \underset{T \in \Im}{Min} \ V(T,F) \qquad (2)$$

or

$$V(T,F_0) \geq V(T_0,F_0) \geq V(T_0,F) \qquad (3)$$

for all $T \in \Im$ and $F \in \wp$. The quantity T_0 is called minimax robust estimate and F_0 the least favorable distribution. Obviously, the choice $F = F_0$ guarantees that the asymptotic variance or covariance matrix is minimal among all the possible candidates.

From the above definition, robust identification of system parameters can be defined as an inverse problem of Eq.(2) for the least favorable distribution F_0, which belongs to sets $\wp$. That is

$$T_0 = \underset{T \in \mathfrak{I}}{Arg\, Min}\, V(T, F_0) \tag{4}$$

where $V(.,.)$ is the appropriate cost function and T_0 is the minimax estimates of system parameters or expansion coefficients.

I.C.   Survey of Robust Identification

Although the robust estimation technique has been well established since the early sixties, applications to state and parameter estimation have been considered only in the last fifteen years.

To the best of the authors' knowledge, the first work on robust identification of systems is due to Poljak and Tsypkin [3,4]. Huber's minimax principle has been used for the simplest identification problem of parameter estimation in a static model. Note that Huber's results have been extended to identification of nonlinear regression in those papers. A somewhat different methodology has been used relying on the Cramer-Rao inequality and minimization of Fisher information criterion rather than asymptotic variance. Starting from the Cramer-Rao inequality, the worst distribution F_0 has been selected from a class of distributions $\wp$. The estimates obtained in the two papers are asymptotically optimal in the minimax sense on the class $\wp$ and coincide with those by Huber. However, one important and serious restriction of the approach is that it can only be applied to identification of static models.

Robust Bayesian estimation and robustified Kalman filter have been proposed by Masreliez and Martin [5]. The modified Kalman filter is essentially a robustified Kalman filter, which is desensitized to the influence of heavy-tailed distributions of either the state or the observation error. In fact, robustness of the proposed method has been achieved by introducing influence functions of a monotone variety rather than a nonmonotonically "redescending" variety. It should be noted that the term "influence function" has been used in the paper. The definition of influence functions is different from the one in robust statistics [1]. However, it is proportional to the definition used in robust statistics. In fact, the definition used in this paper has been widely accepted in many application-oriented papers.

Given a linear, single input-single output ARMAX model, Basu and Vandelinde [6] have tried to solve the robust identification problem using the robust crosscorrelation method.

A class of robust recursive algorithms of stochastic approximation type has been developed by Stanković and Kovacević in many papers for identification of linear discrete-time dynamic systems. A typical example is [7]. In fact, approaches of this type are extensions of previous work in [4] and [5]. Since the optimal solution cannot be achieved, several sub-optimal robust identification procedures have been derived on the basis of suitable approximations.

Some significant work on robust identification of linear systems has been done by Puthenpura and Sinha [8,9,10,11,12]. The methods proposed in those papers can be put into two classes. The first is the class of off-line identification methods and has been described in [10] where the modified maximum likelihood method has been employed for robust identification of linear single input-single output systems. In addition, some important properties of the modified maximum likelihood method have been summarized in that paper. The methods in the other class are based mainly on robust recursive identification of linear systems. Like most other cases, the robustness of these methods results from introducing an appropriate influence function. In those papers, the influence function has been generated by minimizing a class of convex functions that are chosen as the criterion.

Instead of utilizing the technique developed in robust statistics, a distribution-free identification criterion, based on sign changes in the residuals, has been presented in [13]. Using membership set theory, an alternative approach for obtaining the range of possible values of the parameter vectors has been proposed in [14].

Recently, there has been significant progress in robust estimation of states and parameters of dynamic systems [15]. Several robust recursive methods have been proposed by the authors for state and parameter estimation of bilinear systems [16,17,18,19]. The basic idea in these methods is to minimize a convex function so that the effect of outliers is reduced or even eliminated by the so-called influence functions. The proofs of convergence for the recursive algorithms have been developed in those papers. In this chapter, we shall not only summarize the results in those papers but also present some new approaches for robust recursive identification of bilinear systems.

It should be mentioned that there are several other off-line methods proposed by the authors for robust identification of dynamic systems. Since this chapter is dedicated to robust recursive identification of bilinear systems, those methods will be discussed in detail in the other chapter.

II. PROBLEM FORMULATION

In general, referring to section I.B, robust identification can be defined as an inverse problem of Eq.(2) for the selected least favorable distribution F_0. It, however, provides only a hint on how to select a robust procedure suited for a particular estimation problem. To cope with problems associated with conventional identification (finite sampling points and multi-parameters), one commonly used approach is to utilize an asymptotic performance criterion (like asymptotic variance). Statisticians have shown that asymptotic minimax estimates minimizing the corresponding asymptotic criterion are finite-sample minimax estimates under some conditions. Fortunately, the conditions of a high degree of symmetry of criteria and a sufficient number of observations are generally satisfied in most cases where estimation problems often occur. Consequently, robust estimates, instead of the general definition as in Eq.(4) can be obtained by minimizing an asymptotic performance criterion, such as asymptotic variance. In particular,to obtain such robust estimates from finite-samples, a class of convex functions $\rho(.)$ has been introduced [1]. They will be utilized for deriving the robust recursive algorithms in this chapter. Further discussions about a class of general convex functions will be given in detail in the next chapter.

Here, to develop the robust methods, three classes of discrete-time models of single-input single-output bilinear systems are considered. In fact, it is not difficult to extend the methods to be developed in this chapter to multi-input multi-output bilinear systems by using their canonical forms. In addition, it is necessary that knowledge about the order and the structure of bilinear systems be known beforehand. The input signal is assumed to be independent of the noise sequence contaminating the output.

The state-space model of bilinear systems under consideration is assumed to be in the observability canonical form shown below:

$$\left.\begin{array}{l} x(t+1) = Px(t) + Qx(t)u(t) + Ru(t) \\ w(t) = Sx(t) \\ y(t) = w(t) + e(t) \end{array}\right\}$$

(5.a)

where we have $x(t) \in \mathfrak{R}^n$, $u(t)$, $w(t)$, and $y(t) \in \mathfrak{R}$. The matrices and vectors included in the above equation are

$$P = \begin{bmatrix} 0 & I \\ a_n & \cdots & a_1 \end{bmatrix}; \qquad Q = \begin{bmatrix} 0 \\ b_n & \cdots & b_1 \end{bmatrix};$$

$$R^T = \begin{bmatrix} c_1' & \cdots & c_n' \end{bmatrix}; \qquad S = \begin{bmatrix} 1 & 0 & \cdots & 0 \end{bmatrix}$$

(5.b)

and the quantity n is the order of the discrete-time bilinear systems. The sequence $\{e(t)\}$ of additive noise usually contains certain amount of outliers. The $e(t)$s' distributions are assumed to belong to the "ϵ-contaminated family" $\wp_e$ [1]

$$\wp_e = \{ F | F = (1-\epsilon)G + \epsilon H, \qquad 1 \geq \epsilon \geq 0 \}.$$

(6)

Usually, G is assumed to be a normal distribution. Corresponding to the distribution of outliers, H is assumed to be unknown but belonging to some classes of symmetric distributions with zero mean and finite variance. Without losing generality, both G and H are defined as normal density $N_1(. | 0, \sigma_1^2)$ and $N_2(. | 0, \sigma_2^2)$, respectively. There should not be much difference if we choose H as other symmetric distributions rather than normal distributions. To simulate the behaviour of outliers, we may choose $\sigma_2^2 \gg \sigma_1^2$. The quantity ϵ is the probability of occurrence of outliers in the assumed Gaussian distribution $N_1(. | 0, \sigma_1^2)$.

To fit the problem into the framework of recursive identification, the input-output representation of the state-space model (5) should be derived. It is very straightforward to derive the following input-output relation from Eq.(5). That is

$$\{A(z^{-1}) + u(t-n)B(z^{-1})\}y(t) = \{C(z^{-1}) + u(t-n)D(z^{-1})\}u(t)$$

$$+ \{A(z^{-1}) + u(t-n)B(z^{-1})\}e(t)$$

(7.a)

where

$$\left.\begin{array}{l} A(z^{-1}) = 1 - a_1 z^{-1} - \cdots - a_n z^{-n} \\ B(z^{-1}) = -b_1 z^{-1} - \cdots - b_n z^{-n} \\ C(z^{-1}) = c_1 z^{-1} + \cdots + c_n z^{-n} \\ D(z^{-1}) = d_1 z^{-1} + \cdots + d_n z^{-n} \end{array}\right\}$$

(7.b)

and z^{-1} is the commonly used backward shift operator. As to the coefficients $\{c_i, i=1 \dots n\}$ and $\{d_i, i=1 \dots n\}$ in Eq.(7), they can be obtained by using the following equation, that is

$$
\begin{bmatrix} c_1 \\ c_2 \\ \vdots \\ c_n \end{bmatrix} = \begin{bmatrix} 1 & & & \\ -a_1 & 1 & & \\ \vdots & \vdots & \ddots & \\ -a_{n-1} & \cdots & -a_1 & 1 \end{bmatrix} \begin{bmatrix} c_1' \\ c_2' \\ \vdots \\ c_n' \end{bmatrix} \tag{8.a}
$$

$$
\begin{bmatrix} d_1 \\ d_2 \\ \vdots \\ d_n \end{bmatrix} = \begin{bmatrix} 0 & & & \\ -b_1 & 0 & & \\ \vdots & \vdots & \ddots & \\ -b_{n-1} & \cdots & -b_1 & 0 \end{bmatrix} \begin{bmatrix} c_1' \\ c_2' \\ \vdots \\ c_n' \end{bmatrix}. \tag{8.b}
$$

For the sake of brevity, Eq.(7) can be rewritten as

$$
y(t) = \varphi_1^T(t)\theta_0 + \nu(t) \tag{9}
$$

where $\varphi_1(t)$ and θ_0 are the observation vector and the true parameter vector, respectively. They are defined as

$$
\varphi_1^T(t) = \left[Y^T(t-1), Y^T(t-1)u(t-n), U^T(t-1), U^T(t-1)u(t-n) \right] \tag{10}
$$

with

$$
U^T(t-1) = \left[u(t-1) \cdots u(t-n) \right]
$$

$$
Y^T(t-1) = \left[y(t-1) \cdots y(t-n) \right]
$$

and

$$
\theta_0^T = \left[a^T, b^T, c^T, d^T \right] \tag{11}
$$

with

$$
a^T = \left[a_1 \cdots a_n \right] \qquad b^T = \left[b_1 \cdots b_n \right]
$$

$$
c^T = \left[c_1 \cdots c_n \right] \qquad d^T = \left[d_1 \cdots d_n \right].
$$

And the residual $\nu(t)$ is given as

$$
\nu(t) = e(t) - a^T E(t-1) - b^T E(t-1)u(t-n) \tag{12}
$$

with

$$
E^T(t-1) = \left[e(t-1) \cdots e(t-n) \right].
$$

It should be noted that the above input-output model in Eq.(9) will be used in sections III, IV, and V of this chapter. In section VI, since we shall develop a robust recursive method, the robust bootstrap method, for joint estimation of states and parameters of bilinear systems, another kind of models will be considered. Similar to Eq.(5), the state-space model is assumed to have the following observability canonical form:

$$x(t+1) = Px(t) + Qx(t)u(t) + Ru(t) + Le(t)$$
$$w(t) = Sx(t) \tag{13}$$
$$y(t) = w(t) + e(t).$$

The only difference between Eqs. (5) and (13) is that the noise term has been included in the state-space equation. While matrices P and Q and vector S in (13) are same as those in (5.b), vectors R and L are defined as

$$R^T = \begin{bmatrix} c_1 \cdots c_n \end{bmatrix}$$
$$L^T = \begin{bmatrix} d_1 \cdots d_n \end{bmatrix}. \tag{14}$$

Again, the variable $e(t)$ is additive noise, having probability density function belonging to the "ϵ-contaminated family" $\wp_e$. To develop the bootstrap method, an input-output relation of Eq.(13) is preferred. That is

$$y(t) = \varphi_2^T(t)\theta_0 + e(t) \tag{15}$$

where $\varphi_2(t)$ and θ_0 are the observation vector and true parameter vector defined as

$$\varphi_2^T(t) = \begin{bmatrix} X^T(t-n), X^T(t-n)u(t-n), U^T(t-1), E^T(t-1) \end{bmatrix} \tag{16}$$
$$U^T(t-1) = \begin{bmatrix} u(t-1) \cdots u(t-n) \end{bmatrix}$$
$$X^T(t-n) = \begin{bmatrix} x_1(t-n) \cdots x_n(t-n) \end{bmatrix}$$
$$E^T(t-1) = \begin{bmatrix} e(t-1) \cdots e(t-n) \end{bmatrix}$$

and

$$\theta_0^T = \begin{bmatrix} a^T, b^T, c^T, d^T \end{bmatrix} \tag{17}$$
$$a^T = \begin{bmatrix} a_n \cdots a_1 \end{bmatrix} \qquad b^T = \begin{bmatrix} b_n \cdots b_1 \end{bmatrix}$$
$$c^T = \begin{bmatrix} c_1 \cdots c_n \end{bmatrix} \qquad d^T = \begin{bmatrix} d_1 \cdots d_n \end{bmatrix}.$$

To develop two other methods in sections VII and VIII, we shall consider a simple input-output equation of bilinear systems rather than the input-output models described in Eqs.(9) and (15). That is

$$\{A(z^{-1}) + u(t-n)B(z^{-1})\}y(t) = \{C(z^{-1}) + u(t-n)D(z^{-1})\}u(t) + K(z^{-1})e(t) \tag{18}$$

where $A(z^{-1})$, $B(z^{-1})$, $C(z^{-1})$, and $D(z^{-1})$ are given by Eq.(7.b). The polynomial $K(z^{-1})$ is defined as

$$K(z^{-1}) = 1 + k_1 z^{-1} + \cdots + k_n z^{-n}. \tag{19}$$

For brevity, Eq.(18) can be rewritten as

$$y(t) = \varphi_3^T(t)\theta_0 + e(t). \tag{20}$$

Here, $e(t)$ is additive noise, having probability density function belonging to the "ϵ-

contaminated family" $\wp_e$ and $\varphi_3(t)$ and θ_0 are defined as

$$\varphi_3^T(t) = \left[Y^T(t-1), Y^T(t-1)u(t-n), U^T(t-1), U^T(t-1)u(t-n), E^T(t-1) \right] \quad (21)$$

with

$$U^T(t-1) = \left[u(t-1) \cdots u(t-n) \right] \qquad Y^T(t-1) = \left[y(t-1) \cdots y(t-n) \right]$$

$$E^T(t-1) = \left[e(t-1) \cdots e(t-n) \right]$$

and

$$\theta_0^T = \left[a^T, b^T, c^T, d^T, k^T \right] \quad (22)$$

with

$$a^T = \left[a_1 \cdots a_n \right] \qquad b^T = \left[b_1 \cdots b_n \right]$$

$$c^T = \left[c_1 \cdots c_n \right] \qquad d^T = \left[d_1 \cdots d_n \right] \qquad k^T = \left[k_1 \cdots k_n \right].$$

For the models described above, conventional identification methods, i.e. non-robust methods, cease to work in the presence of outliers contaminating the input-output data. Therefore, the objective of robust identification of bilinear systems is to estimate the parameter vector θ or/and states in a manner that the estimates are distorted very little by outliers. In the following sections, we shall present six robust recursive estimation methods to carry out the task.

III. ROBUST RECURSIVE LEAST SQUARES METHOD WITH MODIFIED WEIGHTS

It is known that the recursive least squares methods for parameter estimation has been well developed [20,21]. Utilizing a convex function $\rho(x)$, a robust recursive least squares method with modified weights for bilinear system identification is presented in this section [16][1]. In fact, it is an extension of the robust version by Puthenpura et al. [8].

III.A. Robust Recursive Least Squares Method

Based on Huber's minimax principle, we can define an asymptotic criterion as

$$J_1(\theta) = \frac{1}{N} \sum_{t=1}^{N} \rho\{y(t) - \phi^T(t)\,\theta\} \quad (23)$$

where the observation vector $\phi^T(t)$ is taken as $\varphi_1^T(t)$ defined in Eq.(10) and $\rho(.)$ is a piecewise continuously differentiable convex function. The derivative of the convex function $\rho(.)$ is the influence function given as

$$\psi(\nu) = \frac{d\rho(\nu)}{d\nu}. \quad (24)$$

The optimal estimates of parameters should be, then, the solution of the following equation

---

[1]   Figure 1, lemma 3 and theorem 1 have been reproduced from reference 16, by kind permission of IEE.

$$\frac{1}{N}\sum_{t=1}^{N}\phi(t)\,\psi(y(t) - \phi^T(t)\,\theta) = 0 \tag{25.a}$$

or,

$$\frac{1}{N}\sum_{t=1}^{N}\alpha(t)\,\phi(t)\,(y(t) - \phi^T(t)\,\theta) = 0. \tag{25.b}$$

In above equation, the ratio $\alpha(t)$ is defined as

$$\alpha(t) = \begin{cases} \dfrac{\psi(y(t) - \phi^T(t)\,\theta)}{y(t) - \phi^T(t)\,\theta} \\[2mm] 1 \qquad\qquad if \quad y(t) - \phi^T(t)\,\theta = 0. \end{cases} \tag{26.a}$$

To calculate $\alpha(t)$, however, the vector θ in Eq.(26.a) should be replaced by an estimate $\hat{\theta}$. Here, the estimate $\hat{\theta}(t\text{-}1)$ is utilized based on the information up to and including time $t\text{-}1$. It leads to

$$\alpha(t) = \begin{cases} \dfrac{\psi(y(t) - \phi^T(t)\,\hat{\theta}(t\text{-}1))}{y(t) - \phi^T(t)\,\hat{\theta}(t\text{-}1)} \\[2mm] 1 \qquad\qquad if \quad y(t) - \phi^T(t)\,\hat{\theta}(t\text{-}1) = 0 \end{cases} \tag{26.b}$$

where the quantity $\phi^T(t)\hat{\theta}(t\text{-}1)$ is the predicted output of systems at the t-th instant. Evidently, instead of residuals $\bar{\varepsilon}(t,\theta(t))$ or $\bar{\varepsilon}(t)$, the argument of the influence function is specified as the prediction error $\varepsilon(t,\theta)$ or $\varepsilon(t)$, that is

$$\varepsilon(t,\theta) = y(t) - \phi^T(t)\,\hat{\theta}(t\text{-}1). \tag{27}$$

Following the normal procedure for deriving a recursive algorithm, the parameter estimate can be given by

$$\hat{\theta}(t) = \hat{\theta}(t\text{-}1) + \frac{P(t\text{-}1)\,\phi(t)\,(y(t) - \phi^T(t)\,\hat{\theta}(t\text{-}1))}{1/\alpha(t) + \phi^T(t)P(t\text{-}1)\phi(t)} \tag{28.a}$$

$$P(t) = P(t\text{-}1) - \frac{P(t\text{-}1)\,\phi(t)\,\phi^T(t)P(t\text{-}1)}{1/\alpha(t) + \phi^T(t)P(t\text{-}1)\phi(t)} \tag{28.b}$$

and

$$P(t) = R^{-1}(t) = \left[\sum_{k=1}^{t}\alpha(k)\,\phi(k)\,\phi^T(k)\right]^{-1}. \tag{28.c}$$

The starting values may be chosen as

$$\hat{\theta}(0) = \emptyset; \qquad\qquad P(0) = kI, \qquad k \geqslant 0. \tag{28.d}$$

Obviously, the algorithm described above, the robust recursive least squares method with modified weights, is very similar to the non-robust version. The main difference is including the weight $\alpha(t)$. From the definition in Eq.(26), $\alpha(t)$ depends upon the prediction error and the influence function. If the prediction error is larger than (or less than) the selected turning point ω (or $-\omega$), then the weight $\alpha(t)$ is less than

1, otherwise it remains 1.

The influence function $\psi(.)$ which is introduced for reducing the effect of outliers on the parameter estimates, may be chosen in several different ways. Usually, it is a nonlinear function of residuals rather than the linear function used in the normal quadratic criterion. Results of simulation, using different influence functions, will be included in section V. The choice of the turning point of the influence function will be briefly discussed later in this section.

III.B. Convergence Analysis

In this sub-section, a theorem proving the convergence of the algorithm is presented. To establish the theorem, several lemmas are given first. Notice that one of the lemmas, namely, lemma 1, has been given in [8].

Lemma 1.

For a given influence function ψ, which is
(i)   odd;
(ii)   Riemann summable over any finite interval;
(iii) $\exists\, M, m,\ 0 < M < \infty,\ m > 1$, so that $\forall\, x \geq M,\ \psi(x) \geq m/x$,
the function $f(x)$, which is defined as

$$f(x) = K exp\left(-\int_0^x \psi(t)\,dt\right) \qquad 0 < K < \infty$$

is a probability density function, i.e.

$$\int_{-\infty}^{+\infty} f(x)\,dx = 1.$$

Furthermore, $f(x)$ is even.
                                                                                            ∎

Proof: A proof of the lemma has been given by Puthenpura et al. in [8].
Note: Assumptions (i) and (iii) are used to assure that the influence function $\psi(.)$ is monotonically non-decreasing. Furthermore, the convexity of the resulting cost function $\rho(.)$ described in Eq.(23) is guaranteed.

Lemma 2.

In addition to the assumptions of lemma 1, let

$$r(t) = \frac{\phi^T(t)\big(P(t-1)\big)^2 \phi(t)}{\big(d(t)\big)^2}$$

where

$$d(t) = \big(1 + \alpha(t)\phi^T(t)P(t-1)\phi(t)\big)^2.$$

The matrix $P(t)$ and the observation vector $\phi(t)$ are described by Eqs.(28.b) and (10), respectively. Assuming that there is $\lambda \in (0,+\infty)$ such that $|\nu(t)| < \lambda$ for all $t\in[0,+\infty)$, then

$$\sum_{t=1}^{\infty} r(t) < \infty.$$

                                                                                            ∎

Proof: See Appendix X.A.

Lemma 3.
   Under the assumptions in lemmas 1 and 2, the following is true

$$2M^2\sigma^2\sum_{t=1}^{\infty}\frac{1}{t}\phi^T(t)P(t)\phi(t) < \infty \qquad\qquad w.p.1$$

if

$$\lim_{N\to\infty}\frac{1}{N}\sum_{t=1}^{N}|\phi(t)|^2 < \infty \qquad\qquad w.p.1$$

■

Proof: See Appendix X.A.
Note: The abbreviation $w.p.1$ means with probability one [21].

Theorem 1.
   Consider algorithm (28), with the bilinear system modeled by Eq.(5). In addition to the assumptions in lemmas 1, 2, and 3, assume that
(i)   input series $\{u(t)\}$ is independent of noise sequence $\{e(t)\}$;
(ii)   the series $\{e(t)\}$ is a sequence of random variables which are not necessarily Gaussian but have zero mean and finite variance, such that

$$E\left(\nu(t)\,|\mathfrak{C}_{t-1}\right) = 0$$

$$E\left(\nu^2(t)\,|\mathfrak{C}_{t-1}\right) = \sigma^2 < \infty$$

where $\mathfrak{C}_{t-1}$ is the σ-algebra generated by $\{\nu(0), ..., \nu(t-1), \varphi_1(0), ..., \varphi_1(t)\}$ and $\nu(t)$ is given in Eq.(12);
(iii)   the difference $\{\varepsilon(t)-\nu(t)\} \in \mathfrak{C}_{t-1}$ with $\varepsilon(t)$ defined as the prediction error.
then

$$\left(\hat{\theta}(t) - \theta_0\right)^T R(t)\left(\hat{\theta}(t) - \theta_0\right) \to 0 \qquad\qquad w.p.1 \; as \; N \to \infty$$

and

$$\frac{1}{N}\sum_{t=1}^{N}\left(\bar{\varepsilon}(t) - e(t)\right)^2 \to 0 \qquad\qquad w.p.1 \; as \; N \to \infty$$

where $\bar{\varepsilon}(t)$ is residual.

■

Proof: See Appendix X.A.
Note: The proof is similar to that given by Neveu [22] and Ljung and Söderström [21] with the difference that the increased σ-algebra $\mathfrak{C}_{t-1}$ is generated by $\{\nu(0), ..., \nu(t-1), \varphi_1(0), ..., \varphi_1(t)\}$ instead of $\{e(0), ..., e(t-1), \varphi_1(0), ..., \varphi_1(t)\}$. Therefore, the proof always holds, irrespective of whether the noise under consideration is white or colored. Actually, $\nu(t)$ is a linear expression of $\{e(0), ..., e(t-1)\}$ related to $u(t-n)$.

III.C. Results of Simulation

   To show the robustness of the proposed algorithm, an example has been provided and compared with non-robust methods.   The bilinear system under consideration is described by (5) with matrices and vectors as

$$P = \begin{bmatrix} 0 & 1 \\ -0.7 & 1.4 \end{bmatrix} \qquad Q = \begin{bmatrix} 0 & 0 \\ 0.2 & 0.33 \end{bmatrix} \qquad \text{(29.a)}$$

$$R^T = \begin{bmatrix} 1.0 & 2.0 \end{bmatrix} \qquad S = \begin{bmatrix} 1 & 0 \end{bmatrix}$$

which gives the parameter vector

$$\theta_0^T = \begin{bmatrix} 1.4 & -0.7 & 0.33 & 0.2 & 1.0 & 0.6 & 0.0 & -0.33 \end{bmatrix}. \qquad \text{(29.b)}$$

Two sets of data are generated. The first set is noise-free. Another contains noise term $e(t)$ which belongs to "ϵ-contaminated family" $\wp_e$ described in Eq.(6). The variances of normal noise and outliers occurring with probability $\epsilon = 0.1$ are chosen as $\sigma_1^2 = 0.01$ and $\sigma_2^2 = 20.0$, respectively. The influence function is

$$\psi(x) = \begin{cases} \omega \, Sign(x) & |x| > \omega \\ x & |x| \leq \omega . \end{cases} \qquad \text{(30)}$$

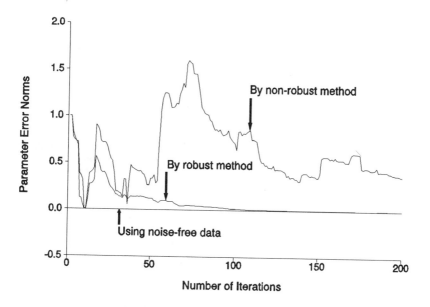

Figure 1   Comparison of parameter error norms by robust and non-robust LS method

In general, it is difficult to select the turning point of the influence function especially when neither variance of normal noise nor variance of outliers is known beforehand. Assuming the known variance of residuals, an empirical formula for calculating the turning point of the influence functions has been suggested by Puthenpura et al. [8]. However, it is too rough to meet requirements in many physical applications. Intuitively, one of the better approaches to solve this problem might be that parameters and noise variance are estimated simultaneously so that the turning point can be determined automatically. Further discussions and a detailed investigation of this approach will be the topic of another chapter. Here, the turning point ω is simply taken as 0.5.

To compare the robustness of the estimated parameters, the following parameter error norm is introduced

$$n(t) = \frac{\|\hat{\theta} - \theta_0\|}{\|\theta_0\|} \tag{31}$$

where the Euclidean norm is used. The error norms of the estimated parameters (robust, non-robust, noise-free) are plotted in Fig. 1. From the results, it is quite evident that the algorithm derived in this section is more powerful when compared to the non-robust least squares method.

IV. ROBUST RECURSIVE INSTRUMENTAL VARIABLE METHOD WITH MODIFIED WEIGHTS

One straightforward extension to the method developed in the last section is the robust recursive instrumental variable (IV) method with modified weights. In fact, a robust version for linear system identification has been proposed in [9]. In this section, a robust instrumental variable method is derived for bilinear system identification which has been proposed by Dai and Sinha in [18][2].

IV.A. Instrumental Variable Method

Note that the least squares method can only produce biased estimates if the contaminating noise is not completely uncorrelated with the observation vector $\varphi_1(t)$, i.e. the noise is not white. The IV method is exclusively designed to solve this problem. The basic idea behind the ordinary least squares and IV, however, is same.

Like the least squares method, parameter estimates by the IV method can be obtained in such a way that the following criterion should be minimized

$$J_2(\theta) = E\left\{\frac{1}{2}(\varepsilon(t,\theta))^2\right\} \tag{32}$$

where $\varepsilon(t,\theta)$ is the prediction error defined in Eq.(27).

Instead of letting the approximate gradient of $\varepsilon(t,\theta)$ as

$$\frac{\partial\varepsilon(t,\theta)}{\partial\theta} \approx \phi(t) \tag{33}$$

with $\phi(t)$ chosen by Eq.(10), the following gradient is chosen

$$\frac{\partial\varepsilon(t,\theta)}{\partial\theta} \approx \zeta(t). \tag{34}$$

The vector $\zeta(t)$ should have desirable properties as below

---

[2]   Figures 2 to 4 have been reproduced from reference 18, by kind permission of IEEE.

$$E\{\varsigma(t)\varepsilon(t,\theta)\} = 0 \qquad (35)$$

and

$$E\{\varsigma(t)\phi^T(t)\} > 0. \qquad (36)$$

Then, the IV estimate can be obtained by solving

$$E\{\varsigma(t)(y(t) - \phi^T(t)\theta)\} = 0. \qquad (37)$$

It has been proved that the IV estimate $\hat{\theta}$ converges to θ_0 as t tends to infinity [21]. The vector $\varsigma(t)$ is called the instrumental variable vector.

IV.B. Robust Instrumental Variable Method

Although the instrumental variable method is especially designed for dealing with colored noise contaminating output sequences, it could only provide very poor estimates of parameters in the presence of outliers. Obviously, one of the effective approaches to solve the problem is to replace (32) by a robust criterion, that is

$$J_3(\theta) = E\{\rho(\varepsilon(t,\theta))\} \qquad (38)$$

with the prediction error described before.

Notice that, for a sufficiently large value of N, Eq.(38) can be rewritten as

$$J_4(\theta) = \frac{1}{N}\sum_{t=1}^{N} \rho(\varepsilon(t,\theta)) \qquad (39)$$

if the disturbances are stationary and ergodic.

Using the IV principle and proceeding on the same lines as that in section III lead to

$$\frac{1}{N}\sum_{t=1}^{N} \varsigma(t)\psi(y(t) - \phi^T(t)\theta) = 0. \qquad (40)$$

Therefore, the robust estimate of θ calculated by the above equation can be obtained recursively from the following relationships

$$\hat{\theta}(t) = \hat{\theta}(t-1) + \frac{P(t-1)\varsigma(t)(y(t) - \phi^T(t)\hat{\theta}(t-1))}{1/\alpha(t) + \phi^T(t)P(t-1)\varsigma(t)} \qquad (41.a)$$

$$P(t) = P(t-1) - \frac{P(t-1)\varsigma(t)\phi^T(t)P(t-1)}{1/\alpha(t) + \phi^T(t)P(t-1)\varsigma(t)} \qquad (41.b)$$

and

$$P(t) = R^{-1}(t) = \left[\sum_{k=1}^{t} \alpha(k)\varsigma(k)\phi^T(k)\right]^{-1}. \qquad (41.c)$$

The quantity $\alpha(t)$ is as described by Eq.(26). The influence function is specified by Eq.(30). The initial values are the same as ones given by Eq.(28.d).

IV.C. Convergence Analysis

The result on convergence of algorithm (41) is stated in the following lemma and theorem.

Lemma 4.
Under the assumptions of lemma 1, the following inequality holds

$$\sum_{t=1}^{\infty} \frac{1}{t} \zeta^T(t) P(t) \zeta(t) < \infty \qquad w.p.1$$

where $P(t)$ is specified in Eq.(41.c), if the following conditions are satisfied:
(i)   There is $\lambda \in (0, +\infty)$ such that $|\nu(t)| < \lambda$ for all $t \in [0, +\infty)$;
(ii) $\forall t \in [0, +\infty)$, the following inequality holds for some $\delta > 0$:

$$\bar{R}(t) = \frac{1}{t} \sum_{k=1}^{t} \zeta(k) \phi^T(k) \geq \delta I$$

(iii)

$$sup \; \|\zeta(t)\|^2 = \|\zeta_0\|^2 < \infty \qquad for \quad t \in [0, +\infty).$$

■

Proof: See Appendix X.B.
Note: Notice that the conditions enforced in this lemma are quite common [21]. The matrix inequality $A \geq B$ means that the matrix A-B is positive semidefinite. In particular, condition (ii) simply implies that $E\{\zeta(t)\phi^T(t)\}$ is positive definite.

Theorem 2.
Consider algorithm (41), corresponding to bilinear systems modeled by (5). In addition to the assumptions in lemmas 1 and 4, assume that
(i)   input series $\{u(t)\}$ is independent of noise $\{e(t)\}$;
(ii)   the series $\{e(t)\}$ is a sequence of random variables which are not necessarily Gaussian but have zero mean and finite variance, such that

$$E\left(\nu(t) \, |\mathfrak{C}_{t-1}\right) = 0$$

$$E\left(\nu^2(t) \, |\mathfrak{C}_{t-1}\right) = \sigma^2 < \infty$$

where $\mathfrak{C}_{t-1}$ is the σ-algebra generated by $\{\nu(0), ..., \nu(t-1), \varphi_1(0), ..., \varphi_1(t)\}$ and $\nu(t)$ is given in Eq.(12);
(iii)   the difference $\{\varepsilon(t)\text{-}\nu(t)\} \in \mathfrak{C}_{t-1}$ with $\varepsilon(t)$ defined as the prediction error.
(iv)   instrumental variable $\{\zeta(t)\}$ and noise $\{e(t)\}$ are independent;
then,

$$\left(\hat{\theta}(t) - \theta_0\right)^T R(t)\left(\hat{\theta}(t) - \theta_0\right) \to 0 \qquad w.p.1 \quad as \quad N \to \infty$$

and

$$\frac{1}{N}\sum_{t=1}^{N} \left(\bar{\varepsilon}(t) - e(t)\right)^2 \to 0 \qquad w.p.1 \quad as \quad N \to \infty$$

where $\bar{\varepsilon}(t)$ is residual.

■

Proof: See Appendix X.B.
The proof of theorem 2 is very similar to the one in Appendix X.A except that some information vector $\phi(t)$ should be replaced by the instrumental variable $\zeta(t)$. As to the robust symmetric IV method which will be discussed in the next section, a proof of convergence may be desired in a straightforward manner by following either of the

approaches given in Appendixes X.A and X.B.

IV.D. Results of Simulation

To illustrate the robustness and convergence of the proposed approach, two examples are presented. The bilinear system under consideration is expressed by (5) with the parameter matrices and vectors

$$P = \begin{bmatrix} 0 & 1 \\ -0.7 & 1.5 \end{bmatrix} \qquad Q = \begin{bmatrix} 0 & 0 \\ 0.2 & 0.33 \end{bmatrix} \tag{42.a}$$

$$R^T = \begin{bmatrix} 1.0 & 0.4 \end{bmatrix} \qquad S = \begin{bmatrix} 1 & 0 \end{bmatrix}.$$

Therefore, the parameter vector is

$$\theta_0^T = \begin{bmatrix} 1.5 & -0.7 & 0.33 & 0.2 & 1.0 & -1.1 & 0.0 & -0.33 \end{bmatrix}. \tag{42.b}$$

Two sets of input-output data are generated by Eq.(5). The noise term $e(t)$, here, belongs to the "ϵ-contaminated family" $\wp_\epsilon$ described in (6). In both cases, the probability ϵ is taken as 0.1 and σ_1^2 as 0.01 except that variances of outliers are different, $\sigma_2^2 = 20.0$ and $\sigma_2^2 = 50.0$, respectively. Correspondingly, the turning points of the influence functions are chosen as 0.3 and 0.5

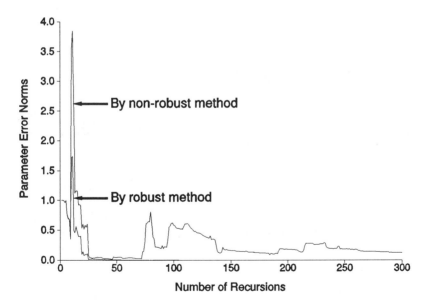

Figure 2   Comparison of parameter error norms by robust and non-robust IV methods (the turning point is 0.3)

The instrumental variable $\zeta(t)$ used in the simulation is the most commonly used one, that is

$$\zeta^T(t) = \left[\hat{Y}^T(t-1), \hat{Y}^T(t-1)u(t-n), U^T(t-1), U^T(t-1)u(t-n) \right] \qquad (43)$$

where $\hat{Y}(t-1)$ is an estimated output vector at $t-1$.

To show the robustness and convergence of the estimated parameters, the parameter error norms (robust and non-robust) defined in Eq.(31) are plotted in Figs. 2 and 3. Again, it is confirmed that the robust method derived in this section is superior to the non-robust one.

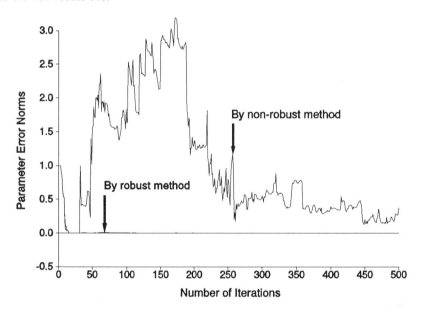

Figure 3   Comparison of parameter error method by robust and non-robust IV methods (the turning point is 0.5)

V.   ROBUST RECURSIVE OUTPUT ERROR METHOD WITH MODIFIED WEIGHTS

One efficient approach for system identification in a recursive manner is output error method [23]. In this section, a robust version of output error method will be developed for bilinear system identification [18,19].

V.A.   Output Error Method

From the point of view of model reference, recursive identification could be formulated as a model tracking problem. This leads to the output error method, which could be less sensitive to properties of noise sequence $\{e(t)\}$ than the normal recursive least squares approach.

Consider a class of bilinear systems expressed by (5). In order that the observation vector $\phi(t)$ does not directly depend on observed y-variables and hence insensitive to the output noise $e(t)$ added to the measurements, the observed y-variables in vector φ_1 are replaced by a sequence of estimated outputs $\{\hat{y}(t)\}$, like Eq.(43) above. That is

$$\phi^T(t) = \left[\hat{Y}^T(t-1), \hat{Y}^T(t-1)u(t-n), U^T(t-1), U^T(t-1)u(t-n) \right] \tag{44}$$

$$U^T(t-1) = \left[u(t-1) \cdots u(t-n) \right]$$

$$\hat{Y}^T(t-1) = \left[\hat{y}(t-1) \cdots \hat{y}(t-n) \right]$$

where $\hat{y}(t-1)$, ..., $\hat{y}(t-n)$, $t=1$, ..., N, can be obtained by the following equations with estimated parameters $\{\hat{a}_i, \hat{b}_i, \hat{c}_i, \hat{d}_i \; i=1,..., n\}$ at time $t-1$, ..., $t-n$, respectively.

$$\{\hat{A}(z^{-1}) + u(t-n)\hat{B}(z^{-1})\}\hat{y}(t) = \{\hat{C}(z^{-1}) + u(t-n)\hat{D}(z^{-1})\}u(t) \tag{45}$$

where

$$\hat{A}(z^{-1}) = 1 - \hat{a}_1 z^{-1} - \cdots - \hat{a}_n z^{-n}$$

$$\hat{B}(z^{-1}) = -\hat{b}_1 z^{-1} - \cdots - \hat{b}_n z^{-n}$$

$$\hat{C}(z^{-1}) = \hat{c}_1 z^{-1} + \cdots + \hat{c}_n z^{-n}$$

$$\hat{D}(z^{-1}) = \hat{d}_1 z^{-1} + \cdots + \hat{d}_n z^{-n}.$$

Obviously, Eq.(45) can be rewritten as

$$\hat{y}(t) = \phi^T(t)\,\hat{\theta}(t) \tag{46}$$

with $\phi(t)$ specified in Eq.(44) and $\hat{\theta}$ given as

$$\hat{\theta}^T(t) = \left[\hat{a}^T, \hat{b}^T, \hat{c}^T, \hat{d}^T \right] \tag{47}$$

$$\hat{a}^T = \left[\hat{a}_1 \cdots \hat{a}_n \right] \qquad \hat{b}^T = \left[\hat{b}_1 \cdots \hat{b}_n \right]$$

$$\hat{c}^T = \left[\hat{c}_1 \cdots \hat{c}_n \right] \qquad \hat{d}^T = \left[\hat{d}_1 \cdots \hat{d}_n \right].$$

Furthermore, natural updating formulas analogous to the recursive least squares algorithm can be easily obtained.

V.B. Robust Recursive Output Error Method

Even though the output error method can produce asymptotic unbiased estimates for noisy measurements, it fails to do so when outliers contaminate the input-output data. Intuitively, a robust version of output error method needs to be derived so that certain robustness of the method can be achieved.

It is very straightforward to develop the robust output error method by following the same procedure as in section III. The only difference is that when forming the observation vector $\varphi_1(t)$ the sequence of estimated outputs $\{\hat{y}(t)\}$ is used rather than the observed output $y(t)$. In other words, the observation vector $\phi(t)$ is specified by Eq.(44), instead of Eq.(10). With the difference described above, the corresponding recursive formulas for the parameter estimate $\hat{\theta}$ and the matrix $P(t)$ can be expressed as

$$\hat{\theta}(t) = \hat{\theta}(t-1) + \frac{P(t-1)\phi(t)\left(y(t) - \phi^T(t)\hat{\theta}(t-1)\right)}{1/\alpha(t) + \phi^T(t)P(t-1)\phi(t)}$$ (48.a)

$$P(t) = P(t-1) - \frac{P(t-1)\phi(t)\phi^T(t)P(t-1)}{1/\alpha(t) + \phi^T(t)P(t-1)\phi(t)}$$ (48.b)

and

$$P(t) = R^{-1}(t) = \left[\sum_{k=1}^{t}\alpha(k)\phi(k)\phi^T(k)\right]^{-1}.$$ (48.c)

It will be seen that algorithm (48) is very similar to algorithm (28) except for a different definition of the observation vector $\phi(t)$. To start the algorithm, the initial values may be chosen as in (28.d).

V.C.   Convergence Analysis

A convergence analysis of the robust output error method is given as below. The theorem to be described is similar to theorem 1 except that a different observation vector $\phi(t)$ is used.

Theorem 3.
Consider algorithm (48), with bilinear systems modeled by (5). In addition to the assumptions in lemmas 1, 2, and 3, assume that
(i)   input series $\{u(t)\}$ is independent of noise sequence $\{e(t)\}$;
(ii)   the series $\{e(t)\}$ is a sequence of random variables which are not necessarily Gaussian but have zero mean and finite variance, such that

$$E\left(\nu(t)\,|\mathfrak{C}_{t-1}\right) = 0$$

$$E\left(\nu^2(t)\,|\mathfrak{C}_{t-1}\right) = \sigma^2 < \infty$$

where $\mathfrak{C}_{t-1}$ is the σ-algebra generated by $\{\nu(0), ..., \nu(t-1), \varphi_1(0), ..., \varphi_1(t)\}$ and $\nu(t)$ is given in Eq.(12);
(iii)   the difference $\{\varepsilon(t)-\nu(t)\} \in \mathfrak{C}_{t-1}$ with $\varepsilon(t)$ defined as the prediction error.
then

$$\left(\hat{\theta}(t) - \theta_0\right)^T R(t)\left(\hat{\theta}(t) - \theta_0\right) \rightarrow 0 \qquad w.p.1 \quad as \quad N \rightarrow \infty$$

and

$$\frac{1}{N}\sum_{t=1}^{N}\left(\bar{\varepsilon}(t) - e(t)\right)^2 \rightarrow 0 \qquad w.p.1 \quad as \quad N \rightarrow \infty$$

where $\bar{\varepsilon}(t)$ is residual.

∎

Proof: See Appendix X.A.
Note that the observation vector $\phi(t)$ defined in Eq.(44) is $\mathfrak{C}_{t-1}$-measurable. Therefore, a different observation vector $\phi(t)$ should not affect the proof of theorems 1 and 3. In addition, it is interesting to note that algorithm (48) coincides with one of the symmetric recursive instrumental variable method if the instrumental variables are

taken as inputs and "model outputs" i.e. estimated outputs. It brings out another way to prove the convergence of the algorithm.

V.D. Results of Simulation

Consider the same bilinear system (5). The true parameter vector is

$$\theta_0^T = \begin{bmatrix} 1.4 & -0.7 & 0.33 & 0.2 & 1.0 & 0.6 & 0.0 & -0.33 \end{bmatrix}. \tag{49}$$

The noise-free output $w(t)$ of the system is perturbed by a sequence of additive noise $e(t)$ generated by Eq.(6). In this example, the same data as in section III are used.

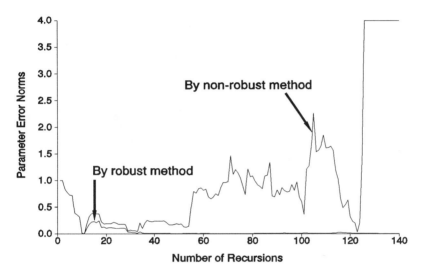

Figure 4   Comparison of parameter error norms by robust and non-robust OE methods

With the parameter error norm defined in Eq.(31), comparison with non-robust output error method is provided in Fig. 4. Here, the influence function is chosen like (30) and the turning point ω is chosen as 0.5. From Fig.4, the robustness of the suggested method is evident. It should be noted that, in Fig.4, when the number of recursions reaches 120, we have put the values of the parameter error norm using the non-robust output error method as the upper boundary 4.0 since the actual values of the norm are too large to be in the same figure.

For the same data, simulation results using different influence functions are showed in Fig. 5. The influence functions are described in Eqs.(30), (50), and (51):

$$\psi(x) = \begin{cases} 1 & |x| > 0 \\ 0 & x = 0 \\ -1 & |x| < 0 \end{cases} \tag{50}$$

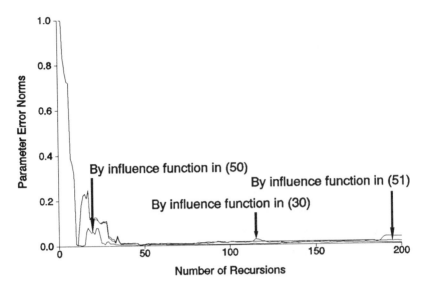

Figure 5   Comparison of parameter error norms by robust OE method with different influence functions

$$\psi(x) = \begin{cases} 0 & |x| > \omega \\ x & |x| \le \omega. \end{cases} \qquad (51)$$

For (51), the turning point ω is set as 6.0. The results show that the turning point for (51) should be about 10 times larger than that for Eq.(30). In addition, the influence function described in (51) does not show us a good convergence because it is hard to get correction for estimated parameters from the incoming data, but the influence function in (30) can do.

VI. ROBUST BOOTSTRAP METHOD FOR COMBINED ESTIMATION OF STATES AND PARAMETERS OF BILINEAR SYSTEMS

Many contributions have been made to the joint estimation of states and parameters [24]. However, most of them are not sufficiently robust to obtain unbiased estimates from data containing outliers, sometimes they even fail to converge. An algorithm, using a Bayesian approach, for robust combined estimation of states and parameters of linear systems in a bootstrap manner has been presented by Puthenpura and Sinha [11]. In this section, a robust bootstrap algorithm is presented to estimate states and parameters of bilinear systems [17]. It is slightly different from the algorithm in [11] and is derived by using the extended least squares approach[3].

VI.A   Bootstrap Method

For simplicity, we assume that the bilinear system under consideration can be modeled by Eq.(13). An input-output form of (13) has been given in Eq.(15).

The bootstrap method for joint estimation of states and parameters of systems is composed of three basic steps. These are:

1. Realization of the given system in a special canonical form;
2. Estimation of the parameters from samples of the input and output and from the nominal values of states;
3. State estimation using estimated parameters and going back to step 2.

Thus steps 2 and 3 are coupled in a bootstrap manner.

VI.B.   Robust Bootstrap Method

Bearing Huber's minimax principle in mind, the robust bootstrap method can be developed by using the extended least squares approach. Denote the estimates of states and noise as $\hat{x}_i(t\text{-}n)$, $i=1, ..., n$ and $\bar{\varepsilon}(t)$ respectively. Then, replacing states and noise in Eq.(15) by the corresponding estimates leads to

$$y(t) = \phi^T(t)\theta + \bar{\varepsilon}(t) \tag{52}$$

where $\phi(t)$ is defined as

$$\phi^T(t) = \left[\hat{X}^T(t\text{-}n), \hat{X}^T(t\text{-}n)u(t\text{-}n), U^T(t\text{-}1), \hat{E}^T(t\text{-}1) \right] \tag{53}$$

$$U^T(t\text{-}1) = \left[u(t\text{-}1) \cdots u(t\text{-}n) \right] \qquad \hat{X}^T(t\text{-}n) = \left[\hat{x}_1(t\text{-}n) \cdots \hat{x}_n(t\text{-}n) \right]$$

$$\hat{E}^T(t\text{-}1) = \left[\bar{\varepsilon}(t\text{-}1) \cdots \bar{\varepsilon}(t\text{-}n) \right].$$

Introduce the robust criterion given by Eq.(23). Then, following almost the same procedure as in section III, the parameter estimate $\hat{\theta}$ of bilinear systems can be obtained recursively by

$$\hat{\theta}(t) = \hat{\theta}(t\text{-}1) + \frac{P(t\text{-}1)\phi(t)\big(y(t) - \phi^T(t)\hat{\theta}(t\text{-}1)\big)}{1/\alpha(t) + \phi^T(t)P(t\text{-}1)\phi(t)} \tag{54.a}$$

$$P(t) = P(t\text{-}1) - \frac{P(t\text{-}1)\phi(t)\phi^T(t)P(t\text{-}1)}{1/\alpha(t) + \phi^T(t)P(t\text{-}1)\phi(t)} \tag{54.b}$$

and

$$P(t) = R^{-1}(t) = \left[\sum_{k=1}^{t} \alpha(k)\phi(k)\phi^T(k) \right]^{-1}. \tag{54.c}$$

The ratio $\alpha(t)$ and the influence function $\psi(x)$ are chosen by Eqs.(26) and (30),

---

[3]   Figures 6 to 8, lemma 5 and theorem 4 have been reproduced from reference 17, by kind permission of Automatica.

respectively. Of course, the turning point ω of the influence function should be appropriately selected beforehand.

After obtaining the estimated parameters, the estimated states and noise can be updated through the following equations:

$$\left.\begin{aligned}
\hat{x}(t+1) &= \hat{P}\hat{x}(t) + \hat{Q}\hat{x}(t)u(t) + \hat{R}u(t) + \hat{L}\bar{\varepsilon}(t) \\
\hat{w}(t) &= S\hat{x}(t) \\
\bar{\varepsilon}(t) &= y(t) - \hat{w}(t)
\end{aligned}\right\} \tag{55}$$

where

$$\hat{P} = \begin{bmatrix} 0 & I \\ \hat{a}_n & \cdots & \hat{a}_1 \end{bmatrix}; \qquad\qquad \hat{Q} = \begin{bmatrix} 0 \\ \hat{b}_n & \cdots & \hat{b}_1 \end{bmatrix}; \tag{56}$$

$$\hat{R}^T = \begin{bmatrix} \hat{c}_1 & \cdots & \hat{c}_n \end{bmatrix}; \qquad \hat{L}^T = \begin{bmatrix} \hat{d}_1 & \cdots & \hat{d}_n \end{bmatrix}; \qquad S = \begin{bmatrix} 1 & 0 & \cdots & 0 \end{bmatrix}.$$

New estimated states can then be used in the estimated observation vector (53) so that the whole procedure can be repeated in a bootstrap manner. To start the procedure, initial values are taken as

$$\hat{\theta}(0) = \phi; \qquad P(0) = kI \qquad k \gg 0; \qquad \hat{x}(0) = \phi. \tag{57}$$

VI.C. Convergence Analysis

A proof of convergence of the robust bootstrap algorithm is provided in this section. Instead of a Martingale convergence proof, the proof will be developed exactly in the way of extended least squares algorithm, a most celebrated example of pseudolinear regression [21].

Lemma 5.

For the algorithm described by Eqs.(54) to (57), denote

$$\tau_i(z^{-1}) = 1 - \sum_{j=1}^{n-i} \left(a_j + b_j u(t-n) \right) z^{-j}$$

$$h = diag \begin{bmatrix} \tau_1(z^{-1}) & \cdots & \tau_{n-1}(z^{-1}) & 1 \end{bmatrix}$$

$$H(z^{-1}) = \frac{1}{1 + K(z^{-1})} \begin{bmatrix} I & 0 \\ & h \\ 0 & h \end{bmatrix}$$

with

$$\begin{bmatrix} k_1 \\ k_2 \\ \vdots \\ k_n \end{bmatrix} = \begin{bmatrix} d_1 \\ d_2 \\ \vdots \\ d_n \end{bmatrix} - \begin{bmatrix} 1 & & & \\ d_1 & 1 & & \\ \vdots & \vdots & \ddots & \\ d_{n-1} & d_{n-2} & \cdots & d_1 & 1 \end{bmatrix} \begin{bmatrix} a_1 + b_1 u(t-n) \\ a_2 + b_2 u(t-n) \\ \vdots \\ a_n + b_n u(t-n) \end{bmatrix}$$

$$K(z^{-1}) = k_1 z^{-1} + k_2 z^{-2} + \cdots + k_n z^{-n}.$$

Then, we have

$$\overline{\varepsilon}(t,\hat{\theta}) = \left(\theta_0 - \hat{\theta}\right)^T \overline{\phi}(t,\hat{\theta}) + e(t)$$

where

$$\overline{\phi}(t,\hat{\theta}) = H(z^{-1})\phi(t,\hat{\theta}).$$

■

Proof: See Appendix X.C.

Note: The notations $\overline{\varepsilon}(t,\hat{\theta})$ and $\overline{\phi}(t,\hat{\theta})$ are introduced respectively to show the dependence of $\overline{\varepsilon}(t)$ and $\overline{\phi}$ (t) on $\hat{\theta}$.

Theorem 4.

Consider the algorithm (54) to (57) for bilinear systems (13), and assume that
(i)   the series $\{e(t)\}$ is a noise sequence with zero mean;
(ii)   the quantities $\phi(t,\hat{\theta})$ and $e(t)$ are independent each other for any $\hat{\theta}$;
(iii)   for every input signal $u(t\text{-}n)$, $t=1 \dots N$, we have

$$Real \left[H(z^{-1}) - \frac{I}{2}\right] > 0.$$

Then, there is

$$Prob \left[\lim_{t \to \infty} \hat{\theta} \to \theta_0\right] = 1 \qquad globally.$$

■

Proof: It is similar to that given by Dugard and Landau [25] and is provided in Appendix X.C.

VI.D. Results of Simulation

To compare the robust bootstrap method with the non-robust version, a simulation example is provided. The bilinear system under consideration is modeled by Eq.(13). The matrices and vectors are given as

$$P = \begin{bmatrix} 0 & 1 \\ -0.7 & 1.5 \end{bmatrix} \qquad Q = \begin{bmatrix} 0 & 0 \\ 0.2 & 0.33 \end{bmatrix} \tag{58}$$

$$R^T = \begin{bmatrix} 0.5 & -1.4 \end{bmatrix} \qquad L^T = \begin{bmatrix} 1.2 & 1.1 \end{bmatrix} \qquad S = \begin{bmatrix} 1 & 0 \end{bmatrix}.$$

The system is disturbed by a sequence of noise $\{e(t)\}$ which belongs to "ϵ-contaminated family" described in (6) with $\sigma_1^2 = 0.01$ and $\sigma_2^2 = 50.0$ respectively. The quantity ϵ is taken as 0.1. For the robust bootstrap method, the turning point ω of the influence function, given in Eq.(30), is taken as 0.2.

The norms of errors in the states and parameters are plotted against the number of samples in Figs. 6, 7, and 8 respectively. From Fig.6 we can see that, using the robust bootstrap method, there is only one large shot of the state error norm at the beginning of recursion. In Fig.7, there are several overshoots of values much larger than 10.0. For convenience, we replace those values of the state error norm by the boundary,

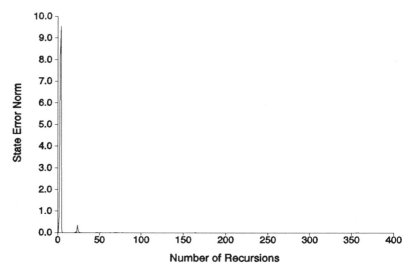

Figure 6   State error norm by robust bootstrap method

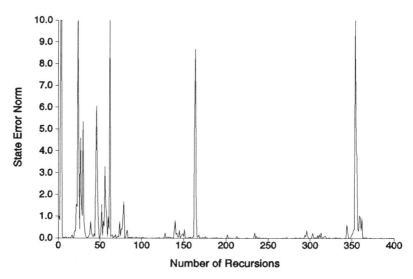

Figure 7   State error norm by non-robust bootstrap method

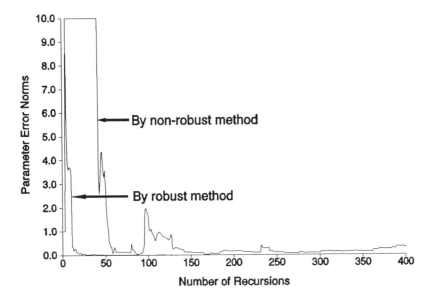

Figure 8   Comparison of parameter error norms by robust and non-robust bootstrap methods

10.0. Similarly, in Fig.8, some values of the parameter error norm using the non-robust bootstrap method have been replaced by the bound 10.0. Obviously, the method proposed in this chapter gives much better results.

VII.   ROBUST EXTENDED LEAST SQUARES METHOD WITH MODIFIED WEIGHTS

One efficient approach to the problem of recursive identification, namely, pseudolinear regression method has been fully developed in [21]. In fact, the output error method investigated in section V is a special case of pseudolinear regression. In this section, a robust version of the extended least squares method, the best known example of a pseudolinear regression method, is proposed.

VII.A. Extended Least Squares Method

To simplify the problem, a particular structure of bilinear system is considered, which has been described in Eq.(18). It is possible, however, to adopt the method to an arbitrary bilinear systems.

The model given in (18) looks like the regression model (9). Therefore, the robust recursive least squares method can be applied for estimating θ if the observation vector $\varphi_3(t)$ is defined by (21). The problem, of course, is that the variables $e(t-1)$, ..., $e(t-n)$ entering φ_3-vector are not measurable, and hence, the recursive least squares method cannot be applied directly. Intuitively, some estimates of the variables $e(t-1)$,

..., $e(t-n)$ are needed to replace the corresponding components. It leads to a combined procedure of estimating θ and reconstructing the unobserved φ_3-components.

Define $\phi(t)$ as

$$\phi^T(t) = \left[Y^T(t-1), Y^T(t-1)u(t-n), U^T(t-1), U^T(t-1)u(t-n), \hat{E}^T(t-1) \right] \qquad (59)$$

$$U^T(t-1) = \left[u(t-1) \cdots u(t-n) \right]$$

$$Y^T(t-1) = \left[y(t-1) \cdots y(t-n) \right]$$

$$\hat{E}^T(t-1) = \left[\overline{\varepsilon}(t-1) \cdots \overline{\varepsilon}(t-n) \right].$$

Here $\overline{\varepsilon}(t)$ is the residual defined as

$$\overline{\varepsilon}(t,\theta) = y(t) - \phi^T(t)\,\hat{\theta}(t). \qquad (60)$$

An obvious algorithm for estimating θ is now obtained from the well known recursive least squares method by defining $\phi(t)$ as Eq.(59). That is the so called extended least squares method.

VII.B. Robust Extended Least Squares Method

Like most non-robust recursive algorithm, the extended least squares method ceases to work due to the effect of outliers contaminating the input-output data. A robust extended least squares method, based on Huber's minimax principle, is developed in this section.

As indicated in the previous sections, various robust algorithms have been obtained by minimizing a robust criterion (23). Bearing this point in mind, the robust extended least squares method can be easily developed by applying the same procedure described in the previous sections to the pseudolinear regression problem. The recursive formulas for calculating robust estimates are:

$$\hat{\theta}(t) = \hat{\theta}(t-1) + \frac{P(t-1)\phi(t)\left(y(t) - \phi^T(t)\,\hat{\theta}(t-1)\right)}{1/\alpha(t) + \phi^T(t)P(t-1)\phi(t)} \qquad (61.a)$$

$$P(t) = P(t-1) - \frac{P(t-1)\phi(t)\phi^T(t)P(t-1)}{1/\alpha(t) + \phi^T(t)P(t-1)\phi(t)} \qquad (61.b)$$

and

$$P(t) = R^{-1}(t) = \left[\sum_{k=1}^{t} \alpha(k)\phi(k)\phi^T(k) \right]^{-1}. \qquad (61.c)$$

The initial values may be taken as those in (28.d).

Note that the observation vector $\phi(t)$ is given by Eq.(59). As to the ratio $\alpha(t)$, it should follow the definition in Eq.(26).

VII.C. Convergence Analysis

A proof of convergence of the robust extended least squares method is given in this section. In fact, the proof is very similar to that by Dugard and Landau [25] and Ljung and Söderström [21].

Lemma 6.

For bilinear systems modeled by Eq.(18), it is true that

$$K(z^{-1})\left(\overline{\varepsilon}(t) - e(t)\right) = \phi^T(t)\left(\theta_0 - \hat{\theta}(t)\right)$$

with $\phi(t)$ defined by Eq.(59).

■

Proof: Considering Eqs.(18) and (20) and using the expression in Eq.(60) give

$$\overline{\varepsilon}(t) = \theta_0^T\left(\varphi_3(t) - \phi(t)\right) + \left(\theta_0 - \hat{\theta}(t)\right)^T\phi(t) + e(t).$$

But

$$\theta_0^T\left(\varphi_3(t) - \phi(t)\right) = \left(K(z^{-1}) - 1\right)\left(e(t) - \overline{\varepsilon}(t)\right).$$

according to the definitions of $K(z^{-1})$ and $\varphi_3(t)$, hence

$$K(z^{-1})\left(\overline{\varepsilon}(t) - e(t)\right) = \phi^T(t)\left(\theta_0 - \hat{\theta}(t)\right).$$

■

Theorem 5.

Consider the algorithm (61) of bilinear systems modeled by Eq.(18), and assume that

(i)   the series $\{e(t)\}$ is a noise sequence with zero mean;

(ii)   the quantities $\phi(t,\hat{\theta})$ and $e(t)$ are independent each other for any $\hat{\theta}$, again $\phi(t,\theta)$ denotes $\phi(t)$ defined by Eq.(59);

(iii)

$$Real\left[\frac{1}{K(z^{-1})} - \frac{I}{2}\right] > 0$$

where $K(z^{-1})$ is defined in Eq.(19).

Then,

$$Prob\left[\lim_{t\to\infty}\hat{\theta} \to \theta_0\right] = 1 \qquad globally.$$

■

Proof: The proof is similar to that given in [25].

Note that if let

$$H(z^{-1}) = \frac{1}{K(z^{-1})}$$

the proof of theorem 5 should follow the line of theorem 4 exactly except some minor and obvious changes.

VII.D Results of Simulation

To illustrate the robustness and convergence of the developed method, a simulation example is provided along with comparison to the corresponding non-robust method. The true system is modeled by Eq.(18) with the parameter vector θ_0 as follows

$$\theta_0^T = \begin{bmatrix} 1.5 & -0.7 & 0.33 & 0.2 & 1.0 & -1.1 & 0.0 & -0.33 & -0.7 & 0.2 \end{bmatrix}. \qquad (62)$$

The noise sequence is generated by Eq.(6), where variances σ_1^2 and σ_2^2 are selected as

0.01 and 20.0, respectively. The probability for an outliers to occur in the given normal noise sequence is 0.1. For the generated data, both non-robust and robust extended least squares methods are used for estimating the parameter vector $\hat{\theta}$. The influence function is chosen by (30) where the turning point ω is selected as 0.3. The results of simulation are shown by Fig.9 in which the error norms, defined in Eq.(31), of both robust and non-robust parameter estimates are plotted. Again there is a couple of points where the parameter error norm values are beyond the bound 4.0 and replaced by 4.0 in order to show the difference more clearly. It is quite obvious to see the advantage of the derived method over the non-robust one.

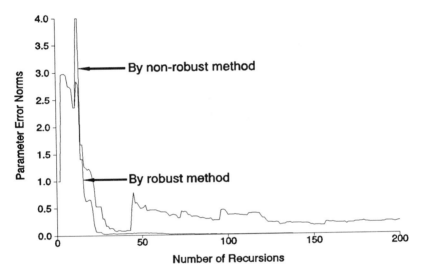

Figure 9   Comparison of parameter error norms by robust and non-robust ELS methods

VIII. ROBUST RECURSIVE PREDICTION ERROR METHOD WITH MODIFIED WEIGHTS

It is well known that the least squares criterion can be interpreted as minimization of the error between the "predicted" and the measured outputs. A similar interpretation could be given to some other methods. The family of these methods go under the name, prediction error identification methods. A general discussion of these methods are given in [21]. In this section, a strengthened prediction error method is presented for robust system identification.

VIII.A. Recursive Prediction Error Method

With a striking formal similarity with the recursive least squares algorithm, the recursive prediction error method has been well developed. It has been also extended to recursive identification of bilinear systems [26]. For the sake of simplicity, the attention will be focused on a particular class of bilinear systems as modeled by Eq.(18).

Based on observations of input $u(\tau)$ and output $y(\tau)$, $0 \le \tau \le t\text{-}1$, assume that the prediction $\hat{y}(t \mid \theta)$ is given as

$$\hat{y}(t \mid \theta) = \phi^T(t)\,\theta \qquad (63)$$

where $\phi(t)$ is given as

$$\phi^T(t) = \left[Y^T(t\text{-}1), Y^T(t\text{-}1)u(t\text{-}n), U^T(t\text{-}1), U^T(t\text{-}1)u(t\text{-}n), E^T(t\text{-}1,\theta) \right] \qquad (64)$$

$$U^T(t\text{-}1) = \left[u(t\text{-}1) \cdots u(t\text{-}n) \right]$$

$$Y^T(t\text{-}1) = \left[y(t\text{-}1) \cdots y(t\text{-}n) \right]$$

$$E^T(t\text{-}1,\theta) = \left[\varepsilon(t\text{-}1,\theta) \cdots \varepsilon(t\text{-}n,\theta) \right]$$

and θ is specified in Eq.(22). The prediction error $\varepsilon(t,\theta)$ at t-th instant is defined as

$$\varepsilon(t,\theta) = y(t) - \hat{y}(t \mid \theta). \qquad (65)$$

A reasonable criterion of how well the model performs is the sum of squared prediction errors

$$J_S(\theta) = \frac{1}{N}\sum_{t=1}^{N} \left(\varepsilon(t,\theta) \right)^2. \qquad (66)$$

Denote the negative derivative of $\varepsilon(t,\theta)$ with respect to θ by

$$\xi(t,\theta) = \left[-\frac{d\,\varepsilon(t,\theta)}{d\theta} \right]^T. \qquad (67)$$

The prediction error method then can be put into a recursive form only at the price of certain approximations. The recursive formulas are

$$\hat{\theta}(t) = \hat{\theta}(t\text{-}1) + \frac{P(t\text{-}1)\xi(t)\,\varepsilon(t)}{1 + \xi^T(t)P(t\text{-}1)\xi(t)} \qquad (68.\text{a})$$

$$P(t) = P(t\text{-}1) - \frac{P(t\text{-}1)\xi(t)\xi^T(t)P(t\text{-}1)}{1 + \xi^T(t)P(t\text{-}1)\xi(t)} \qquad (68.\text{b})$$

and

$$P(t) = R^{-1}(t) = \left[\sum_{k=1}^{t} \xi(k)\,\xi^T(k) \right]^{-1} \qquad (68.\text{c})$$

with

$$\varepsilon(t) = y(t) - \phi^T(t)\,\hat{\theta}(t\text{-}1) \qquad (68.\text{d})$$

$$\xi(t) = \left(1 - \hat{K}(z^{-1}) \right)\xi(t) + \phi(t) \qquad (68.\text{e})$$

$$\hat{K}(z^{-1}) = 1 + \hat{k}_1 z^{-1} + \cdots + \hat{k}_n z^{-n}. \qquad (68.\text{f})$$

The vector $\phi(t)$ is updated by Eq.(64) with $\varepsilon(t,\theta)$ replaced by $\varepsilon(t)$.

VIII.B. Robust Recursive Prediction Error Method

As we know, the basic idea of the recursive prediction error method is to minimize a sum of squared prediction errors. Therefore, the estimate generated by the algorithm (68) could be affected deadly by outliers contaminating the observations of systems. Based on Huber's minimax principle, a strengthened version of the recursive prediction error method can be developed for robust identification of both linear and bilinear systems.

Similar to the asymptotic criterion $J_1(\theta)$ in Eq.(23), a robust criterion is introduced to derive the robust recursive prediction error method, that is

$$J_N(\theta) = J_6(\theta) = \sum_{t=1}^{N} \rho\big(\varepsilon(t,\theta)\big). \tag{69}$$

Again, $\rho(.)$ is a convex function suggested earlier, whose derivative about its argument is the so called influence function $\psi(.)$. The $\varepsilon(t,\theta)$ is the prediction error defined in Eq.(65).

Let $\hat{\theta}(t\text{-}1)$ be an estimate at t-1. To recursively obtain an estimate $\hat{\theta}(t)$ that approximately minimizes $J_t(\theta)$ in Eq.(69), the Taylor expansion of $J_t(\theta)$ around $\hat{\theta}(t\text{-}1)$ should be minimized. It leads to

$$\hat{\theta}(t) = \hat{\theta}(t\text{-}1) - \left\{\ddot{J}_t\big(\hat{\theta}(t\text{-}1)\big)\right\}^{-1}\left\{\dot{J}_t\big(\hat{\theta}(t\text{-}1)\big)\right\}^{T} + o(|\hat{\theta}(t) - \hat{\theta}(t\text{-}1)|) \tag{70}$$

where the $\dot{J}$ and $\ddot{J}$ denote the first and second differentiations with respect to θ, and $o(x)$ denotes a function such that $o(x) \rightarrow 0$ as $|x| \rightarrow 0$.

In order to evaluate the above equation, some approximations have to be made:

1.   The estimate $\hat{\theta}(t)$ is assumed to be found in a small neighbourhood of $\hat{\theta}(t\text{-}1)$, then,

$$o\big(|\hat{\theta}(t) - \hat{\theta}(t\text{-}1)|\big) \approx 0 \tag{71.a}$$

and

$$\ddot{J}_t\big(\hat{\theta}(t)\big) \approx \ddot{J}_t\big(\hat{\theta}(t\text{-}1)\big). \tag{71.b}$$

2.   The estimate $\hat{\theta}(t\text{-}1)$ is indeed the optimal estimate at time t-1, so that

$$\dot{J}_{t-1}\big(\hat{\theta}(t\text{-}1)\big) \approx 0. \tag{72}$$

3.   Assume

$$\ddot{\varepsilon}(t,\theta)\psi(t,\theta) \approx 0 \tag{73}$$

where $\psi(t,\theta)$ denotes the influence function at time t for given θ.

4.   The influence function is such chosen that

$$\frac{d\psi(t,\theta)}{d\varepsilon(t,\theta)} = 1 \tag{74}$$

when $|\varepsilon(t,\theta)|$ is small enough.

Considering the notation in Eq.(67) and assumption 4, $\dot{J}_t(\hat{\theta}(t))$ could be expressed as

$$\big(\dot{J}_t(\hat{\theta})\big)^{T} = \dot{J}_{t-1}(\hat{\theta}) - \alpha(t)\xi(t,\theta)\,\varepsilon(t,\theta). \tag{75}$$

Again $\alpha(t)$ denotes the ratio of $\psi(t,\theta)$ to $\varepsilon(t,\theta)$ and has been described in Eq.(26).

Furthermore, the second derivative of $J_t(\theta)$ with respect to θ can be given by

$$\ddot{J}_t(\theta) = \ddot{J}_{t-1}(\theta) + \alpha(t)\xi(t,\theta)\xi^{T}(t,\theta) + \ddot{\varepsilon}(t,\theta)\psi(t,\theta). \tag{76}$$

Substituting Eqs.(71.b) and (73) into Eq.(76) leads to

$$\ddot{J}_t(\hat{\theta}(t)) = \ddot{J}_{t-1}(\hat{\theta}(t-1)) + \alpha(t)\xi(t,\hat{\theta}(t-1))\xi^T(t,\hat{\theta}(t-1)). \tag{77}$$

With assumption 2 inserted into Eq.(75), the first derivative of $J_t(\theta)$ can be given as

$$\{\dot{J}_t(\hat{\theta}(t))\}^T = -\alpha(t)\xi(t,\hat{\theta}(t-1))\varepsilon(t,\hat{\theta}(t-1)). \tag{78}$$

Put

$$R(t) = \ddot{J}_t((\hat{\theta})); \qquad P(t) = R^{-1}(t) \tag{79}$$

and then, substituting Eqs.(71.a), (79), and (78) into Eq.(70) gives

$$\hat{\theta}(t) = \hat{\theta}(t-1) + \alpha(t)P(t)\xi(t,\hat{\theta}(t-1))\varepsilon(t,\hat{\theta}(t-1)) \tag{80.a}$$

$$R(t) = R(t-1) + \alpha(t)\xi(t,\hat{\theta}(t-1))\xi^T(t,\hat{\theta}(t-1)). \tag{80.b}$$

Straightforwardly, a commonly used recursive form of Eq.(80) can be obtained by employing the famous matrix inverse lemma, that is

$$\hat{\theta}(t) = \hat{\theta}(t-1) + \alpha(t)P(t)\xi(t,\hat{\theta}(t-1))\varepsilon(t,\hat{\theta}(t-1)) \tag{81.a}$$

$$P(t) = P(t-1) - \frac{P(t-1)\xi(t,\hat{\theta}(t-1))\xi^T(t,\hat{\theta}(t-1))P(t-1)}{1/\alpha(t) + \xi^T(t,\hat{\theta}(t-1))P(t-1)\xi(t,\hat{\theta}(t-1))}. \tag{81.b}$$

To implement the algorithm (81), however, the vector $\xi(t,\hat{\theta}(t-1))$ and prediction error $\varepsilon(t,\hat{\theta}(t-1))$ should be somehow calculated first.

Consider the model of bilinear systems (18). From the definition of the prediction $\hat{y}(t\,|\,\theta)$ in Eq.(63), it is quite straightforward to obtain

$$\hat{y}(t\,|\,\theta) + k_1\hat{y}(t-1\,|\,\theta) + \cdots + k_n\hat{y}(t-n\,|\,\theta)$$
$$= \sum_{i=1}^{n}(a_i + k_i)y(t-i) + \sum_{i=1}^{n}b_iy(t-i)u(t-n) + \sum_{i=1}^{n}(c_i + d_iu(t-n))u(t-i). \tag{82}$$

Recalling the definition of $\xi(t,\theta)$ in Eq.(67), taking the derivatives of $\hat{y}(t-i\,|\,\theta)$ $i=0, 1,...,$ n with respect to θ leads to

$$\xi(t,\theta) + k_1\xi(t-1,\theta) + \cdots + k_n\xi(t-n,\theta) = \phi(t) \tag{83}$$

with $\phi(t)$ defined by Eq.(64).

Naturally, an approximation of Eqs.(65) and (83) is to make only one time iteration with the current estimates and use previous values of ε and ξ as initial values. This implies $\varepsilon(t,\hat{\theta}(t-1))$ is approximated by $\varepsilon(t)$, calculated according to

$$\varepsilon(t) = y(t) - \phi^T(t)\hat{\theta}(t-1). \tag{84.a}$$

Similarly, a natural approximation of $\xi(t,\hat{\theta}(t-1))$ is $\xi(t)$, computed by means of

$$\xi(t) + k_1\xi(t-1) + \cdots + k_n\xi(t-n) = \phi(t). \tag{84.b}$$

Using $\varepsilon(t)$ and $\xi(t)$ defined in Eq.(84) produces the robust recursive prediction error method as below

$$\hat{\theta}(t) = \hat{\theta}(t-1) + \frac{P(t-1)\xi(t)\varepsilon(t)}{1/\alpha(t) + \xi^T(t)P(t-1)\xi(t)} \tag{85.a}$$

$$P(t) = P(t-1) - \frac{P(t-1)\xi(t)\xi^T(t)P(t-1)}{1/\alpha(t) + \xi^T(t)P(t-1)\xi(t)} \tag{85.b}$$

and

$$P(t) = R^{-1}(t) = \left[\sum_{k=1}^{t} \alpha(k)\,\xi(k)\,\xi^{T}(k) \right]^{-1}. \tag{85.c}$$

The initial values could be chosen as in Eq.(28.d).

It should be quite clear from the algorithm (84) and (85), how the corresponding algorithm for robust identification of linear systems is constructed. In fact, the problem of linear systems is simply a particular case of bilinear systems.

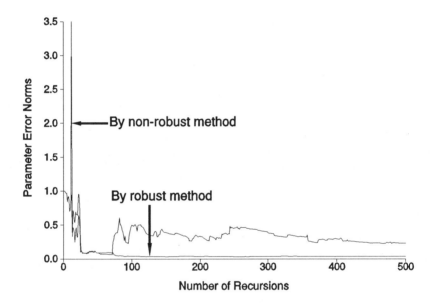

Figure 10   Comparison of parameter error norms by robust and non-robust PE methods

VIII.C. Results of Simulation

In order to demonstrate the robustness of the developed method, a simulation example is included. The bilinear system under consideration is described by Eq.(18) and the parameter vector θ_0 follows the specification in Eq.(62). For the normal noise, variance σ_1^2 is chosen as 0.01. To simulate the behaviour of outliers, variance σ_2^2 is introduced and taken as the value 20.0. The quantity ϵ, which indicates the probability of outliers occurring in the given normal noise series, is simply selected as 0.1. Given the generated data, the robust recursive prediction error method, described by Eqs.(84) and (85), is utilized to estimate the parameter vector. The influence function used here satisfies the definition in Eq.(30) with the turning point selected as 0.3.

The traditional recursive prediction error method, given by Eq.(68), is used to process the same set of data. To illustrate the advantage of the robust algorithm over the non-robust one, the error norm, defined in Eq.(31) is used. The error norm of the estimated parameters (robust and non-robust) are plotted in Fig.10. From the figure, the robustness of the proposed method can be easily verified.

IX. CONCLUSIONS

The contributions of this chapter are the six robust identification methods for bilinear systems. Three of these are derived without knowing corresponding robust version for linear system identification. The others are extensions of previous work to bilinear systems, but the proofs of convergence are different and more accurate. In addition, the robust bootstrap method developed in section VI has been derived by using the extended least squares approach instead of a Bayesian approach utilized in [11].

The basic idea of developing these methods, however, is the same as suggested in [8], that is, to utilize a kind of convex functions rather than the traditional least squares criterion so that an influence function can be introduced to generate a class of robust recursive algorithms. As a matter of fact, the class of robust algorithms is very similar to the recursive weighted least squares algorithm. The weights used in these algorithms are defined as a ratio of the influence function to the prediction error. When the absolute value of prediction errors is equal to or smaller than a pre-specified value, called the turning point, the weights are identity. Otherwise, the larger the absolute value of prediction errors, the smaller are the weights. This is the way to reduce the effect of outliers on estimates and hence to achieve robustness of the algorithms.

One of the drawbacks of these robust algorithms, however, is that the turning point of influence functions should be somehow pre-determined. As a result, it leads to another difficult task especially when there is no information available about noise. Even though certain typical information, for example mean or variance, is known beforehand, some empirical formulas for describing the relationship between the turning point and noise variance are required to calculate the appropriate turning point of influence functions. Obviously, this could limit physical applications in a variety of real-life situations. To solve the problem, a new and more powerful approach is needed. This is the topic of the next chapter.

X. APPENDIXES

X.A. A Martingale Convergence Proof of Theorem 1

X.A.1. Proof of Lemma 2

Since $\alpha(t) > 0$ and $P(t-1)$ positive definite, it follows that $d(t) > 1$. From (28.b), we have $P(t)\phi(t) = P(t-1)\phi(t)/d(t)$. Therefore

$$\phi^T(t)\left(P(t)\right)^2\phi(t) = \frac{\phi^T(t)P(t)P(t-1)\phi(t)}{d(t)}$$

$$< \frac{tr\left(P(t-1) - P(t)\right)}{\alpha(t)}.$$

Also, from (28.b), the following equation is true

$$\phi^T(t)(P(t))^2\phi(t) = \frac{\phi^T(t)P(t-1)P(t-1)\phi(t)}{(d(t))^2}.$$

Therefore

$$r(t) < \frac{tr(P(t-1) - P(t))}{\alpha(t)}.$$

Without lose of generality, the estimate $\hat{\nu}(t)$ of $\nu(t) > 0$ is assumed for $t \in [0, +\infty)$. Then from lemma 1, the following inequality holds

$$\frac{1}{\alpha(t)} = \frac{\hat{\nu}(t)}{\psi(\hat{\nu}(t))} \le \frac{(\hat{\nu}(t))^2}{m}$$

i.e.

$$\frac{1}{\alpha(t)} < \frac{\lambda^2}{m} \qquad \text{for } t \in [0, +\infty). \tag{A.1}$$

Since $d(t) \gg 1$ for $t \in [0, +\infty)$, then

$$r(t) < \frac{\lambda^2}{m} tr(P(t-1) - P(t)).$$

Furthermore, it comes to the conclusion of the lemma, that is

$$\sum_{t=1}^{\infty} r(t) < \frac{\lambda^2}{m} tr(P(0) - P(\infty)) < \infty.$$

∎

X.A.2. Proof of Lemma 3
From lemma 2, it is straightforward to have

$$\sum_{t=1}^{\infty} \phi^T(t)(P(t))^2\phi(t) < \infty.$$

Since

$$\phi^T(t)P(t)\phi(t) \le \|P^{-1}(t)\| \, \phi^T(t)(P(t))^2\phi(t)$$

where $\|P^{-1}(t)\|$ is the operator norm of $P^{-1}(t)$, then

$$\|P^{-1}(t)\| \le trP^{-1}(t) = \sum_{k=1}^{t} \alpha(k) |\phi(k)|^2.$$

Under the assumptions in lemma 1, we may obtain

$$\alpha(t) < M \tag{A.2}$$

for $t \in [0, +\infty)$, we say M is a bound of $\alpha(t)$. Consequently, we have

$$\sum_{t=1}^{\infty} \frac{1}{t} \phi^T(t)P(t)\phi(t) \le \sum_{t=1}^{\infty} \frac{trP^{-1}(t)}{t} \phi^T(t)(P(t))^2\phi(t)$$

$$= \sum_{t=1}^{\infty} \frac{1}{t} \sum_{k=1}^{t} \alpha(k) |\phi(k)|^2 \phi^T(t)(P(t))^2\phi(t).$$

With (A.2) and the assumption in this lemma

$$\lim_{N \to \infty} Sup \frac{1}{N} \sum_{t=1}^{N} |\phi(t)|^2 < \infty \qquad w.p.1$$

the following inequality can be established

$$\sum_{t=1}^{\infty} \frac{1}{t} \phi^T(t) P(t) \phi(t) \leq M \sum_{t=1}^{\infty} \frac{1}{t} \sum_{k=1}^{t} |\phi(k)|^2 \phi^T(t) (P(t))^2 \phi(t)$$

$$\leq M \lim_{N \to \infty} Sup \frac{1}{N} \sum_{k=1}^{N} |\phi(k)|^2 \sum_{t=1}^{\infty} \phi^T(t) (P(t))^2 \phi(t)$$

$$< \infty. \qquad w.p.1$$

That is,

$$\sum_{t=1}^{\infty} \frac{1}{t} \phi^T(t) P(t) \phi(t) < \infty. \qquad w.p.1$$

This leads to the end of proof of lemma 3.

■

X.A.3. Proof of Theorem 1

To prove the theorem, the algorithm (28) may be rewritten

$$\hat{\theta}(t) = \hat{\theta}(t-1) + \alpha(t) P(t-1) \phi(t) \bar{\varepsilon}(t) \qquad \text{(A.3a)}$$

$$P^{-1}(t) = P^{-1}(t-1) + \alpha(t) \phi(t) \phi^T(t) \qquad \text{(A.3b)}$$

where $\bar{\varepsilon}(t)$ is the residuals at the t-th instant. Subtracting θ_0 from both sides of (A.3a), denoting $\tilde{\theta}(t) = \hat{\theta}(t) - \theta_0$, leads to

$$\tilde{\theta}(t) = \tilde{\theta}(t-1) + \alpha(t) P(t-1) \phi(t) \bar{\varepsilon}(t) \qquad \text{(A.4a)}$$

or,

$$\tilde{\theta}(t) = \tilde{\theta}(t-1) + \alpha(t) P(t) \phi(t) \varepsilon(t) \qquad \text{(A.4b)}$$

where $\varepsilon(t)$ is the prediction error at the t-th instant. Premultiplying both sides of (A.4a) by $\tilde{\theta}^T(t) P^{-1}(t-1)$ gives

$$\tilde{\theta}^T(t) P^{-1}(t-1) \tilde{\theta}(t)$$
$$= \tilde{\theta}^T(t) P^{-1}(t-1) \tilde{\theta}(t-1) + \alpha(t) \tilde{\theta}^T(t) P^{-1}(t-1) P(t-1) \phi(t) \bar{\varepsilon}(t)$$
$$= \tilde{\theta}^T(t) P^{-1}(t-1) \tilde{\theta}(t-1) + \alpha(t) \tilde{\theta}^T(t) \phi(t) \bar{\varepsilon}(t).$$

Using (A.3b) yields

$$\tilde{\theta}^T(t) P^{-1}(t) \tilde{\theta}(t) = \tilde{\theta}^T(t) P^{-1}(t-1) \tilde{\theta}(t-1) + \alpha(t) \tilde{\theta}^T(t) \phi(t) \bar{\varepsilon}(t) + \alpha(t) \left(\tilde{\theta}^T(t) \phi(t) \right)^2.$$

Introducing $T(t) = \tilde{\theta}^T(t) P^{-1}(t) \tilde{\theta}(t)$ and substituting (A.4a) into the above equation give

$$T(t) = T(t-1) + \alpha(t) \tilde{\theta}^T(t-1) \phi(t) \bar{\varepsilon}(t) + \alpha(t) \tilde{\theta}^T(t) \phi(t) \bar{\varepsilon}(t) + \alpha(t) \left(\tilde{\theta}^T(t) \phi(t) \right)^2.$$

Substituting (A.4a) again leads to

$$T(t) = T(t-1) - \phi^T(t) P(t-1) \phi(t) \left(\alpha(t) \bar{\varepsilon}(t) \right)^2 \qquad \text{(A.5)}$$
$$+ 2\alpha(t) \tilde{\theta}^T(t) \phi(t) \bar{\varepsilon}(t) + \alpha(t) \left(\tilde{\theta}^T(t) \phi(t) \right)^2.$$

The last two terms in (A.5) are

$$2\alpha(t)\tilde{\theta}^T(t)\phi(t)\bar{\varepsilon}(t) + \alpha(t)\big(\tilde{\theta}^T(t)\phi(t)\big)^2$$
$$= \alpha(t)\phi^T(t)\tilde{\theta}(t)\big\{\phi^T(t)\tilde{\theta}(t) + 2\big(\bar{\varepsilon}(t)-\nu(t)\big)\big\} + 2\alpha(t)\nu(t)\phi^T(t)\tilde{\theta}(t). \tag{A.6}$$

Since

$$\bar{\varepsilon}(t) = \nu(t) - \phi^T(t)\tilde{\theta}(t) \tag{A.7}$$

then, (A.6) can be written as

$$2\alpha(t)\tilde{\theta}^T(t)\phi(t)\bar{\varepsilon}(t) + \alpha(t)\big(\tilde{\theta}^T(t)\phi(t)\big)^2$$
$$= \alpha(t)\phi^T(t)\tilde{\theta}(t)\big(-\phi^T(t)\tilde{\theta}(t)\big) + 2\alpha(t)\nu(t)\phi^T(t)\tilde{\theta}(t). \tag{A.8}$$

Taking into account of (A.4b), the conditional expectation of the last term in (A.8) is

$$E\big(\alpha(t)\nu(t)\phi^T(t)\tilde{\theta}(t) \,|\, \mathfrak{C}_{t-1}\big)$$
$$= E\big(\alpha(t)\nu(t)\phi^T(t)\tilde{\theta}(t-1) \,|\, \mathfrak{C}_{t-1}\big) + E\big(\nu(t)\phi^T(t)P(t)\phi(t)\alpha^2(t)\,\varepsilon(t) \,|\, \mathfrak{C}_{t-1}\big)$$
$$= \phi^T(t)P(t)\phi(t)E\big\{\alpha^2(t)\,[\nu^2(t)+\nu(t)\big(\varepsilon(t)-\nu(t)\big)]\,\big|\,\mathfrak{C}_{t-1}\big\}$$
$$\quad + \phi^T(t)\tilde{\theta}(t-1)E\big(\alpha(t)\nu(t) \,|\, \mathfrak{C}_{t-1}\big) \tag{A.9}$$

Considering assumption (ii) in the theorem, we have

$$E\big\{\nu^2(t) + \nu(t)\big(\varepsilon(t)-\nu(t)\big) \,|\, \mathfrak{C}_{t-1}\big\} = \sigma^2 > 0$$

$$E\big\{\nu(t) \,|\, \mathfrak{C}_{t-1}\big\} = 0$$

and

$$E\big\{\alpha^2(t) \,|\, \mathfrak{C}_{t-1}\big\} \le M^2 > 0 \qquad\qquad \forall\; t \in [0,+\infty).$$

then,

$$E\big\{\alpha^2(t)\,[\,\nu^2(t) + \nu(t)\big(\varepsilon(t)-\nu(t)\big)]\,|\,\mathfrak{C}_{t-1}\big\}$$
$$\le E\big\{\nu^2(t) + \nu(t)\big(\varepsilon(t)-\nu(t)\big) \,|\, \mathfrak{C}_{t-1}\big\}E\big\{\alpha^2(t) \,|\, \mathfrak{C}_{t-1}\big\}$$
$$\le M^2\sigma^2.$$

Therefore,

$$E\big(\alpha(t)\nu(t)\phi^T(t)\tilde{\theta}(t) \,|\, \mathfrak{C}_{t-1}\big) \le M^2\sigma^2\phi^T(t)P(t)\phi(t). \tag{A.10}$$

Take the conditional expectation on the both sides of (A.5) and substitute (A.10) into (A.5), we may obtain

$$E\big(T(t) \,|\, \mathfrak{C}_{t-1}\big) \le T(t-1) + 2M^2\sigma^2\phi^T(t)P(t)\phi(t)$$
$$\qquad\qquad - E\big(\alpha(t)\phi^T(t)\tilde{\theta}(t)\phi^T(t)\tilde{\theta}(t) \,|\, \mathfrak{C}_{t-1}\big)$$
$$\qquad\qquad - \phi^T(t)P(t-1)\phi(t)E\big(\alpha^2(t)\bar{\varepsilon}^2(t) \,|\, \mathfrak{C}_{t-1}\big)$$
$$\qquad \le T(t-1) + 2M^2\sigma^2\phi^T(t)P(t)\phi(t)$$
$$\qquad\qquad - E\big(\alpha(t)\phi^T(t)\tilde{\theta}(t)\phi^T(t)\tilde{\theta}(t) \,|\, \mathfrak{C}_{t-1}\big)$$

i.e.

$$E\big\{T(t) + \alpha(t)\big(\phi^T(t)\tilde{\theta}(t)\big)^2 \,|\, \mathfrak{C}_{t-1}\big\} \le T(t-1) + 2M^2\sigma^2\phi^T(t)P(t)\phi(t).$$

Introducing

$$\tilde{T}(t) = \frac{1}{t}\left[T(t) + \sum_{k=1}^{t}\alpha(k)\left(\phi^T(k)\,\tilde{\theta}(k)\right)^2\right]$$

(obviously $\tilde{T}(t)>0$) and noting that $\phi^T(k)\tilde{\theta}(k) \in \mathfrak{C}_{t-1}$ for $k\leq t-1$, we have

$$E\left(t\tilde{T}(t)\,|\mathfrak{C}_{t-1}\right)$$

$$= E\left\{T(t) + \alpha(t)\left(\phi^T(t)\,\tilde{\theta}(t)\right)^2\,|\mathfrak{C}_{t-1}\right\} + \sum_{k=1}^{t-1}\alpha(k)\left(\phi^T(k)\,\tilde{\theta}(k)\right)^2$$

$$\leq T(t-1) + 2M^2\sigma^2\phi^T(t)P(t)\phi(t) + \sum_{k=1}^{t-1}\alpha(k)\left(\phi^T(k)\,\tilde{\theta}(k)\right)^2$$

$$= (t-1)\tilde{T}(t-1) + 2M^2\sigma^2\phi^T(t)P(t)\phi(t)$$

i.e.

$$E\left(\tilde{T}(t)\,|\mathfrak{C}_{t-1}\right) \leq \tilde{T}(t-1) - \frac{1}{t}\tilde{T}(t-1) + \frac{1}{t}2M^2\sigma^2\phi^T(t)P(t)\phi(t). \qquad \text{(A.11)}$$

Using the lemmas above and the theorem given by Neveu [22] leads to the final results. That is

$$\tilde{\theta}^T(t)P^{-1}(t)\,\tilde{\theta}(t) \to 0 \qquad\qquad w.p.1 \quad as \quad t \to \infty$$

and

$$\frac{1}{N}\sum_{t=1}^{N}\left(\bar{\varepsilon}(t) - \nu(t)\right)^2 \qquad\qquad w.p.1 \quad as \quad N \to \infty.$$

It ends the proof of the theorem.

∎

X.B.   A Martingale Convergence Proof of Theorem 2

X.B.1.   Proof of Lemma 4
Let $R^0(t)$ denotes the matrix as below

$$R^0(t) = \frac{1}{t}\sum_{k=1}^{t}\alpha(k)\varsigma(k)\phi^T(k). \qquad \text{(A.12)}$$

Again, from lemma 1 and assumption (i) in this lemma, Eq.(A.1) in Appendix A is true. Therefore, considering assumption (ii) in this lemma leads to, $\forall\, t \in [0,+\infty)$

$$R^0(t) > \frac{m}{\lambda^2}\frac{1}{t}\sum_{k=1}^{t}\varsigma(k)\phi^T(k)$$

$$\geq \frac{m}{\lambda^2}\delta I > 0. \qquad \text{(A.13)}$$

Furthermore, $\forall\, t \in [0,+\infty)$, it comes to

$$\left(R^0(t)\right)^{-1} < \Gamma I \qquad\qquad \text{(A.14)}$$

for $0<\Gamma<\infty$.
Using (A.14), it is straightforward to have

$$\varsigma^T(t)P(t)\varsigma(t) = \frac{1}{t}\varsigma^T(t)\big(R^0(t)\big)^{-1}\varsigma(t)$$

$$< \frac{\Gamma}{t}\varsigma^T(t)\varsigma(t). \tag{A.15}$$

Considering Eq.(A.15) along with assumption (iii) in this lemma gives the conclusion of this lemma, that is

$$\sum_{t=1}^{\infty}\frac{1}{t}\varsigma^T(t)P(t)\varsigma(t) < \sum_{t=1}^{\infty}\frac{\Gamma}{t}\varsigma^T(t)\varsigma(t)$$

$$< \Gamma\|\varsigma_0\|^2\sum_{t=1}^{\infty}\frac{1}{t^2} < \infty. \qquad w.p.1 \tag{A.16}$$

That ends the proof of lemma 4.

∎

X.B.2. Proof of Theorem 2
For the sake of convenience, the algorithm (41) is rewritten

$$\hat{\theta}(t) = \hat{\theta}(t-1) + \alpha(t)P(t-1)\varsigma(t)\overline{\varepsilon}(t) \tag{A.17a}$$

$$P^{-1}(t) = P^{-1}(t-1) + \alpha(t)\varsigma(t)\phi^T(t) \tag{A.17b}$$

where $\overline{\varepsilon}$ is the residual at the t-th instant. Subtracting θ_0 from both sides of (A.17a), denoting $\tilde{\theta}(t) = \hat{\theta}(t)-\theta_0$, leads to

$$\tilde{\theta}(t) = \tilde{\theta}(t-1) + \alpha(t)P(t-1)\varsigma(t)\overline{\varepsilon}(t) \tag{A.18a}$$

or,

$$\tilde{\theta}(t) = \tilde{\theta}(t-1) + \alpha(t)P(t)\varsigma(t)\varepsilon(t) \tag{A.18b}$$

where $\varepsilon(t)$ is the prediction error at the t-th instant. Premultiplying both sides of (A.18a) by $\tilde{\theta}^T(t)P^{-1}(t-1)$ gives

$$\tilde{\theta}^T(t)P^{-1}(t-1)\tilde{\theta}(t) = \tilde{\theta}^T(t)P^{-1}(t-1)\tilde{\theta}(t-1) + \alpha(t)\tilde{\theta}^T(t)\varsigma(t)\overline{\varepsilon}(t).$$

Substituting (A.17b) into the above equation leads to

$$\tilde{\theta}^T(t)P^{-1}(t)\tilde{\theta}(t) = \tilde{\theta}^T(t)P^{-1}(t-1)\tilde{\theta}(t-1) + \alpha(t)\tilde{\theta}^T(t)\varsigma(t)\overline{\varepsilon}(t) + \alpha(t)\tilde{\theta}^T(t)\varsigma(t)\phi^T(t)\tilde{\theta}(t).$$

Introducing $T(t)=\tilde{\theta}^T(t)P^{-1}(t)\tilde{\theta}(t)$ and substituting (A.18a) into the above equation give

$$T(t) = T(t-1) + \alpha(t)\tilde{\theta}^T(t-1)\varsigma(t)\overline{\varepsilon}(t)$$
$$+ \alpha(t)\tilde{\theta}^T(t)\varsigma(t)\overline{\varepsilon}(t) + \alpha(t)\tilde{\theta}^T(t)\varsigma(t)\phi^T(t)\tilde{\theta}(t).$$

Substituting (A.18a) again yields

$$T(t) = T(t-1) - \varsigma^T(t)P(t-1)\varsigma(t)\big(\alpha(t)\overline{\varepsilon}(t)\big)^2$$
$$+ 2\alpha(t)\tilde{\theta}^T(t)\varsigma(t)\overline{\varepsilon}(t) + \alpha(t)\tilde{\theta}^T(t)\varsigma(t)\phi^T(t)\tilde{\theta}(t). \tag{A.19}$$

Since

$$\overline{\varepsilon}(t) = \nu(t) - \phi^T(t)\tilde{\theta}(t) \tag{A.20}$$

the last two terms in (A.19) can be rewritten

$$2\alpha(t)\tilde{\theta}^T(t)\zeta(t)\overline{\varepsilon}(t) + \alpha(t)\tilde{\theta}^T(t)\zeta(t)\phi^T(t)\tilde{\theta}(t)$$
$$= \alpha(t)\zeta^T(t)\tilde{\theta}(t)\left\{\phi^T(t)\tilde{\theta}(t) + 2\big(\overline{\varepsilon}(t) - \nu(t)\big)\right\} + 2\alpha(t)\nu(t)\zeta^T(t)\tilde{\theta}(t) \qquad \text{(A.21)}$$
$$= \alpha(t)\tilde{\theta}^T(t)\zeta(t)\big(-\phi^T(t)\tilde{\theta}(t)\big) + 2\alpha(t)\nu(t)\zeta^T(t)\tilde{\theta}(t).$$

Considering (A.18b), the conditional expectation of the last term in (A.21) is

$$E\big(\alpha(t)\nu(t)\zeta^T(t)\tilde{\theta}(t)\,|\mathbb{C}_{t-1}\big)$$
$$= E\big(\alpha(t)\nu(t)\zeta^T(t)\tilde{\theta}(t-1)\,|\mathbb{C}_{t-1}\big) + E\big(\nu(t)\zeta^T(t)P(t)\zeta(t)\alpha^2(t)\varepsilon(t)\,|\mathbb{C}_{t-1}\big)$$
$$= \zeta^T(t)P(t)\zeta(t)E\left\{\alpha^2(t)\left[\nu^2(t) + \nu(t)\big(\varepsilon(t) - \nu(t)\big)\right]\,|\mathbb{C}_{t-1}\right\}$$
$$\qquad + \zeta^T(t)\tilde{\theta}(t-1)E\big(\alpha(t)\nu(t)\,|\mathbb{C}_{t-1}\big) \qquad \text{(A.22)}$$

From assumption (ii) in the theorem, we have

$$E\left\{\nu^2(t) + \nu(t)\big(\varepsilon(t) - \nu(t)\big)\,|\mathbb{C}_{t-1}\right\} = \sigma^2 > 0$$

$$E\{\nu(t)\,|\mathbb{C}_{t-1}\} = 0$$

and
$$E\left\{\alpha^2(t)\,|\mathbb{C}_{t-1}\right\} \leq M^2 > 0 \qquad \forall\ t \in [0, +\infty).$$

Then, using Schwarcz inequality leads to

$$E\left\{\alpha^2(t)\left[\nu^2(t) + \nu(t)\big(\varepsilon(t) - \nu(t)\big)\right]\,|\mathbb{C}_{t-1}\right\}$$
$$\leq E\left\{\nu^2(t) + \nu(t)\big(\varepsilon(t) - \nu(t)\big)\,|\mathbb{C}_{t-1}\right\}E\left\{\alpha^2(t)\,|\mathbb{C}_{t-1}\right\}$$
$$\leq M^2\sigma^2.$$

It simply implies

$$E\big(\alpha(t)\nu(t)\zeta^T(t)\tilde{\theta}(t)\,|\mathbb{C}_{t-1}\big) \leq M^2\sigma^2\zeta^T(t)P(t)\zeta(t). \qquad \text{(A.23)}$$

Take the conditional expectation on the both sides of (A.19) and then substitute (A.21) and (A.23) in (A.19), we may obtain

$$E\big(T(t)\,|\mathbb{C}_{t-1}\big) \leq T(t-1) + 2M^2\sigma^2\zeta^T(t)P(t)\zeta(t)$$
$$\qquad - E\big(\alpha(t)\zeta^T(t)\tilde{\theta}(t)\phi^T(t)\tilde{\theta}(t)\,|\mathbb{C}_{t-1}\big)$$
$$\qquad - \zeta^T(t)P(t-1)\zeta(t)E\big(\alpha^2(t)\overline{\varepsilon}^2(t)\,|\mathbb{C}_{t-1}\big)$$
$$\leq T(t-1) + 2M^2\sigma^2\zeta^T(t)P(t)\zeta(t)$$
$$\qquad - E\big(\alpha(t)\zeta^T(t)\tilde{\theta}(t)\phi^T(t)\tilde{\theta}(t)\,|\mathbb{C}_{t-1}\big)$$

i.e.

$$E\left\{T(t) + \alpha(t)\zeta^T(t)\tilde{\theta}(t)\phi^T(t)\tilde{\theta}(t)\,|\mathbb{C}_{t-1}\right\} \leq T(t-1) + 2M^2\sigma^2\zeta^T(t)P(t)\zeta(t).$$

Introducing

$$\tilde{T}(t) = \frac{1}{t}\left[T(t) + \sum_{k=1}^{t}\alpha(k)\tilde{\theta}^T(k)\zeta(k)\phi^T(k)\tilde{\theta}(k)\right] \qquad \text{(A.24)}$$

(obviously $\tilde{T}(t) > 0$) and noting that $\tilde{\theta}^T(k)\zeta(k)\phi^T(k)\tilde{\theta}(k) \in \mathbb{C}_{t-1}$ for $k \leq t-1$, we have

$$E\left(t\,\tilde{T}(t)\mid \mathfrak{C}_{t-1}\right)$$

$$= E\left(T(t) + \alpha(t)\,\tilde{\theta}^T(t)\,\varsigma(t)\,\phi^T(t)\,\tilde{\theta}(t)\mid \mathfrak{C}_{t-1}\right) + \sum_{k=1}^{t-1}\alpha(k)\,\tilde{\theta}^T(k)\,\varsigma(k)\,\phi^T(k)\,\tilde{\theta}(k)$$

$$\le T(t-1) + 2M^2\sigma^2\varsigma^T(t)P(t)\varsigma(t) + \sum_{k=1}^{t-1}\alpha(k)\,\tilde{\theta}^T(k)\,\varsigma(k)\,\phi^T(k)\,\tilde{\theta}(k)$$

$$= (t-1)\tilde{T}(t-1) + 2M^2\sigma^2\varsigma^T(t)P(t)\varsigma(t)$$

i.e.

$$E\left(\tilde{T}(t)\mid \mathfrak{C}_{t-1}\right) \le \tilde{T}(t-1) - \frac{1}{t}\tilde{T}(t-1) + \frac{1}{t}2M^2\sigma^2\varsigma^T(t)P(t)\varsigma(t). \qquad \text{(A.25)}$$

Using lemma 4 and the theorem given by Neveu [22] leads to the final conclusions:

$$\tilde{\theta}^T(t)P^{-1}(t)\,\tilde{\theta}(t) \to 0 \qquad\qquad w.p.1 \;\; as \;\; t \to \infty$$

and

$$\frac{1}{N}\sum_{t=1}^{N}\left(\bar{\varepsilon}(t) - \nu(t)\right)^2 \qquad\qquad w.p.1 \;\; as \;\; N \to \infty.$$

It ends the proof of the theorem.   ∎

X.C. A Convergence Proof Using the Ordinary Differential Equation Approach Associated with Robust Bootstrap Method

X.C.1. Proof of Lemma 5

First, let us introduce $\tilde{\theta}(t) = \hat{\theta}(t)-\theta_0$ and $\bar{\varepsilon}(t) = y(t)-\hat{w}(t)$, then

$$\bar{\varepsilon}(t) = -\hat{X}^T(t-n)\left(\bar{a} + \bar{b}u(t-n)\right) - \hat{E}^T(t-1)\bar{d}$$
$$\qquad\quad - U^T(t-1)\bar{c} + \hat{X}^T(t-n)\left(a + bu(t-n)\right) + \tilde{E}^T d + e(t) \qquad \text{(A.26)}$$

Based on Eq.(13), $X(t-n)$ can be expressed as

$$\left.\begin{aligned}
x_1(t-n) &= y(t-n) - e(t-n)\\
x_2(t-n) &= y(t-n+1) - c_1 u(t-n) - e(t-n+1) - d_1 e(t-n)\\
&\;\;\vdots\\
x_n(t-n) &= y(t-1) - \sum_{i=1}^{n-1}c_i u(t-i-1) - \sum_{i=1}^{n-1}d_i e(t-i-1) - e(t-1)
\end{aligned}\right\} \qquad \text{(A.27a)}$$

Similarly, $\hat{X}(t-n)$ is given by

$$\left.\begin{aligned}
\hat{x}_1(t-n) &= y(t-n) - \bar{\varepsilon}(t-n)\\
\hat{x}_2(t-n) &= y(t-n+1) - \hat{c}_1 u(t-n) - \bar{\varepsilon}(t-n+1) - \hat{d}_1\bar{\varepsilon}(t-n)\\
&\;\;\vdots\\
\hat{x}_n(t-n) &= y(t-1) - \sum_{i=1}^{n-1}\hat{c}_i u(t-i-1) - \sum_{i=1}^{n-1}\hat{d}_i\bar{\varepsilon}(t-i-1) - \bar{\varepsilon}(t-1)
\end{aligned}\right\} \qquad \text{(A.27b)}$$

So,

$$\tilde{X}^T(t-n)\big(a + bu(t-n)\big)$$

$$= \sum_{i=1}^{n}\big(\overline{\varepsilon}(t-i) - e(t-i)\big)\big(a_i + b_i u(t-n)\big)$$

$$+ \sum_{j=1}^{n-1}\big(a_{n-j} + b_{n-j}u(t-n)\big)\sum_{i=1}^{j}\big[d_i\big(\overline{\varepsilon}(t-n+j-i) - e(t-n+j-i)\big)\big] \qquad \text{(A.28)}$$

$$+ \sum_{j=1}^{n-1}\big(a_{n-j} + b_{n-j}u(t-n)\big)\sum_{i=1}^{j}\tilde{c}_i u(t-n+j-i)$$

$$+ \sum_{j=1}^{n-1}\big(a_{n-j} + b_{n-j}u(t-n)\big)\sum_{i=1}^{j}\tilde{d}_i\overline{\varepsilon}(t-n+j-i).$$

Substituting (A.28) into (A.26) and rearranging each term produce

$$\overline{\varepsilon}(t) - e(t)$$

$$= -\hat{X}^T(t-n)\big(\bar{a} + \bar{b}u(t-n)\big) - \sum_{i=1}^{n}\tilde{c}_i\left[1 - \sum_{j=1}^{n-i}\big(a_j + b_j u(t-n)\big)z^{-j}\right]u(t-i) \qquad \text{(A.29)}$$

$$- \sum_{i=1}^{n}\tilde{d}_i\left[1 - \sum_{j=1}^{n-i}\big(a_j + b_j u(t-n)\big)z^{-j}\right]\overline{\varepsilon}(t-i) + \big(E^T(t-1) - \hat{E}^T(t-1)\big)\big(d - f\big)$$

where

$$f = \begin{bmatrix} \big(a_1 + b_1 u(t-n)\big) \\ \big(a_2 + b_2 u(t-n)\big) + \big(a_1 + b_1 u(t-n)\big)d_1 \\ \vdots \\ \big(a_n + b_n u(t-n)\big) + \sum_{i=1}^{n-1}\big(a_i + b_i u(t-n)\big)d_{n-i} \end{bmatrix}. \qquad \text{(A.30)}$$

That is

$$\big(1 + K(z^{-1})\big)\big(\overline{\varepsilon}(t) - e(t)\big) = -\hat{X}^T(t-n)\big(\bar{a} + \bar{b}u(t-n)\big)$$
$$- \sum_{i=1}^{n}\tilde{c}_i\tau_i(z^{-1})u(t-i) - \sum_{i=1}^{n}\tilde{d}_i\tau_i(z^{-1})\overline{\varepsilon}(t-i). \qquad \text{(A.31)}$$

Following the notations in lemma 5, it is easy to prove the lemma. ∎

X.C.2. Proof of Theorem 4

The differential equations associated with algorithm (54) to (57) are as follows:

$$\frac{d\theta(t)}{dt} = R^{-1}(t)f\big(\theta(t)\big)$$
$$\frac{dR(t)}{dt} = \overline{G}\big(\theta(t)\big) - R(t) \qquad \text{(A.32)}$$

where

$$f(\hat{\theta}(t)) = E(\phi(t,\hat{\theta})\alpha(t)\overline{\varepsilon}(t))$$
$$\overline{G}(\hat{\theta}(t)) = E(\phi(t,\hat{\theta})\alpha(t)\phi^T(t,\hat{\theta})).$$

(A.33)

From lemma 5, the first equation in (A.33) can be rewritten as

$$f(\hat{\theta}(t)) = E\left\{\phi(t,\hat{\theta})\alpha(t)\left[\left(\theta_0 - \hat{\theta}\right)^T\overline{\phi}(t,\hat{\theta}) + e(t)\right]\right\}.$$

Again,

$$\overline{\phi}(t,\hat{\theta}) = H(z^{-1})\phi(t,\hat{\theta})$$

corresponds to the particular input signal $u(t\text{-}n)$ in $H(z^{-1})$. Based on assumptions (i) and (ii)

$$f(\hat{\theta}(t)) = E\left[\alpha(t)\phi(t,\hat{\theta})\overline{\phi}^T(t,\hat{\theta})\left(\theta_0 - \hat{\theta}(t)\right) + E(\phi(t,\hat{\theta})\alpha(t)e(t))\right]$$
$$= -\alpha(t)\tilde{G}(\hat{\theta}(t))\tilde{\theta}(t)$$

where

$$\tilde{G}(\hat{\theta}(t)) = E\left(\phi(t,\hat{\theta})\overline{\phi}^T(t,\hat{\theta})\right).$$

Define the following Lyapunov function

$$V(\tilde{\theta}(t),R) = \tilde{\theta}^T(t)R(t)\tilde{\theta}(t)$$

(A.34)

then along the trajectories of (A.32), we have

$$\frac{dV(\tilde{\theta}(t),R)}{dt} = -\alpha(t)\tilde{\theta}^T(t)\left[\tilde{G}^T(\hat{\theta}(t)) + \tilde{G}(\hat{\theta}(t)) - G(\hat{\theta}(t)) + R(t)\right]\tilde{\theta}(t)$$

where

$$G(\hat{\theta}(t)) = E\left(\phi(t,\hat{\theta})\phi^T(t,\hat{\theta})\right).$$

For the particular input signal $u(t\text{-}n)$, we have

$$\tilde{G}^T(\hat{\theta}(t)) + \tilde{G}(\hat{\theta}(t)) - G(\hat{\theta}(t)) = \frac{1}{2\pi}\int_{-\pi}^{\pi}\left(\Gamma(\omega)H^T(e^{j\omega}) + H(e^{j\omega})\Gamma^T(\omega) - I\Gamma(\omega)\right)d\omega$$

where $\Gamma(\omega)$ is the spectral density matrix of the stationary process $\phi(t,\hat{\theta})$. Since $\Gamma(\omega)$ is real and symmetric and $H(e^{j\omega})$ is real and diagonal

$$\tilde{G}^T(\hat{\theta}(t)) + \tilde{G}(\hat{\theta}(t)) - G(\hat{\theta}(t)) = \frac{1}{2\pi}\int_{-\pi}^{\pi}2Re\left[H(e^{j\omega}) - \frac{I}{2}\right]\Gamma(\omega)\,d\omega.$$

It is clear that the matrix on left side of above equation is positive definite if $H(e^{j\omega})\text{-}I/2$ is s.p.r. for the particular input $u(t\text{-}n)$, $t = 1, ..., N$. Therefore, we have

$$\frac{dV(\hat{\theta}(t),R)}{dt} \leq 0$$

i.e.

$$Prob\left[\lim_{t\to\infty}\hat{\theta} \to \theta_0\right] = 1 \qquad globally.$$

This completes the proof of theorem 4.   ∎

XI. REFERENCES

1.   P.J. Huber, *Robust Statistics*, John Wiley & Sons, New York (1981).

2.   P.J. Huber, "Robust estimation of a location parameter", *Ann. Math. Stat.*, 35, pp.73-101 (1964).

3.   B.T. Poljak and Ja.Z. Tsypkin, "Robust identification", *Proc. 4th IFAC Symp. Ident. Syst. Par. Est.*, pp.203-224, Tbilisi (1976).

4.   B.T. Poljak and Ja.Z. Tsypkin, "Robust identification", *Automatica*, 16(1), pp.53-63 (1980).

5.   C.I. Masreliez and R.D. Martin, "Robust Bayesian estimation for a linear model and robustifying the Kalman filter", *IEEE Trans. Automat. Contr.*, AC-22(3), pp.361-371 (1977).

6.   S. Basu and D. Vandelinde, "Robust identification of a linear system", *Proc. Fifteenth Ann. Allertion Conf. on Communication, Control and Computing*, pp.221-230 (1977).

7.   S.S. Stanković and B.D. Kovacević, "Analysis of robust stochastic approximation algorithms for process identification", *Automatica*, 22(4), pp.483-488 (1986).

8.   S.C. Puthenpura, N.K. Sinha, and O.P. Vidal, "Application of M-estimation i robust recursive system identification", *IFAC Symp. on Stochastic Control*, pp.23-30 (1985).

9.   S.C. Puthenpura and N.K. Sinha, "Robust instrumental variables method for system identification", *Control Theory and Advanced Technology*, 1(3), pp.175-188 (1985).

10. S.C. Puthenpura and N.K. Sinha, "Modified maximum likelihood method for the robust estimation of system parameters from very noisy data", *Automatica*, 22(2), pp.231-235 (1986).

11. S.C. Puthenpura and N.K. Sinha, "Robust bootstrap method for joint estimation of states and parameters of a linear system", *ASME J. of Dynamic. Syst., Measu., and Control*, 108(3), pp.255-263 (1986).

12. S.C. Puthenpura and N.K. Sinha, "Robust identification from impulse and step response", *IEEE Trans. on Industrial Electronics*, IE-34(3), pp.366-370 (1987).

13. A. Venot, L. Pronzato, E. Ealter, and J.-F. Lebruchec, "A distribution-free criterion for robust identification, with applications in system modelling and image processing", *Automatica*, 22(1), pp.105-109 (1986).

14.   E. Walter and H. Piet-Lahanier, "Robust nonlinear parameter estimation in the bounded noise case", *Proc. 25th IEEE Conf. Decision and Control*, **2** pp.1037-1042 (1986).

15.   H. Dai, *Robust Identification of Bilinear Systems*, Ph.D. Thesis, McMaster University, Hamilton, Ontario (1990).

16.   H. Dai and N.K. Sinha, "Robust recursive least squares method with modified weights for bilinear system identification", *IEE Proceedings-D Control Theory and Applications*, **136**(3), pp.122-126 (1989).

17.   H. Dai and N.K. Sinha, "Robust combined estimation of states and parameters of bilinear systems", *Automatica*, **25**(4), pp.613-616 (1989).

18.   H. Dai and N.K. Sinha, "Robust recursive instrumental variable method with modified weights for bilinear system identification", *IEEE Trans. on Industrial Electronics*, **38**(1), pp.1-7 (1991).

19.   H. Dai and N.K. Sinha, "Robust recursive output error method for bilinear system identification", *Proc. 8th IFAC Symp. Ident. Syst. Par. Est.*, **2**, pp.1141-1146 (1988).

20.   N.K. Sinha and B. Kuszta, *Modeling and Identification of Dynamic Systems*, Van Nostrand Reinhold Company, New York (1983).

21.   L. Ljung and T. Söderström, *Theory and Practice of Recursive Identification*, The MIT Press (1983).

22.   J. Neveu, *Discrete Parameter Martingales*, North Holland Publishing Company, Amsterdam (1975).

23.   I.D. Landau, "Unbiased recursive identification using model reference techniques", *IEEE Trans. Automat. Contr.*, **AC-21**(2), pp.194-202 (1976).

24.   H. El-Sherief and N.K. Sinha, "Bootstrap estimation of parameters and states of linear multivariable systems", *Automatica*, **16**(2), pp.340-343 (1979).

25.   L. Dugard and I.D. Landau, "Recursive output error identification algorithm theory and evaluation", *Automatica*, **16**(5), pp.443-462 (1980).

26.   F. Fnaiech and L. Ljung, "Recursive identification of bilinear systems", *Int. J. Control*, **45**(2), pp.453-470 (1987).

Robust Controller Design: A Bounded-Input, Bounded-Output Worst-Case Approach

Munther A. Dahleh*
Department of Electrical Engineering and Computer Science
Massachuasetts Institute of Technology
Cambridge, MA 02139

Abstract

This chapter introduces a general framework for designing robust control systems in the presence of uncertainty when the specifications are posed in the time-domain. Necessary and sufficient conditions for performance robustness and a synthesis methodology utilizing linear programming are presented. Since many of the past work involved quadratic performance objectives, this chapter outlines a set of new tools needed to handle Peak-to-Peak specifications. The objective in here is to present many of the central results in this discipline in an intuitive fashion rather than a complete rigorous treatment of the subject.

*This research was supported by Wright Patterson AFB under grant F33615-90-C-3608, C.S. Draper Laboratory under grant DL-H-441636 and NSF under grant 9157306-ECS.

1 Introduction

Feedback controller design is primarily concerned with designing control systems that can deliver high quality performance despite the presence of uncertainty. Eventhough a real system is not uncertain, it is desirable to think of it as such to reflect our imprecise or partial knowledge of its dynamics. On the other hand, uncertainty in the noise and disturbances can be cast under "real uncertainties", as it is practically impossible to provide exact models of such inputs. The description of this *uncertainty*, plant or input uncertainty, depends on the particular physical system and the operation environment and thus it is problem-dependent. To aid the designer in this process, analytical theory has been developed to provide a design methodology for synthesizing controllers to achieve "certain" performance objectives in the presence of a specific class of uncertainty. It is understood that a particular design problem may not fit the assumptions needed in any of these problems, nevertheless such methods can provide at least three important pieces of information to the designer (one can think of many more): The first is providing a starting point for design, maybe in terms of achieving robust stability and nominal performance, but possibly also achieving robust performance. The second is highlighting the tradeoffs in designing the controller for the system given. As intuitively expected, robustifying the system to handle larger classes of uncertainty results in more conservative design and hence in the loss of some performance. The third is capturing in a quantitative way the fundamental limitations and capabilities of feedback design. This is possibly the most important of all since one would like to know whether certain performance specifications can be achieved, and if so, with what controllers. Examples of such analytical theory are the H_2, H_∞ and the ℓ_1 theory.

The ℓ_1 theory is primarily a time-domain theory in which performance specifications as well as input and plant uncertainty are posed in the time-domain in terms of the maximum norm (ℓ_∞-norm). This is a natural set-up since most performance specifications are magnitude-specifications such as tracking errors, maximum deviations from a nominal point, limits on actuator authority and so on. Disturbances in general have the property of persistence and boundedness and are best described in terms of the ℓ_∞ norm. Examples of such disturbances are wind gusts, friction and so on.

In this chapter, an overview of the ℓ_1 theory is given starting from simple nominal performance problems and leading all the way to a general theory that allows for synthesizing controllers to achieve robust performance. The basic ideas involved in concepts of robust stabilization are introduced in a simple intuitive way, without obscuring these ideas through rigorous proofs. The synthesis of ℓ_1 controllers is explained through an analogy with standard linear programs, the difference being only in that the ℓ_1 problem has an infinite number of variables. Most of the exact proofs are referred to as they are needed.

2   Preliminaries

First, some notation regarding standard concepts for input/output systems. For more details, consult [1, 2] and references therein.

$\ell_{\infty,e}$ denotes the extended space of sequences in $\mathbb{R}^N$, $f = \{f_0, f_1, f_2, \ldots\}$. ℓ_∞ denotes the set of all $f \in \ell_{\infty,e}$ such that

$$\|f\|_{\ell_\infty} = \sup_k |f(k)|_\infty < \infty$$

where $|f(k)|_\infty$ is the standard ℓ_∞ norm on vectors. $\ell_{\infty,e} \backslash \ell_\infty$ denotes the set $\{f : f \in \ell_{\infty,e} \text{ and } f \notin \ell_\infty\}$. $\ell_p, p \in [1, \infty)$, denotes the set of all sequences, $f = \{f_0, f_1, f_2, \ldots\}$ in R^N such that

$$\|f\|_{\ell_p} = \left(\sum_k |f(k)|_p^p \right)^{1/p} < \infty.$$

c_0 denotes the subspace of ℓ_∞ in which every function x satisfies

$$\lim_{k \to \infty} x(k) = 0.$$

S denotes the standard shift operator.

P_k denotes the kth-truncation operator on $\ell_{\infty,e}$:

$$P_k : \{f_0, f_1, f_2, \ldots\} \longrightarrow \{f_0, \ldots, f_k, 0, \ldots\}$$

Let $H : \ell_{\infty,e} \longrightarrow \ell_{\infty,e}$ be a nonlinear operator. H is called *causal* if

$$P_k H f = P_k H P_k f, \quad \forall k = 0, 1, 2, \ldots,$$

H is called *strictly causal* if

$$P_k H f = P_k H P_{k-1} f, \quad \forall k = 0, 1, 2, \ldots$$

H is called *time-invariant* if it commutes with the shift operator:

$$H S = S H.$$

Finally, H is called ℓ_p *stable* if

$$\|H\| = \sup_k \sup_{\substack{f \in \ell_{p,e} \\ P_k f \neq 0}} \frac{\|P_k H f\|_{\ell_p}}{\|P_k f\|_{\ell_p}} < \infty.$$

The quantity $\|H\|$ is called the *induced operator norm over ℓ_p*.
$\mathcal{L}_{TV}$ denotes the set of all *linear* causal ℓ_∞-stable operators. This space is characterized by infinite block lower triangular matrices of the form:

$$\begin{pmatrix} H_{00} & & 0 \\ H_{10} & H_{11} & \\ \vdots & \vdots & \ddots \end{pmatrix}$$

where H_{ij} is a $p \times q$ matrix. This infinite matrix representation of H acts on elements of ℓ_∞^q by multiplication, i.e. if $u \in \ell_\infty^q$, then $y := Hu \in \ell_\infty^p$ where $y(k) = \sum_{j=0}^k H_{kj} u(j) \in \mathbb{R}^p$. The induced norm of such an operator is given by:

$$\|H\|_{\mathcal{L}_{TV}} = \sup_i |(H_{i1} \ldots H_{ii})|_1$$

where $|A|_1 = \max_i \sum_j |a_{ij}|$. $\mathcal{L}_{TI}$ denotes the set of all $H \in \mathcal{L}_{TV}$ which are *time-invariant*. It is well known that $\mathcal{L}_{TI}$ is isomorphic to ℓ_1 and the matrix representation of the operator has a Toeplitz structure. Every element in $\mathcal{L}_{TI}$ is associated with a λ-transform defined as:

$$\hat{H}(\lambda) = \sum_{k=0}^{\infty} H(k) \lambda^k$$

The collection of all such transforms is usually denoted by $\mathbf{A}$, which will be equipped with the same norm as the ℓ_1 norm.

Throughout this paper, systems are thought of as operators. So, the composition of two operators G, H is denoted as GH, if both are time-invariant then $GH \in \ell_1$ (or $\mathcal{L}_{TI}$), and the induced norm is denoted by $\|GH\|_1$. When the λ-transform is referred to specifically, we use the notation $\hat{H}$ for the transform of H. Also, all operator spaces are matrix-valued functions whose dimensions will be suppressed in general whenever understood from the context.

3   The ℓ_1 Norm

Let T be a linear time-invariant system given by

$$z(t) = (Tw)(t) = \sum_{k=0}^{t} T(k)w(t-k).$$

The inputs and outputs of the system are measured by their maximum amplitude over all time; otherwise known as the ℓ_∞ norm, i.e.

$$\|w\|_\infty = \max_j \sup_k |w_j(k)|.$$

The ℓ_1 norm of the system T is precisely equal to the maximum amplification the system exerts on bounded inputs. This measure defined on the system T is known as the induced operator norm and is mathematically defined as follows:

$$\|T\| = \sup_{\|w\|_\infty \leq 1} \|Tw\|_\infty = \|T\|_1$$

where $\|T\|_1$ is the ℓ_1-norm of the pulse response and is given by

$$\|T\|_1 = \max_i \sum_j \sum_k |t_{ij}(k)|.$$

A system is said to be ℓ_∞- stable if it has a bounded ℓ_1 norm and the space of all such system will be denoted by ℓ_1. From this definition, it is clear that the system attenuates inputs if its ℓ_1 norm is strictly less than unity. For control systems applications, if the objective involves minimizing maximum deviations, rejection of bounded disturbances, tracking uniformly in time, or meeting certain objectives with hard constraints on the signals, a problem with an ℓ_1 norm criterion arises.

In the case where the inputs and outputs of the linear system are measured by the ℓ_2 norm, then the gain of the system is given by the H_∞ norm and is given by:

$$\|\hat{T}\|_\infty = \sigma_{max}(\hat{T}(e^{i\theta})).$$

The two induced norms are related as follows:

$$\|\hat{T}\|_\infty \leq C\|T\|_1$$

where C is a constant depending only on the dimension of the matrix T. In other words, every system inside ℓ_1 is also inside H_∞, however the converse is not true. This means that there exists ℓ_2 stable linear time-invariant systems that are not ℓ_∞ stable; an example is the function with the λ-transform given by [3];

$$\hat{T}(\lambda) = e^{\frac{1}{1-\lambda}}.$$

Thus, for LTI systems, minimizing the ℓ_1 norm of a systems guarantees that the H_∞ norm is bounded. This means that this system will have good ℓ_2-disturbance rejection properties as well as ℓ_∞-disturbance rejection properties. Also, the ℓ_1 norm is more tied-up to BIBO stability notions and hence quite desirable to work with. The disadvantage in working with the ℓ_1 norm is the fact that it is a Banach space of operators operating on a Banach space, not a Hilbert space. Many of the standard tools are not usable, however, this overview will present new techniques for handling problems of this kind.

4   Stability and Parameterization

Consider the system G described as in Figure 1. G is a 2×2 Block matrix, u is the control input, y is the measured output, w is the exogenous input, z is the regulated output, and K is the feedback controller. Let $H(G, K)$

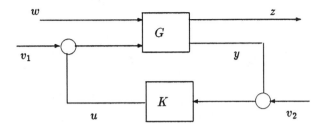

Figure 1: Closed Loop System

denote the map

$$H(G,K) = \begin{pmatrix} w \\ v_1 \\ v_2 \end{pmatrix}, \longrightarrow \begin{pmatrix} z \\ u \\ y \end{pmatrix}$$

The closed loop system is ℓ_∞-stable if and only if $H(G,K) \in \ell_1$. The map of interest is the map between w to z, denoted by T_{zw}:

$$T_{zw} = G_{11} + G_{12}K(I - G_{22}K)^{-1}G_{21}$$

This particular mapping captures the *performance objectives*. If $w \in \ell_\infty$ with $\|w\|_\infty \leq 1$ but other than that is arbitrary, then the nominal performance problem is defined as:

$$\inf_{K \text{ stabilizing}} (\sup_w \|T_{zw}w\|_\infty) = \inf_{K \text{ stabilizing}} \|T_{zw}\|_1.$$

4.1   Controller Parameterization

There are several ways for arriving to the parameterization of all stabilizing controllers, one of which is via coprime factorization [4, 5, 6]. In the 2-input 2-output set-up, the stabilizability of the system G is equivalent to the stabilizability of G_{22}. Let G_{22} have the bi-coprime factorization with $G_{22} = NM^{-1} = \tilde{M}^{-1}\tilde{N}$:

$$\begin{bmatrix} \tilde{X} & -\tilde{Y} \\ -\tilde{N} & \tilde{M} \end{bmatrix} \begin{bmatrix} M & Y \\ N & X \end{bmatrix} = I$$

where all quantities above are stable. All stabilizing controllers are then parametrized as:

$$K = (Y - MQ)(X - NQ)^{-1} = (\check{X} - Q\check{N})^{-1}(\check{Y} - Q\check{M}) \qquad Q \text{ stable.}$$

This parameterization has two major advantages: the first is that it furnishes the space of all stabilizing controllers (including time-varying and nonlinear) in terms of one parameter in a vector space, eventhough the space of all stabilizing controllers is not in itself a vector space. The second advantage, which is a great surprise, is that it transforms the complicated mapping T_{zw} which is a nonlinear expression in the controller, to an affine linear function in the parameter Q. By simple manipulations, T_{zw} is given by:

$$T_{zw} = T_1 - T_2 Q T_3$$

where

$$T_1 = G_{11} + G_{12} M \check{Y} G_{21}$$

$$T_2 = G_{12} M$$

$$T_3 = \check{M} G_{21}.$$

Define the feasible space $\mathcal{S}$ as follows:

$$\mathcal{S} = \{R \in \ell_1 | R = T_2 Q T_3, \ Q \in \ell_1\}.$$

The ℓ_1 optimal control problem is defined as follows:

$$\inf_{R \in \mathcal{S}} \|T_1 - R\|_1.$$

5   Examples of Nominal Performace Objectives

In this section, a few examples are presented to illustrate this formulation. These examples will show in a very simple way the advantages of using the ℓ_∞-norm on signals to reflect realistic time-domain specifications.

5.1   Disturbance Rejection

In many real world applications, output disturbance and/or noise is persistent, i.e. continues acting on the system as long as the system is in operation. This implies that such inputs have infinite energy, and thus one cannot model them as "bounded-energy signals". Also, disturbances can be correlated with the inputs to the plant in a nonlinear fashion that makes it difficult to get accurate information about its statistics. Nevertheless, one can get a good estimate on the maximum amplitude of such inputs. In general, we will assume that the disturbance is the output of a linear-time invariant filter subjected to signals of magnitude less than or equal to one, i.e

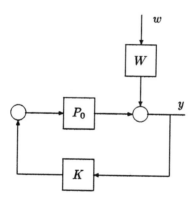

Figure 2: Disturbance Rejection Problem

$$d = Ww, \quad ||w||_\infty \leq 1.$$

The disturbance rejection problem is defined as follows: Find a feedback controller that minimizes the maximum amplitude of the output over all possible disturbances. The two-input two-output system shown in Figure 2 is given by

$$z = P_0 u + Ww$$

$$y = P_0 u + W w$$

5.2 Command Following with Saturations

The command following problem is equivalent to the disturbance rejection problem and is shown in Figure 3. In here, we will show how to pose this problem, in the presence of saturation nonlinearities at the input of the plant, as an ℓ_1 -optimal control problem. Define the function

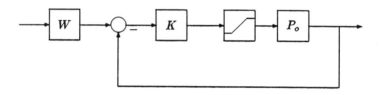

Figure 3: Command Following with Input Saturation

$$Sat(u) = \begin{cases} u & |u| \le U_{\max} \\ U_{\max} & |u| \ge U_{\max} \end{cases}$$

Let the plant be described as

$$P_u = P_0 Sat(u)$$

where P_0 is LTI. Let the commands be modeled as

$$r = Ww, \ ||w||_\infty \le 1.$$

The objective is to find a controller K such that y follows r uniformly in time. Keeping in mind the saturation function, it is clear that the allowable inputs have to have $||u||_\infty \le U_{\max}$. Let γ be a performance level desired, and define

$$z = \begin{bmatrix} \frac{1}{\gamma}(y - r) \\ \frac{1}{U_{\max}} u \end{bmatrix}$$

with

$$y = P_0 u$$

The problem is equivalent to finding a controller such that

$$\left(\sup_{w} \|z\|_\infty\right) < 1$$

which is an ℓ_1 -optimal control problem.

Comment It is known to many that for such a problem, one will introduce some nonlinear function to minimize the effects of saturations on the plant. The formulation of this problem as a ℓ_1 minimization problem is an example of the usage of this theory to highlight the fundamental limitations of linear controller design.

5.3   Robust Stability

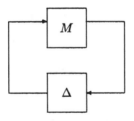

Figure 4: Stability Robustness Problem

Underlying most of the stability robustness results is the small gain theorem. Basically the theorem guarantees the stability of a feedback system consisting of an interconnection of two stable systems as in Figure 4 if the product of their gains is less than unity. This theorem is quite general and applies to nonlinear time-varying systems with any notion of ℓ_p-stability. The small gain condition is in general not necessary for stability, however, it can be necessary if one of the systems in the feedback was arbitrary. Most of these results were previously treated in the literature for the case of ℓ_2-stability

[14], but were not studied in the case of ℓ_∞-stability. The next theorem is a surprising result in this sense and is due to Dahleh and Ohta [7].

Theorem 1 *Let M be a linear time-invariant system and Δ be a linear (possibly time-varying) stable system. The closed loop system shown in Figure 4 is ℓ_∞-stable for all Δ with gain $\sup_{\|f\|_\infty \leq 1} \|\Delta f\|_\infty < 1$ if and only if $\|M\|_1 \leq 1$.*

The theorem implies the following: if $\|M\|_1 > 1$, then there exists a stable time-varying perturbation with gain less than unity such that the closed loop system is unstable. It is important that the perturbations Δ are allowed to be time-varying, otherwise the theorem may not be satisfied.

Example 1 *Let*

$$\hat{M}(\lambda) = 0.9\frac{\lambda - 0.5}{1 - .5\lambda}.$$

The ℓ_1 norm of M is equal to 1.35 and the H_∞ norm is equal to 0.9. Any destabilizing linear time-invariant perturbation satisfies:

$$\|\Delta\|_1 \geq \|\hat{\Delta}\|_\infty \geq \frac{1}{0.9} > 1.$$

However, a time-varying perturbation exists with a gain smaller than one.

As a consequence of the small gain theorem, it is possible to provide stability robustness conditions for some classes of perturbed systems.

5.4 Unstructured Multiplicative Perturbations

Consider the case where the system has input uncertainty in a multiplicative form as in Figure 5, i.e.

$$\Omega = \{P|P = P_0(I + W_1\Delta W_2), \Delta \text{ is time-varying with}\|\Delta\| < 1\}.$$

If a controller is designed to stabilize P_0, under what conditions will it stabilize the whole set Ω? By simple manipulations of the closed loop system, the problem is equivalent to the stability robustness of the feedback system in Figure 4, with $M = W_2P_0K(I - P_0K)^{-1}W_1$. A necessary and sufficient

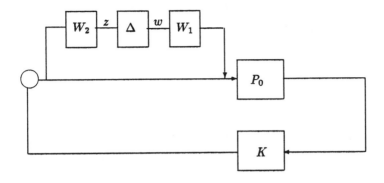

Figure 5: Multiplicative Perturbations

condition for robust stability is then given by $\|M\|_1 \leq 1$. The resulting 2-input 2-output description is given by:

$$y = P_0 u + P_0 W_1 w$$

$$z = W_2 u$$

5.5   Robustness in the Presence of Stable Coprime Factor Perturbations

Let P_0 be a linear time invariant, finite dimensional plant. As usual, $P_0 = G_{22}$. The graph of P_0 over the space ℓ_q is given by [5]:

$$G^q(P_0) = G_{P_0}\ell_q \; where \; G_{P_0} = \begin{bmatrix} M \\ N \end{bmatrix}$$

Define the following class of plants as in Figure 6:

$$\Omega_q = \{P|G_P = \begin{bmatrix} M + \Delta_1 \\ N + \Delta_2 \end{bmatrix} \; and \; \| \begin{bmatrix} \Delta_1 \\ \Delta_2 \end{bmatrix} \| \leq 1\}$$

where Δ_i's are ℓ_q-stable linear time-varying systems, and the norm is the induced ℓ_q norm. The next theorem gives a necessary and sufficient condition for a controller that stabilizes P_0 to stabilize all $P \in \Omega_\infty$. A similar result in the case of $P \in \Omega_2$ was proved in [8]. It is evident that Ω_∞ contains time varying plants which will be essential for the proof of the next theorem capturing the stability robustness conditions [9]:

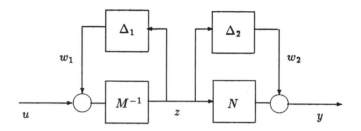

Figure 6: Coprime Factor Perturbations

Theorem 2 *If K stabilizes P_0, then K stabilizes all $P \in \Omega_\infty$ if and only if*

$$\left\| \begin{bmatrix} \tilde{X} - Q\tilde{N} & -\tilde{Y} + Q\tilde{M} \end{bmatrix} \right\|_1 \leq 1.$$

It is interesting to note that this class of perturbations is a more natural one in the case of unstable plants. It amounts to perturbing the graph of the operator, rather than the operator directly. This class of perturbations allow unstable perturbations and can result in changing the locations and the number of the unstable poles of the system. A description of the coprime perturbations of a system can be derived in a natural way from parameteric identification techniques in which a fixed order polynomial is identified for both M, N. The incorporation of such robustness results in adaptive controllers has proved quite effective as shown in [10, 11].

If the above condition is not satisfied, then the theorem asserts that there exists an admissible time-varying plant which the controller does not stabilize. The 2-input 2-output description that results in the above criterion is given by:

$$y = P_0 w_1 + w_2 + P_0 u$$

$$z = M^{-1} w_1 + M^{-1} u$$

Comment There is a subtle difference between the construction of the plant which is destabilized by the controller in this case and the plant in the case of multiplicative perturbations. In the multiplicative perturbation case, if the stability robustness condition is not satisfied, the constructed destabilizing perturbation results in a plant with an internal unstable cancellation, thus cannot be stabilized by any controller. In the case of coprime factor perturbations, such plants can be ruled out from the set, and the destabilizing perturbation results in a stabilizable plant, however, not stabilized with the specific controller used [9].

5.6   How big is this class of perturbation?

As mentioned earlier, the small gain condition is applicable for general nonlinear time-varying systems, and hence tends to be a conservative condition for stability robustness. Nevertheless, it is a powerful tool for representing realistic classes of uncertainty such as unmodelled dynamics, ignored nonlinearities, time delays and so on.

Since the approach presented above is the same for the H_∞ problem, it is worthwhile comparing the class of perturbations that have gain less than unity over ℓ_2 with the class of perturbations that have gain less than unity over ℓ_∞. If the perturbations are restricted to time-invariant ones, the ℓ_∞-stable perturbations with gain less than unity lie inside the unit ball of ℓ_2-stable perturbations (for the multivariable case, the unit ball will be scaled by a constant). This follows directly from the norm inequality between ℓ_1 and H_∞. If the perturbations are allowed to be time-varying, then the two sets are not comparable. Earlier, an example was presented that shows that the H_∞ ball is larger than the ℓ_1 ball. The next example furnishes a time varying operator which is ℓ_∞-stable but not ℓ_2 stable.

Example 2 *Define Δ as follows:*

$$(\Delta f)(k) = f(0)$$

Clearly, this operator is ℓ_∞ stable but not ℓ_2 stable.

5.7   Duality between Stability and Performance

In the previous examples, each of the robust stabilization problems was shown to be equivalent to some performance problem where a fictitious disturbance is injected at the output of the perturbation, and an error is measured at the input of the perturbation. So the transfer function to be minimized is simply the function *seen* by the perturbations. This says that robust stability is equivalent to some nominal performance problem. In the next section, the dual of this idea will be used: Performance will be equivalent to a robust stability problem in the presence of some fictitious perturbation. This will make the derivation of robust performance conditions a tractable problem.

6   A Unified Approach for Stability and Performance Robustness

In this section, it is shown that a general class of plant uncertainty can be described by linear fractional transformations. The robust performance problem will be posed as a robust stability problem in the presence of structured perturbations.

6.1   Linear Fractional Transformation

Given a 2×2 Block matrix transfer function G, we can define linear fractional transformations as

$$F_\ell(G, K) = G_{11} + G_{12}K(I - G_{22}K)^{-1}G_{21}$$

$$F_u(G, \Delta) = G_{22} + G_{21}\Delta(I - G_{11}\Delta)^{-1}G_{12}$$

Consider the 3×3 system matrix G shown in Figure 7 mapping

$$\begin{pmatrix} v \\ w \\ u \end{pmatrix} \longrightarrow \begin{pmatrix} r \\ z \\ y \end{pmatrix}$$

where $v = \Delta r$, $\Delta \in \mathcal{D}$, $u = Ky$. w and z denote the exogenous inputs and regulated outputs respectively, and $\mathcal{D}$ denotes the set of admissible perturbations.

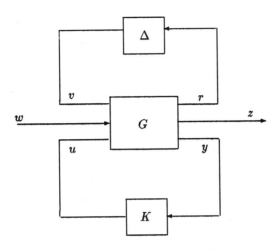

Figure 7: Perturbations Through LFT

$F_\ell(G, K)$ is the transfer matrix mapping

$$\begin{pmatrix} v \\ w \end{pmatrix} \longrightarrow \begin{pmatrix} r \\ z \end{pmatrix}$$

with $u = Ky$. $F_\ell(G, K)$ is the lower linear fractional transformation corresponding to G partitioned conformally with

$$\begin{pmatrix} v \\ w \\ \cdots \\ u \end{pmatrix} \longrightarrow \begin{pmatrix} r \\ z \\ \cdots \\ y \end{pmatrix}$$

$F_u(G, \Delta)$ is the transfer matrix mapping

$$\begin{pmatrix} w \\ u \end{pmatrix} \longrightarrow \begin{pmatrix} z \\ y \end{pmatrix}$$

with $v = \Delta r$. $F_u(G, \Delta)$ is the upper linear fractional transformation corresponding to G partitioned conformally with

$$\begin{pmatrix} v \\ \cdots \\ w \\ u \end{pmatrix} \longrightarrow \begin{pmatrix} r \\ \cdots \\ z \\ y \end{pmatrix}$$

Consider the class of plants $\Omega(G, \mathcal{D})$ in Figure 7 described as:

$$\Omega(G, \mathcal{D}) = \{P | P = F_u(G, \Delta) \text{ for some } \Delta \in \mathcal{D}\}.$$

It will be shown later on that this represent a wide class of plant uncertainty.

6.2 A General Class of Structured Uncertainty

We now formally set up the stability and performance robustness problem for a class of structured uncertainty. The configuration we shall use in the setup of the robustness problem is shown in Figure 7. The 3×3 system matrix G represents the particular structure of interconnection of the nominal plant and the perturbations Δ, and is therefore linear, time-invariant, and stable. The perturbation Δ has the form

$$\Delta = \begin{pmatrix} \Delta_1 & 0 & \dots & 0 \\ 0 & \Delta_2 & & \vdots \\ \vdots & & \ddots & 0 \\ 0 & \dots & 0 & \Delta_n \end{pmatrix}.$$

Each Δ_i represents the perturbation between two points in the system, and has norm less than or equal to one. Of course there is no loss of generality in assuming that the chosen bound on the norms of each of the Δ_i's is one, since any other set of numbers could be absorbed in G. We will restrict the Δ_i's to be *strictly* causal in order to guarantee the well posedness of the system. This is not a serious restriction and can be removed if it is known that the perturbation/nominal system connection is well-posed. Accordingly we can define the classes of perturbations to which the $\Delta_i's$ belong. Assuming the perturbations enter at n places, and that each has p_i inputs and q_i outputs we have

$$\Delta_i \in \mathbf{\Delta}(p_i, q_i)$$

where $\mathbf{\Delta}(p_i, q_i) := \left\{ \Delta \in \mathcal{L}_{TV}^{p_i \times q_i} | \ \Delta \text{ is strictly causal and } \|\Delta\| \leq 1 \right\}$

Note that Δ_i is not dependent in any way on Δ_j when $j \neq i$. The only restriction is that Δ_i belongs to $\mathbf{\Delta}(p_i, q_i)$ for each i. Next let $p = \sum_i p_i$, and $q = \sum_i q_i$. By $\mathcal{D}[(p_1, q_1); \cdots; (p_n, q_n)]$ we mean the set of all operators mapping ℓ^q_∞ to ℓ^p_∞ of the form:

$$\Delta = \begin{pmatrix} \Delta_1 & 0 & \ldots & 0 \\ 0 & \Delta_2 & & \vdots \\ \vdots & & \ddots & 0 \\ 0 & \ldots & 0 & \Delta_n \end{pmatrix}$$

where Δ_i belongs to $\Delta(p_i, q_i)$. When the pairs (p_i, q_i) are known, they will be dropped from the notation and $\mathcal{D}$ will be understood to mean the above set. We will say the system in Figure 7 achieves robust stability if the system is stable for all $\Delta \in \mathcal{D}[(p_1, q_1); \ldots; (p_n, q_n)]$.

Stability Robustness Problem. *Find necessary and sufficient conditions for the controller K to stabilize the class of plants*

$$\{P|P = F_u(G, \Delta)|\Delta \in \mathcal{D}\}.$$

On the other hand, we will say the system in Figure 7 achieves robust performance if the system is stable, and the effect of the exogenous inputs w on the regulated output z is attenuated for all $\Delta \in \mathcal{D}[(p_1, q_1); \ldots; (p_n, q_n)]$.

Performance Robustness Problem. *Find necessary and sufficient conditions for the controller K to robustly stabilize the class of plants*

$$\{P|P = F_u(G, \Delta)|\Delta \in \mathcal{D}\}.$$

and satisfy

$$\|T_{zw}\|_{\mathcal{L}_{TV}} < 1 \quad \forall \Delta \in \mathcal{D}.$$

When the controller is connected, the closed loop system of the plant and controller is given by $F_l(G, K)$ and in the next section will be denoted by M. The perturbation Δ will be connected to M as a feedback system.

6.3   Example

Consider the case where the system has both input uncertainty and output uncertainty as shown in Figure 8. Define new variables v_1, v_2, r_1, r_2,

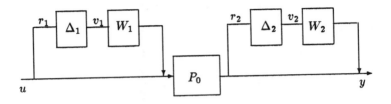

Figure 8: Input and Output Perturbations

where v_i is the output of Δ_i and r_i is the input to Δ_i. Let G be the transfer matrix from

$$\begin{pmatrix} v_1 \\ v_2 \\ u \end{pmatrix} \longrightarrow \begin{pmatrix} r_1 \\ r_2 \\ y \end{pmatrix}$$

and is given by:

$$G = \begin{pmatrix} 0 & 0 & I \\ P_0 W_1 & 0 & P_0 \\ P_0 W_1 & W_2 & P_0 \end{pmatrix}$$

Then, this class of uncertainty is represented as

$$\Omega = \{ P = F_u(G, \Delta) | \; \Delta \; \text{diagonal}, \; \|\Delta\| < 1 \}.$$

6.4   Performance Robustness Versus Stability Robustness

In this section, we will establish a useful relationship between stability and performance robustness, that will be used later in the solution of our problem. This is achieved in the theorem below proved by Khammash and Pearson [12]. It states that performance robustness in one system is equivalent to stability robustness in another one formed by adding a fictitious perturbation. A similar result has been shown to hold when the perturbations are linear time-invariant and when the 2-norm is used to characterize the perturbation class. The same proof does not apply here though, due to the assumed time-varying nature of the perturbations. The usefulness of this theorem stems from the fact that we can now concentrate on finding conditions for achieving stability robustness alone. Once we do, performance robustness comes for free.

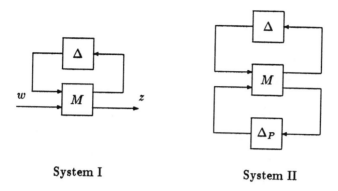

System I   System II

Figure 9: Stability Robustness vs. Performance Robustness

Consider the two systems shown in Figure 9, where $M \in \ell_1^{p \times q}$ and $\Delta_i \in \Delta(p_i, q_i)$. In system II, w is an input vector of size $\tilde{p}$ and z is an output vector of size $\tilde{q}$. In system I, $\Delta_p \in \Delta(\tilde{p}, \tilde{q})$. It follows that $p = \tilde{p} + \sum_i p_i$ and $q = \tilde{q} + \sum_i q_i$. Subdivide M in the following manner:

$$M = \begin{pmatrix} \tilde{M}_{11} & \tilde{M}_{12} \\ \tilde{M}_{21} & \tilde{M}_{22} \end{pmatrix}$$

where $\tilde{M}_{11} \in \ell_1^{\tilde{p} \times \tilde{q}}$.

We now state the following theorem establishing the relation between System I and System II.

Theorem 3 *The following four statements are equivalent:*

i) System I achieves robust stability.

ii) $(I - M\tilde{\Delta})^{-1}$ *is* ℓ_∞*- stable for all* $\tilde{\Delta} \in \mathcal{D}[(p_1, q_1); \cdots; (p_n, q_n); (\tilde{p}, \tilde{q})].$

iii) $(I - \tilde{M}_{11}\Delta)^{-1}$ *is* ℓ_∞*- stable and* $\|\tilde{M}_{22} + \tilde{M}_{21}\Delta(I - \tilde{M}_{11}\Delta)^{-1}\tilde{M}_{12}\| < 1$, *for all* Δ *belonging to* $\mathcal{D}[(p_1, q_1); \cdots; (p_n, q_n)].$

iv) System II achieves robust performance.

The theorem basically says the following: to achieve robust performance, wrap a fictitious Δ_p mapping z to w, and design a controller to achieve robust stability for the new system.

7 Conditions for Stability Robustness

In the previous sections, we have shown that the general Robust Performance problem is equivalent to a Robust Stability problem in the presence of structured perturbations. For any stabilizing controller, the conditions for Robust Stability are equivalent to necessary and sufficient conditions for the invertibility of $(I - \Delta M)$, where M is a stable transfer function and Δ is block diagonal, i.e. $\Delta \in \mathcal{D}$. The next section will present these conditions in an exact form, which are due to Khammash and Pearson [12].

7.1 Structured Small Gain Theorem

Our Interest is to derive non-conservative conditions to guarantee the invertibility of $(I - \Delta M)$, where $\Delta \in \mathcal{D}$. From the small Gain Theorem, one condition is given by

$$\|M\|_1 \leq 1$$

This condition is clearly conservative, i.e. if $\|M\|_1 > 1$ then the destabilizing perturbation may not have the diagonal structure shown. We can reduce the conservatism of the small gain theorem by introducing diagonal scalings that commute with the class of perturbations $\mathcal{D}$. In the sequel, we will treat the case of scalar $\Delta'_i s$. The MIMO case follows in the same way.

Let $\mathcal{X}$ denote the class of all diagonal matrices with positive elements, $X \in \mathcal{X}$ if $X = diag(x_1, \ldots, x_n)$ and $x_i > 0$. It is evident that

$$X \Delta X^{-1} = \Delta \quad \forall \Delta \in \mathcal{D}, X \in \mathcal{X}$$

Hence, $(I - \Delta M)$ is given by $X^{-1}(I - \Delta X M X^{-1})X$, and is stable if

$$\|X M X^{-1}\|_1 \leq 1$$

In otherwords, if the above equation is valid for some X, then $(I - \Delta M)$ has a stable inverse. The least conservative choice will be

$$\inf_{X \in \mathcal{X}} ||X M X^{-1}||_1 \leq 1.$$

While the above condition is less conservative, it is not clear how far it is from being necessary. The surprising fact is that the above condition is both necessary and sufficient, the proof of necessity will be highlighted below.

Denote by $\bar{M}$ the matrix $(||m_{ij}||)$, i.e. the matrix constructed by taking the norms of the ij^{th} entry. It is straightforward to show that

$$||M|| = |\bar{M}|_1$$

where $|A|_1 = \max_i \sum_j |a_{ij}|$. The next theorem is a standard result from linear algebra known as the Perone-Frobenious theorem [13], and applies to all positive matrices. For simplicity, it is assumed that $\bar{M}$ has strictly positive elements.

Theorem 4 *Given any matrix M, let $\bar{M}$ be defined as above, then the following hold:*

i) $\inf_{X \in \mathcal{X}} ||X M X^{-1}||_1 = \lambda_{max}(\bar{M})$.

ii) $\exists \tilde{x} \in \mathbb{R}^n, \tilde{x}_i > 0$ *such that* $\bar{M}\tilde{x} = \lambda_{max}(\bar{M})\tilde{x}$.

iii) *Define* $\tilde{X} = diag(\tilde{x}_1, \ldots, \tilde{x}_n)$, *then*

$$\inf_{X \in \mathcal{X}} ||X M X^{-1}||_1 = ||\tilde{X} M \tilde{X}^{-1}||_1.$$

It is interesting to note that $\tilde{X}$ gives the optimal scaling of both inputs and outputs to incorporate the directional information in the least conservative way. The theorem shows that the computation of $\tilde{X}$ is straightforward for every fixed matrix M.

To summarize, a sufficient condition for robust stability in the presence of structured perturbations is given by:

$$\inf_{X \in \mathcal{X}} ||X M X^{-1}||_1 = \lambda_{max}(\bar{M}) \leq 1.$$

In the sequel, it is shown that the above condition is also necessary. To do that, we will present two key lemmas from which the proof will follow immediately.

Lemma 1 *Given any* $G \in \ell_1^{n \times n}$ *such that* $||G_i||_1 = g_i > 1$, *where* G_i *is the* i^{th} *row of* G. *Then,* $\exists f \in \ell_{\infty,e}^n \backslash \ell_\infty^n$ *(unbounded),* $n^* > 0$ *and* $m > 1$ *such that*

$$\frac{||P_{k-1}G_i f||_\infty}{||P_k f||_\infty} \geq m \quad \forall k \geq n^* \; \forall i$$

In words, this Lemma says that if the norm of each row of a matrix function G is greater than 1, then there exits an unbounded signal f which gets amplified (in the ℓ_∞ sense) at each output channel after some fixed time n^*. This signal is then used to generate a time varying Δ that results in an unstable $(I - \Delta M)^{-1}$.

Lemma 2 *Given* $G_i \in \ell_1^{1 \times n}$ *such that* $||G_i||_1 = g_i > 1$. *There exists a* $\Delta_i \in \mathcal{L}_{TV}^{1 \times 1}, ||\Delta_i|| < 1$ *s.t.*

$$f_i - \Delta_i G_i f \in \ell_\infty$$

Equivalently

$$(I - \Delta G)f \in \ell_\infty \text{ with } \Delta = \text{ diag } (\Delta_1, \ldots, \Delta_n).$$

The idea of the construction is quite straightforward: The signal Gf has components that are amplified in comparison to each component of f, after some time n^*. Thus it is possible to map back $(Gf)_i$ to f_i through a linear operator with norm less than 1, for all time greater than n^*. With this, we are in a position to prove that condition (1) is both necessary and sufficient.

Theorem 5 *Let* $M \in \ell_1^{n \times n}$, $\Delta \in \mathcal{D}$. *A necessary and sufficient condition for the inverse of* $(I - \Delta M)$ *to be* ℓ_∞ *-stable is given by*

$$\inf_{X \in \mathcal{X}} ||XMX^{-1}||_1 = \lambda_{max}(\bar{M}) \leq 1$$

where $\bar{M}$ *is the matrix of norms of* M.

Proof. : Suppose $\lambda_{max}(\bar{M}) > 1$, then there exists a positive vector $x \in \mathbb{R}^n$ s.t.

$$\bar{M}x = \lambda_{\max}x$$

Define $X = \text{diag}\,(x_1 \ldots, x_n)$, and M_i be the ith row of M. We have

$$||M_iX|| = |\bar{M}_ix| = \lambda_{\max}x_i$$

or equivalently

$$||\frac{1}{x_i}M_iX|| = \lambda_{\max} > 1.$$

Let $G = X^{-1}MX$, then $G_i = \frac{1}{x_i}M_iX$ and $||G_i|| > 1$. By Lemma 2, $\exists$ a $\Delta \in \mathcal{D}$ s.t $(I - \Delta G)^{-1}$ is not ℓ_∞-stable. However, $(1 - \Delta G)^{-1} = X(I - \Delta M)^{-1}X^{-1}$. Hence $(I - \Delta M)^{-1}$ is not ℓ_∞-stable.   ■.

Another interpretation of X is simply the direction for the worst input, i.e. if we redefine the input f in Lemma 1 as $\tilde{f} = Xf$, then M exerts the maximum amplification in that direction, the amplification is given by $\lambda_{\max}$.

Finally, it is worth noting that the problem of Stability Robustness and Performance Robustness of structured perturbations were originally formulated and developed by Doyle [14, 15] in what is now known as the μ theory. Similarities and contrasts between the μ theory and the method presented here are discussed in [16].

8   Synthesis of ℓ_1-Optimal Controllers

The basic synthesis problem can be stated as follows: Find a controller K such that

$$\inf_{X \in \mathcal{X}} ||XMX^{-1}||_1 = \lambda_{max}(\bar{M}) \leq 1$$

Incorporating the Q parameterization, the problem is equivalently stated as:

$$\inf_{Q \in \ell_1} \inf_{X \in \mathcal{X}} ||X(T_1 - T_2QT_3)X^{-1}||_1.$$

Exact synthesis by simultaneously minimizing over Q and X is still an open problem. The best known method is an iterative method in which optimization in each of these variables is done independently. The infimization in $\mathcal{X}$ can be done exactly using Perone-Frobenious theorem for every fixed Q. Infimization over Q is the main topic of the ℓ_1 synthesis problem and will be described in the sequel. It is interesting to note that the problem is not jointly convex in both of these parameters and hence this scheme is guaranteed to converge only to local minima.

The problem of ℓ_1 minimization was formulated by Vidyasagar in [17] as a minimax criterion that parallels the standard H_∞ problem [18] and was solved by Dahleh and Pearson in [19, 20]. Much work has been done on this problem afterwards, some of which can be found in [9, 16, 21, 22, 23, 24, 25, 26]. The solution of Dahleh and Pearson hinged on the duality theory of minimum distance problems. In the sequel we will show that solutions can be obtained using only standard linear programming ideas, and the duality theory of Linear Programs.

8.1 Characterization of the subspace $\mathcal{S}$

Recall the ℓ_1 minimization problem:

$$\inf_{R \in \mathcal{S}} \|T_1 - R\|_1 \qquad\qquad (OPT)$$

where the subspace $\mathcal{S}$ is given by:

$$\mathcal{S} = \{R \in \ell_1 | R = T_2 Q T_3, \ Q \in \ell_1\}$$

The subspace $\mathcal{S}$ is in general limited by many factors such as zeros of $\hat{T}_2$ and $\hat{T}_3$ inside the unit disc, the rank of both of these systems and so on. There are two cases that arise, the first we term the *good rank* case in which the only constraints on $\mathcal{S}$ are the zeros of $\hat{T}_2, \hat{T}_3$ in the unit disc, and the second we term the *bad rank* case in which rank constraints also exists. One should bear in mind that the basic idea behind this characterization is the solvability of the equation

$$\hat{R} = \hat{T}_2 \hat{Q} \hat{T}_3$$

for a stable Q.

8.2   Good Rank Case

In here, it is assumed that $\hat{T}_2$ has full row rank $(= m)$ and $\hat{T}_3$ has full column rank $(= n)$. The reason for the term good rank is that the characterization of the subspace S is obtained by finitely many equations and thus the optimization problem over S will have an exact solution. For simplicity, it is assumed that $\hat{T}_2, \hat{T}_3$ have full rank for all $|\lambda| = 1$.

The good rank problem usually arises in situation where one is interested in only one error function. An example of this is the sensitivity minimization problem discussed earlier. It is worthwhile noting that most of the interesting control problems violate these conditions, however, solutions for this class of problems offer a great insight into the solution of the general problem. In the next few sections, the presentation is essentially identical to [24], and is presented here for completeness.

Consider the Smith McMillan form decomposition of $\hat{T}_2, \hat{T}_3$:

$$\hat{T}_2 = \hat{L}_2 \hat{M}_2 \hat{R}_2$$
$$\hat{T}_3 = \hat{L}_3 \hat{M}_3 \hat{R}_3 \tag{3.1}$$

where L_2, R_2, L_3, and R_3 are (polynomial) unimodular matrices and M_2, M_3 are rational matrices which have the familiar diagonal forms:

$$\hat{M}_2 = \begin{pmatrix} \frac{\epsilon_1(\lambda)}{\psi_1(\lambda)} & & & 0 & \cdots & 0 \\ & \ddots & & \vdots & \ddots & \vdots \\ & & \frac{\epsilon_m(\lambda)}{\psi_m(\lambda)} & 0 & \cdots & 0 \end{pmatrix} ; \quad \hat{M}_3 = \begin{pmatrix} \frac{\epsilon_1'(\lambda)}{\psi_1'(\lambda)} & & \\ & \ddots & \\ & & \frac{\epsilon_n'(\lambda)}{\psi_n'(\lambda)} \\ 0 & \cdots & 0 \\ \vdots & \ddots & \vdots \\ 0 & \cdots & 0 \end{pmatrix}$$

Let $\mathcal{Z}_{23}$ denote the set of all $\lambda \in \bar{D}$ which are zeros of either $\hat{T}_2$ or $\hat{T}_3$. Then for each $\lambda_0 \in \mathcal{Z}_{23}$ we can define a non-decreasing sequence of non-negative integers $\Sigma_2(\lambda_0)$ corresponding to the multiplicities with which the term $(\lambda - \lambda_0)$ appears on the diagonal of $\hat{M}_2$. That is:

$$\Sigma_2(\lambda_0) := (\sigma_2^i(\lambda_0))_{i=1}^m$$

means:

$$\frac{\epsilon_i(\lambda)}{\psi_i(\lambda)} = (\lambda - \lambda_0)^{\sigma_2^i(\lambda_0)} \hat{g}_i(\lambda) \qquad i = 1, \ldots, m$$

where $\hat{g}_i(\lambda)$ has no poles or zeros at $\lambda = \lambda_0$. We can define similarly a set of sequences $\Sigma_3(\lambda_0)$ for each $\lambda_0 \in \mathcal{Z}_{23}$ which correspond to the multiplicities of the λ_0's on the diagonal of M_3. A sequence $\Sigma_3(\lambda_0)$ is sometimes referred to as the sequence of *structural indices* of λ_0 in $\hat{T}_2$.

We can also define m polynomial row vectors of dimension m and n polynomial column vectors of dimension n as follows:

$$\hat{\alpha}_i(\lambda) = (\hat{L}_2^{-1})_i(\lambda) \qquad\qquad i = 1, \ldots, m$$

$$\hat{\beta}_j(\lambda) = (\hat{R}_3^{-1})^j(\lambda) \qquad\qquad j = 1, \ldots, n$$

where subscript i indicates the i-th row and superscript j indicates the j-th column. Given the above definitions, a precise notion of interpolation which will be used in the sequel is presented.

Definition 1 *Given $\hat{T}_2$ and $\hat{T}_3$ as above and $\hat{R} \in \mathbf{A}^{m \times n}$, we say $\hat{R}$ interpolates $\hat{T}_2$ (from the left) and $\hat{T}_3$ (from the right) if the following condition is satisfied: Given any zero $\lambda_0 \in \mathcal{Z}_{23}$ of $\hat{T}_2$ and/or $\hat{T}_3$ with structural indices $\Sigma_2(\lambda_0)$ and $\Sigma_3(\lambda_0)$ in $\hat{T}_2$ and $\hat{T}_3$, respectively, we have for all $i \in \{1, \ldots, m\}$ and $j \in \{1, \ldots, n\}$:*

i). $(\hat{\alpha}_i \hat{R})^{(k)}(\lambda_0) = 0, \qquad k = 0, \ldots, \sigma_2^i - 1$

ii). $(\hat{R}\hat{\beta}_j)^{(k)}(\lambda_0) = 0, \qquad k = 0, \ldots, \sigma_3^j - 1$

iii). $\sum_{l=0}^{k-\sigma_3^j} \sum_{r=0}^{\sigma_3^j-1} \binom{k}{l}\binom{k-l}{r}[\hat{\alpha}_i^{(l)}\hat{R}^{(k-l-r)}\hat{\beta}_j^{(r)}](\lambda_0) = 0, \quad k = \sigma_3^j, \ldots, \sigma_2^i + \sigma_3^j - 1$

Or

$$\sum_{l=0}^{k-\sigma_2^i} \sum_{r=0}^{\sigma_3^i-1} \binom{k}{l}\binom{k-l}{r}[\hat{\alpha}_i^{(l)}\hat{R}^{(k-l-r)}\hat{\beta}_j^{(r)}](\lambda_0) = 0, \quad k = \sigma_2^i, \ldots, \sigma_3^j + \sigma_2^i - 1$$

where the argument of $\sigma_2^i(\cdot)$ and $\sigma_3^j(\cdot)$ is understood to be λ_0 and superscript (k) indicates the k-th derivative with respect to z.

Note that this condition simplifies greatly in the case of a zero λ_0 which is not common to $\hat{T}_2$ and $\hat{T}_3$; if it is a zero only of $\hat{T}_2$, for example, we have $\Sigma_3(\lambda_0) = (0)_{j=1}^n$ and parts (ii) and (iii) are trivially satisfied for all i and j. The next theorem gives a characterization of the subspace $\mathcal{S}$ in the good rank case. These conditions can be interpreted as a set of bounded linear functionals annihilating elements in $\mathcal{S}$.

Theorem 6 *Let $\hat{T}_2, \hat{T}_3$ satisfy the good rank assumptions, and $\hat{R} \in \mathbf{A}^{m \times n}$. Then there exists $\hat{Q} \in \mathbf{A}$ satisfying $\hat{R} = \hat{T}_2 \hat{Q} \hat{T}_3$ if and only if $\hat{R}$ interpolates $\hat{T}_2$ and $\hat{T}_3$.*

Proof. It is easily shown that for $\hat{R} \in \mathbf{A}^{m \times n}$ there exists $\hat{Q} \in \mathbf{A}^{m \times n}$ if and only if for all $\lambda_0 \in \mathcal{Z}_{23}$, $i \in \{1, \ldots, m\}$ and $j \in \{1, \ldots, n\}$ we have:

$$(\hat{\alpha}_i \hat{R} \hat{\beta}_j)^{(k)}(\lambda_0) = 0, \qquad k = 0, \ldots, \sigma_2^i(\lambda_0) + \sigma_3^j(\lambda_0) - 1$$

The proof of the theorm follows from Weiner's theorm. The detailed proof can be found in [24].

The conditions captured in the above theorem can be viewed as forcing a collection of linear functionals to annihilate the subspace S. The construction of such functionals is straightforward and follows similar to the SISO example shown in the sequel.

8.3   The Bad rank Case

In this case, $\hat{T}_2$ has full column rank $= n_2$ and $\hat{T}_3$ has full row rank $= n_3$. This situation occurs frequently in controller design since one is in general interested in many objectives for minimization. An example of the bad rank problem is the tracking example with saturations presented earlier.

In the sequel, it is assumed that there exists n_2 rows of $\hat{T}_2$ and n_3 columns of $\hat{T}_3$ which are linearly independent for all λ on the unit circle. This assumption simplifies the exposition although it is not necessary. In general, it is enough to assume the above for one point on the unit circle [27]. Under this assumption, $\hat{T}_2$ and $\hat{T}_3$ can be written in the following form without loss of generality (possibly requiring the interchange of inputs and/or outputs):

$$\hat{T}_1 = \begin{pmatrix} \hat{T}_{21} \\ \hat{T}_{22} \end{pmatrix}$$
$$\hat{T}_2 = (\hat{T}_{31} \quad \hat{T}_{32})$$

where $\hat{T}_{21}$ has dimensions $n_2 \times n_2$ and is invertible and $\hat{T}_{31}$ has dimensions $n_3 \times n_3$ and is invertible. Moreover, $\hat{T}_{21}$ and $\hat{T}_{31}$ have no zeros on the unit circle. Thus $\hat{R} = \hat{T}_2 \hat{Q} \hat{T}_3$ can be written:

$$\hat{R} = \begin{pmatrix} \hat{T}_{21} \\ \hat{T}_{22} \end{pmatrix} \hat{Q} (\hat{T}_{31} \quad \hat{T}_{32}) = \begin{pmatrix} \hat{R}_{11} & \hat{R}_{12} \\ \hat{R}_{21} & \hat{R}_{22} \end{pmatrix}$$

Notice that $\hat{T}_{21}$ and $\hat{T}_{31}$ define a good rank sub-problem, forcing $\hat{R}_{11}$ to interpolate their zeros. Nevertheless, this is not enough to characterize all the admissible $\hat{R}$'s. The choice of $\hat{R}_{11}$ determines uniquely $\hat{Q}$. The rest of the elements in the $\hat{R}$ matrix have to be consistent with this solution, which in turn dictates a set of relations between the $\hat{R}_{ij}$. Define the following polynomial coprime factorizations:

$$\hat{T}_{22}\hat{T}_{21}^{-1} = \hat{D}_2^{-1}\hat{N}_2$$
$$\hat{T}_{31}^{-1}\hat{T}_{32} = \hat{N}_3\hat{D}_3^{-1}$$

Using these definitions, we state the following result characterizing the feasible set S for this case.

Theorem 7 *Given $\hat{T}_2, \hat{T}_3$ with the assumption on the bad rank case, and $\hat{R} \in \mathbf{A}$, there exists $\hat{Q} \in \mathbf{A}$ satisfying $\hat{R} = \hat{T}_2\hat{Q}\hat{T}_3$ if and only if:*

$$i). \quad \left(-\hat{N}_2 \quad \hat{D}_2 \right) \begin{pmatrix} \hat{R}_{11} & \hat{R}_{12} \\ \hat{R}_{21} & \hat{R}_{22} \end{pmatrix} = 0$$

$$ii). \quad \left(\hat{R}_{11} \quad \hat{R}_{12} \right) \begin{pmatrix} -\hat{N}_3 \\ \hat{D}_3 \end{pmatrix} = 0$$

$$iii). \quad \hat{R}_{11} \text{ interpolates } \hat{T}_{21} \text{ and } \hat{T}_{31}.$$

The conditions shown in parts i, ii are convolution constraints on the ℓ_1 sequence. The interpolation condition in the last part can be tightened, since only the common zeros of $\hat{T}_{21}$ and $\hat{T}_{22}$ need to be interpolated.

The discussion above shows that the characterization of this subspace can be summerized by defining two operators:

$$\mathcal{V} : \ell_1^{m \times n} \to \mathbb{R}^s$$

and

$$\mathcal{C} : \ell_1^{m \times n} \to \ell_1^r$$

where s and r are some integers. The first operator captures the interpolation constraints, and thus has a finite dimensional range, and the second captures the convolution constraints. These two operators can be constructed in a straightforward fashion, book-keeping being the only difficulty. To overcome this problem, it is helpful to think of R as a vector rather than a matrix. To illustrate this, let the operator $\mathcal{W}$ be a map from $\ell_1^{m \times n}$ to ℓ_1^{mn} defined as follows:

$$(\mathcal{W}R)(k) = \begin{pmatrix} r_{11}(k) \\ \vdots \\ r_{m1}(k) \\ r_{21}(k) \\ \vdots \\ r_{mn}(k) \end{pmatrix}$$

The operator $\mathcal{W}$ is a one-to-one and onto operator, whose inverse is equal to its adjoint (a fact used later). It simply re-arranges the variables in R. The conditions on R presented in the above theorem can be written explicitly in terms of each component of R.

To construct the first operator $\mathcal{V}$, recall that each interpolation condition is interpreted as a bounded linear functional on R. By stacking up these functionals, the operator $\mathcal{V}$ is constructed. The following is an illustrative example.

Example 3 *Suppose $\hat{T}_{21}$ and $\hat{T}_{31}$ are SISO and both have N distinct zeros a_i in the open unit disc. Then the matrix $\mathcal{V}$ is given by $\mathcal{V} = V_\infty \mathcal{W}$ where*

$$V_\infty = \begin{pmatrix} Re(a_i^0) & 0 & 0 & 0 & Re(a_i^1) & 0 & \ldots & Re(a_i^j) & 0 & \ldots \\ Im(a_i^0) & 0 & 0 & 0 & Im(a_i^1) & 0 & \ldots & Im(a_i^j) & 0 & \ldots \end{pmatrix}$$

for $i = 1, \ldots, N$ and $j = 0, 1, 2, \ldots$.

For the second operator $\mathcal{C}$, recall that convolution can be interpreted as a multiplication by a block Toeplitz matrix, in this case with finite memory. By simple rearrangement, the operator is constructed with its image inside ℓ_1^r. Hence $\mathcal{C}$ is given by $\mathcal{C} = T\mathcal{W}$ where T is a block lower triangular matrix. For a detailed example, see [20, 24]. To illustrate the construction of the operator T, consider the following example.

Example 4 *consider the coprime-factor perturbation problem presented earlier for a SISO. The condition for stability robustness is given by [9]:*

$$\left\| \begin{bmatrix} \tilde{V} - Q\tilde{N} & -\tilde{U} + Q\tilde{M} \end{bmatrix} \right\|_1 \leq 1$$

In this case, $T_2 = 1$ *and* $T_3 = (\tilde{N} \quad -\tilde{M})$. *Since* $\tilde{M}^{-1}\tilde{N} = NM^{-1}$ *with* N, M *coprime, the conditions in the above theorem translate to*

$$(R_{11} \quad R_{12}) \begin{pmatrix} M \\ N \end{pmatrix} = 0$$

The matrix T *is then given by (since* $W = I$*):*

$$T = \begin{pmatrix} (m(0) & n(0)) & 0 & 0 & 0 & 0 & \cdots \\ (m(1) & n(1)) & (m(0) & n(0)) & 0 & 0 & 0 & \cdots \\ (m(2) & n(2)) & (m(1) & n(1)) & (m(0) & n(0)) & 0 & 0 & \cdots \\ (m(3) & n(3)) & (m(2) & n(2)) & (m(1) & n(1)) & (m(0) & n(0)) & 0 & \cdots \\ & \vdots & & \vdots & & \vdots & & \vdots & & \vdots & \vdots \end{pmatrix}$$

It is interesting to note that the operator C *captures all the conditions and no interpolation conditions are needed. The conditions presented in the theorem can be redundant, and can be significantly reduced [27]. Generally, the operator* $W \neq I$ *and so the operator* T *will not be exactly a Toeplitz matrix, although it will have a similar structure.*

The subspace S *is then the set of all elements* $R \in \ell_1^{m \times n}$ *such that* $\mathcal{V}R = 0$ *and* $CR = 0$. *Let* $b_1 = \mathcal{V}T_1$ *and* $b_2 = CT_1$. *The* ℓ_1 *optimization problem can be restated as:*

$$\inf \|\Phi\|_1 \qquad \text{subject to} \quad \mathcal{V}\Phi = b_1, \quad C\Phi = b_2 \qquad (OPT).$$

9   Duality in Linear Programming

It is well known that optimization problems minimizing the ℓ_1-norm with linear constraints inside $\mathbb{R}^n$ are equivalent to linear programming (LP) problems. For this reason, it is natural to expect that the ℓ_1 optimal control problem (OPT) is also equivalent to a linear programming problem. The difference is that such problems may have infinitely many variables and constraints, and thus only solvable by approximation. The duality theory in linear programming is primarily motivated by a desire to reduce the number of variables in a problem. It will be quite interesting if the infinite dimensional linear programs can be converted to finite-dimensional ones through some duality argument. In this section, it is shown that the good rank problems can be exactly transformed to finite-dimensional linear programs, however the bad rank problems may not. For the later, only approximate solutions can be obtained.

The following result is standard in linear programming and is known as the duality theorem: Let $x \in \mathbb{R}^n$, A is an $m \times n$ matrix, b is a $1 \times m$ vector and c is a $1 \times n$ vector. The following two problems are equivalent:

$$
\begin{array}{ll}
\min c^T x \quad = & \max b^T y \\
\text{subject to} & \text{subject to} \\
\quad Ax \leq b & \quad A^T y \geq c \\
\quad x \geq 0 & \quad y \geq 0
\end{array}
$$

In words, a minimization problem with n variables and m constraints is equivalent to a maximization problem with m variables and n constraints. The constraint matrix of the second problem is the adjoint matrix (transpose) of the first. In a more mathematical terminology, y is in the range of the adjoint matrix, hence is an element of the dual space of $\mathbb{R}^m$.

9.1   The ℓ_1 problem as a Linear Program

In this section, only the SISO case is treated. The objective is simply to highlight the basic ideas involved in solving such problems not to present the most general solutions. Notice that in this case, the only constraints are due to $\mathcal{V}$ presented in the example above. An equivalent statement of OPT is given by:

$$\min \mu$$

subject to

$$\|\Phi\|_1 \leq \mu$$

$$\mathcal{V}\Phi = b_1$$

$$\Phi \in \ell_1$$

To show the resemblance between OPT and standard linear programs, we can perform the following changes: Φ can be split into negative and positive components, $\Phi = \Phi^1 - \Phi^2$ where $\phi^1(k), \phi^2(k) \geq 0$. This is a standard trick in linear programming that allows one to convert the ℓ_1 norm to a linear objective function. By performing the minimization over Φ^1, Φ^2 the solution lies on the corners of the constraints, and hence at least one of the $\phi^1(k)$, $\phi^2(k)$ will be zero for every k. This shows that OPT is equivalent to:

$$\min \mu$$

subject to

$$\sum_0^\infty \phi^1(k) + \phi^2(k) - \mu \leq 0$$

$$\mathcal{V}(\Phi^1 - \Phi^2) = b_1$$

$$\phi^1(k), \phi^2(k) \geq 0$$

Let $e^T = (1, 1, 1, \ldots)$. To put this problem in a matrix form, define the space $X = \ell_1 \times \ell_1 \times \mathbb{R}$ which can be viewed simply as an infinite sequence of variables. Define the matrix $\mathcal{A}$ decomposed conformally with X, whose range lie inside $\mathbb{R}^{2N+1}$ as follows:

$$\mathcal{A} = \begin{pmatrix} e^T & e^T & -1 \\ \mathcal{V} & -\mathcal{V} & 0 \\ -\mathcal{V} & \mathcal{V} & 0 \end{pmatrix}$$

Also, define the vector $x \in X$, $b \in \mathbb{R}^{2N+1}$ and the infinite vector c (also decomposed conformally with X) as follows:

$$x = \begin{pmatrix} \Phi^1 \\ \Phi^2 \\ \mu \end{pmatrix} \qquad b = \begin{pmatrix} 0 \\ b_1 \\ -b_1 \end{pmatrix} \qquad c = \begin{pmatrix} 0 \\ 0 \\ 1 \end{pmatrix}$$

The matrix $\mathcal{A}$ has $2N + 1$ rows and an infinite number of columns. OPT is equivalent to the following linear program:

$$\min c^T x$$

subject to

$$\mathcal{A}x \leq b$$

$$x \geq 0$$

Using the LP duality theorem, OPT is equivalent to another maximization problem in terms of a vector of dimension $2N + 1$. Let $\beta \in \mathbb{R}^{2N+1}$ be decomposed as:

$$\beta = \begin{pmatrix} \beta_1 \\ \beta_2 \\ \beta_3 \end{pmatrix} \qquad \beta_1 \in \mathbb{R}, \quad \beta_2, \beta_3 \in \mathbb{R}^N$$

The maximization problem is given by:

$$\max(\beta_2 - \beta_3)^T b_1$$

subject to

$$\beta_1 e + (\beta_2 - \beta_3)^T \mathcal{V} \geq 0$$

$$\beta_1 e - (\beta_2 - \beta_3)^T \mathcal{V} \geq 0$$

$$-\beta_1 \geq -1$$

$$\beta_1 \in \mathbb{R}, \quad \beta_2, \beta_3 \in \mathbb{R}^N$$

Finally, substituting $\alpha = \beta_2 - \beta_3$, then OPT is equivalent to the problem:

$$\max \alpha^T b_1$$

subject to

$$|\alpha^T \mathcal{V}| \leq e$$

where the inequality is taken pointwise. This problem is a finite dimensional linear program with infinitely many constraints. However, the matrix $\mathcal{V}$ has coefficients that decay exponentially, and one can show that only a finite number of constraints is needed to obtain an exact solution [19]. An explicit bound on the length of the finite dimensional problem can be derived.

Once a finite-dimensional dual problem has been determined and solved, the solution of the original problem (i.e. solution for Φ) can be obtained directly from the linear program. The process of obtaining the primal solution is known as the alignment problem.

Note that in the above, an inequality formulation of the LP problem was chosen instead of the equality one. Although in the SISO this is not needed, for MIMO problems, the equivalent linear program will always have mixed constraints: equality and inequality constraints, and thus the above formulation is more direct. Similar results follow from the duality theory for Lagrange Multipliers and are reported in [16].

9.2   The General Case

In general, OPT may not be equivalent to a finite dimensional Linear Programming problem. In that case only approximate solutions can be obtained.

We will not treat this case in here, however a good treatment is found in [16, 20, 24, 27]. A straightforward way for solving the problem is to consider only finite length solutions for Φ, and solve a finite dimensional LP as described earlier. As the length of the solution increases, suboptimal solutions to OPT can be obtained. Duality theory can then be invoked to provide estimates for the distance between the actual minimum and the norm of the suboptimal solution.

10   Conclusions

This chapter presented an overview of the ℓ_1 design methodology as a tool to synthesize controllers to achieve good performance in the presence of uncertainty. It was shown through prototype problems that this formulation is well suited for problems where Peak-to-Peak specifications and constraints are required. A general framework that allows incorporating stability robustness and performance robustness was presented from which computable, non-conservative conditions were derived. These conditions were shown to be equivalent to computing the spectral radius of some matrix, which was then simplified tremendously with the utilization of Perone-Frobenious theorem. The synthesis problem involved solving an ℓ_1 optimization problem, which was shown to be intimately related to infinite Linear Programming problems.

The results presented in this chapter were only discrete-time results. The interest in discrete-time systems stems from the fact that in many practical situations, one is interested in designing digital controllers for a continuous-time plant. Such problems are known as sampled-data systems and has recently received a lot of attention in the literature. For hybrid systems with Peak-to-Peak specifications, it is shown in [28, 29, 30] that these specifications can be met by solving a higher dimensional discrete-time ℓ_1 problem. This in turn justifies the body of work on the pure discrete-time case.

There are many related results in the area of ℓ_1 optimal control design. The sampled-data problem mentioned earlier is one area. A related area is the design of controllers for multi-rate sampled systems, or periodic systems [31]. Recently, a lot of attention has been given to the study of the structure of the ℓ_1 controllers for the bad rank case searching for some separation structure similar to that of the H_2 and H_∞ problems [32]. Exact solutions have been constructed for simple *bad rank* problems in [26, 27]. Also, it was shown that optimal solutions can require dynamic controllers eventhough all the states are available [22]. Other properties of optimal ℓ_1 solutions are still under investigation [23]. Good demonstrations of the ℓ_1 theory on practical problems can be found in [33].

In many practical problems, the external inputs include some fixed signals such as reference inputs. In such problems, the design methodology does not fall under the *worst case* paradigm, however still tractable using the above methods. In particular, controllers that will achieve performance objectives in terms of overshoot and settling time can be synthesized by solving linear programming problems and are discussed in details in [34, 35, 21].

There are quite a few directions of research that are needed for the development of this methodology. To mention some of these, the problem of studying the structure of the ℓ_1 controllers is quite important. Apart from providing a better insight into the design, the knowledge of the structure of the optimal controller can simplify the computations involved in a non-trivial fashion. Another research problem is concerned with model reduction: How can the order of the controller be reduced so that both stability and a level of performance is maintained. Also, the exact synthesis problem which involves a minimization of a spectral radius objective function is still an open problem.

References

[1] C.A. Desoer and M. Vidyasagar, "Feedback Systems: Input-Output Properties," Academic Press, Inc, N.Y., 1975.

[2] J. C. Willems, "The Analysis of Feedback Systems." MIT Press, Cambridge, 1971.

[3] S.P. Boyd and J.C. Doyle, "Comparison of Peak and RMS Gains for Discrete-Time Systems," *Syst. and Contr. Lett.* **9**, 1-6, (1987).

[4] B.A. Francis, "A Course in H_∞ Control Theory, Springer-Verlag, 1987.

[5] M. Vidyasagar, "Control Systems Synthesis: A Factorization Approach," MIT press, 1985.

[6] D.C. Youla, H.A. Jabr, and J.J. Bongiorno, "Modern Wiener-Hopf Design of Optimal Controllers–part 2: The Multivariable Case," *IEEE-Trans. A-C* **21**, 319-338, (1976).

[7] M.A. Dahleh and Y. Ohta, "A Necessary and Sufficient Condition for Robust BIBO Stability," *Syst. and Contr. Lett.* **11**, 271-275, (1988).

[8] M. Vidyasagar and H. Kimura, "Robust Controllers for Uncertain Linear Multivariable Systems," *Automatica* **22**, 85-94, (1986).

[9] M.A. Dahleh, "BIBO Stability Robustness for Coprime Factor Perturbations," *to appear.*

[10] M. Dahleh and M.A. Dahleh, "Optimal Rejection of Persistent and Bounded Disturbances: Continuity Properties and Adaptation," *IEEE Trans. A-C* **35**, 687-696, (1990).

[11] P. Voulgaris M.A. Dahleh and Lena Valavani," Slowly-Varying Systems: ℓ^∞ to ℓ^∞ Performance and Implications to Adaptive Control," *submitted.*

[12] M. Khammash and J. B. Pearson, "Performance Robustness of Discrete-Time Systems with Structured Uncertainty," *IEEE Trans. A-C* **36**, 398-412, (1991).

[13] R. Horn and C. Johnson, "Matrix Analysis," Cambridge University Press, 1985.

[14] J.C. Doyle and G. Stein, "Multivariable Feedback Design: Concepts for a Classical/Modern Synthesis," *IEEE-Trans. A-C* **26**, 4-16, (1981).

[15] J. C. Doyle, "Analysis of Feedback Systems with Structured Uncertainty," *IEEE Proceedings* **129**, 242-250, (1982).

[16] M.A. Dahleh and M.H. Khammash, "Stability and Performance Robustness in the Presence of Bounded But Unknown Inputs," to appear in *Automatica* special issue on robust control.

[17] M. Vidyasagar "Optimal Rejection of Persistent Bounded Disturbances," *IEEE-Trans. A-C* **31**, 527-534, (1986).

[18] G. Zames, "Feedback and Optimal Sensitivity: Model Reference Transformations, Multiplicative Seminorms, and Approximate Inverses," *IEEE-Trans. A-C* **26**, 301-320, (1981).

[19] M.A. Dahleh and J.B. Pearson, "l^1 Optimal Feedback Controllers for MIMO Discrete-Time Systems," *IEEE Trans. A-C* **32**, 314-322, (1987).

[20] M.A. Dahleh and J.B. Pearson, "Optimal Rejection of Persistent Disturbances, Robust Stability and Mixed Sensitivity Minimization," *IEEE Trans. A-C* **33**, 722-731, (1988).

[21] G. Deodhare and M. Vidyasasgar. "Some Results on ℓ_1-Optimality of Feedback Control Systems: The SISO Discrete-Time Case", *IEEE Trans. A-C*, **35**, 1082-1085, (1990).

[22] I. Diaz-Bobillo and M.A. Dahleh, "State Feedback ℓ_1 Optimal Controllers Can be Dynamic," LIDS, MIT, Report No. LIDS-P-2051, (1991).

[23] I. Diaz-Bobillo and M.A. Dahleh, "On the Solution of the ℓ_1 Optimal Multi-Block Problem,", MIT Report in preparation.

[24] J.S. McDonald and J.B. Pearson, "ℓ_1-Optimal Control of Multivariable Systems with Output Norm Constraints," *Automatica* **27**, 317-329, (1991).

[25] M.A. Mendlovitz, "A Simple Solution to the ℓ_1 Optimization Problem," *Syst. and Contr. Lett.* **12**, 461-463, (1989).

[26] O.J. Staffans, "Mixed Sensitivity Minimization Problems with Rational ℓ_1-optimal solutions", *Journal of Optimization Theory and Applications* **70**, 173-189, (1991).

[27] O.J. Staffans, "On the Four-block Model Matching Problem in ℓ_1," Helsinki University of Technology, Espoo, Report A289, (1990).

[28] B.A. Bamieh, M.A. Dahleh and J.B. Pearson, "Minimization of the L^∞-Induced Norm for Sampled-Data Systems," *submitted*.

[29] G. Dullerud and B.A. Francis, "$\mathcal{L}^1$ Performance in Sampled-Data Systems," *submitted*.

[30] M.H. Khammash. "Necessary and Sufficient Conditions for the Robustness of Time-Varying Systems with Applications to Sampled-Data Systems," *submitted*.

[31] M.A. Dahleh, P. Voulgaris and L. Valavani, "Optimal and Robust Controllers for Periodic and Multi-Rate Systems," *to appear*.

[32] J.C. Doyle, K. Glover, P.P. Khargonekar and B.A. Francis, "State Space Solutions to Standard H_2 and H_∞ Control Problems," *IEEE Trans. A-C* **34**, 831-847, (1989).

[33] M.A. Dahleh and D. Richards, "Application of Modern Control Theory on a Model of the X-29 Aircraft," LIDS, MIT, report No. LIDS-P-1932 (1989).

[34] S.P. Boyd and C.H. Barratt, "Linear Controller Design: Limits of Performance," Prentice Hall, N.J., 1991.

[35] M.A. Dahleh and J.B. Pearson. "Minimization of a Regulated Response to a Fixed Input," *IEEE Trans. A-C* **33**, 924-930,(1988).

TECHNIQUES IN ROBUST STATE ESTIMATION
THEORY WITH APPLICATIONS

Mehrdad Saif
saif@cs.sfu.ca
School of Engineering Science
Simon Fraser University
Burnaby, British Columbia V5A 1S6
CANADA

I. INTRODUCTION

The problem of estimating the state of a linear stationary dynamical system subject to unknown time varying plant (and possibly output) disturbances is considered in this article. Along with approaches for robust estimation, several applications of the proposed estimators are also illustrated. Design of controllers to achieve output regulation for the above problem have been considered by a number of authors (e.g. [1-4]), and in particular by Davison [5-8] who refers to the problem as robust (industrial) servomechanism problem. Since in many practical industrial systems the entire state is not accessible for feedback purposes, the state estimation in the robust servomechanism problem is an important topic that needs to be addressed.

In Section II, two new approaches for designing reduced order robust state estimators are presented. These estimators are referred to as the unknown input observer (UIO), and proportional integral observer (PIO). Necessary and sufficient conditions for existence of these observers are stated along with appropriate design procedures. In section III, we will illustrate how various uncertainties within the plant's model can mathematically be modeled as unknown disturbances acting on the system. This will be of interest and use when designing a robust controller/estimator. In Section IV a number of application domains are considered and the solution approach is illustrated through examples. In particular, the robust estimator is extended to deal with problems in 1) Totally decentralized state estimation in large scale systems, 2) Optimal modal servomechanism with some inaccessible state variables, and 3) Fault tolerant control system design. Finally, the conclusions and closing remarks are presented in Section V.

II. THE ROBUST ESTIMATION PROBLEM

In this section two reduced order state estimators are presented. The first estimator referred to as the unknown input observer (UIO) is capable of estimating the state of the system in spite of the time varying disturbances acting in the plant. The second estimator referred to as the proportional integral observer (PIO) is capable of estimating both the state as well as constant disturbance which may be acting on both the plant and the system's output.

A. Unknown Input Observer (UIO)

The problem of estimating the state of a linear time invariant dynamical system subject to both known and unknown inputs (e.g. disturbances) has been considered by a number of authors [11-21]. Notions of controlled and conditioned invariance along with their properties were applied to the problem of reconstruction and observation of linear systems with unknown inputs in the original work of [9]. Further geometric results on the observability of the systems with unknown inputs were recently reported in [10]. As for constructing an estimator capable of estimating the state of the system under consideration, there has basically been two approaches to the problem. One approach is based on formulation of the unknown inputs as the output of a known dynamical system or the approximation of the unknown disturbances by a polynomial. The estimation problem is then solved through the application of standard observer theory by augmenting the disturbance dynamics to that of the plant. Clearly, these schemes increase the dimension of the problem by the order of the dynamical system generating the disturbance, and existence of the estimator will be dependent on the observability of the augmented system. In addition, they require some a priori information regarding the nature of the disturbances. References [11-13] belong to this class of approaches. The other approach does not make any assumption regarding the nature of the unknown inputs. References [14-22] are representative of the second approach to the problem. Here techniques ranging from trial and error to more systematic schemes based on multivariable system inverse, geometric, and direct algebraic concepts have been proposed for the estimation problem. The approach presented here is based on [22] and its extensions. The estimator design technique is believed to be more direct and easier to comprehend than the other previously proposed techniques.

Consider a linear stationary dynamical system driven by exogenous disturbances represented as

$$\dot{x} = Ax + Bu + Dv \tag{1}$$

$$y = Cx = [0 \quad I]x \tag{2}$$

where $x \in \Re^n$ is the state, $u \in \Re^q$ is the control, $v \in \Re^m$ unknown and time varying disturbance input, and $y \in \Re^p$ is the output of the system. It is assumed that the measurement equation is in the form given in Eq. (2), or the system has been transformed via a similarity transformation (see [24]) so that the output is given by that particular structure. In addition, it is assumed that the disturbance distribution matrix D is of full rank. If this is not the case, then this matrix can be decomposed (see Section III) into the product of two full rank matrices, and by redefining the disturbance vector and the distribution matrix one can arrive at a full rank substitute of $\mathbf{D}$.

It is desired to design an estimator capable of estimating the (n-p) unavailable state variables of the above system. To do this, consider the following structure for the estimator

$$\dot{\mathbf{w}} = \mathbf{Fw} + \mathbf{Ey} + \mathbf{Lu} \tag{3}$$

$$\hat{\mathbf{x}} = \mathbf{Rw} + \mathbf{Ty} \tag{4}$$

The following result states the necessary condition for existence of such an estimator.

Theorem 1 - There exist an UIO for the system in Eq. (1-2) if the following necessary condition is satisfied

$$rank(\mathbf{D}) = rank\begin{pmatrix} \mathbf{D}_1 \\ \mathbf{D}_2 \\ \mathbf{D}_3 \end{pmatrix} = rank\begin{pmatrix} \mathbf{D}_2 \\ \mathbf{D}_3 \end{pmatrix} = m$$

where $\mathbf{D}_1 \in \mathfrak{R}^{(n-p)\times m}$, $\mathbf{D}_2 \in \mathfrak{R}^{(p-m)\times m}$, $\mathbf{D}_3 \in \mathfrak{R}^{m\times m}$.

The above can be easily proved by noting that if an asymptotically stable estimator is to exist, then the estimation error ought to go to zero independent of inputs, outputs, and the disturbance. Therefore, by defining the estimator's error and using the above fact we arrive at the following necessary conditions

$$(\mathbf{I} - \mathbf{TC})\mathbf{D} = 0$$

$$(\mathbf{I} - \mathbf{TC})\mathbf{B} = \mathbf{RL}$$

The first condition in the above implies that

$$\mathbf{T}\begin{bmatrix} \mathbf{D}_2 \\ \mathbf{D}_3 \end{bmatrix} = \mathbf{D}$$

and this implies that

$$rank\begin{pmatrix} \mathbf{D}_2 \\ \mathbf{D}_3 \end{pmatrix} = m$$

The above result leads immediately to the following

Corollary 1 - A necessary condition for existence of the UIO is that the number of unknown inputs (m) be less than or equal to the number of outputs (p).

Without any loss of generality assume that the matrix D_3 is nonsingular (if this is not the case the state variables can be relabled). Assuming that m<p, the following theorem states the necessary and sufficient conditions for existence of the estimator.

Theorem 2 - The dynamical system in Eq. (3), with its parameters defined in the following, is an observer for the system in Eq. (1-2) if and only if the pair $\{\overline{A}_{21}, \overline{A}_{11}\}$ is completely observable. In addition, the observer's gain M can be selected such that the closed loop eigenspectrum of the estimator are assigned to any arbitrary locations provided that the complex conjugacy hold.

$$H = (\overline{A}_{13} - M\overline{A}_{23}) + (\overline{A}_{11} - M\overline{A}_{21})(D_1 - MD_2)D_3^{-1} \tag{5}$$

$$G = (\overline{A}_{12} - M\overline{A}_{22}) + (\overline{A}_{11} - M\overline{A}_{21})M \tag{6}$$

$$F = (\overline{A}_{11} - M\overline{A}_{21}) \tag{7}$$

$$L = (\overline{B}_1 - M\overline{B}_2) \tag{8}$$

$$E = [G \quad H] \tag{9}$$

$$\overline{A}_i = A_i - D_i D_3^{-1} A_3 = [\overline{A}_{i1} \quad \overline{A}_{i2} \quad \overline{A}_{i3}] \quad \text{for} \quad i = 1,2 \tag{10}$$

$$\overline{B}_i = B_i - D_i D_3^{-1} B_3 \quad \text{for} \quad i = 1,2 \tag{11}$$

$$R = \begin{pmatrix} I \\ 0 \end{pmatrix} \quad T = \begin{pmatrix} [M & (D_1 - MD_2)D_3^{-1}] \\ I \end{pmatrix} \tag{12}$$

where the following decomposition of Eq. (1) is used, and the estimate of the state is given by Eq. (4).

$$\dot{x} = \begin{pmatrix} A_1 \\ A_2 \\ A_3 \end{pmatrix} x + \begin{pmatrix} B_1 \\ B_2 \\ B_3 \end{pmatrix} u + \begin{pmatrix} D_1 \\ D_2 \\ D_3 \end{pmatrix} v$$

For a straight forward proof of this theorem see [22].

In such cases where the number of the unknown inputs and the output are the same (p=m), we have the following result.

Theorem 3 - If the number of unknown inputs are equal to those of outputs, then an asymptotically stable estimator of the form given in Eq. (3-4) with fixed gain will exist if and only if all the resulting closed loop poles of the estimator are in the left hand plane.

Proof. Partition the dynamical system in Eq. (1-2) in the following way

$$\begin{pmatrix} \dot{x}_1 \\ \dot{y} \end{pmatrix} = \begin{pmatrix} A_{11} & A_{12} \\ A_{21} & A_{22} \end{pmatrix} \begin{pmatrix} x_1 \\ y \end{pmatrix} + \begin{pmatrix} B_1 \\ B_2 \end{pmatrix} u + \begin{pmatrix} D_1 \\ D_2 \end{pmatrix} v \qquad (13)$$

Notice that from Theorem 1, D_2 above is invertible. In addition, let us partition matrices R and T in Eq. (4) as follows

$$R = \begin{pmatrix} R_1 \\ 0 \end{pmatrix} \quad , \quad T = \begin{pmatrix} T_1 \\ I \end{pmatrix} \qquad (14)$$

Define the estimation error as

$$e = x_1 - \hat{x}_1$$

Suppose that an asymptotically stable estimator exists. In that case, the estimation error must be governed by the following dynamical equation

$$\dot{e} = \Omega e$$

where the matrix Ω has all of its eigenvalues in the left hand plane. Using Eqs. (3,4,13,14) in the above, we will arrive at the following

$$\Omega = A_{11} - T_1 A_{21} \qquad (15)$$

$$\Omega R_1 = R_1 F \qquad (16)$$

$$A_{12} - R_1 E - T_1 A_{22} + \Omega T_1 = 0 \qquad (17)$$

$$B_1 - T_1 B_2 = R_1 L \qquad (18)$$

$$D_1 = T_1 D_2 \qquad (19)$$

Eq. (19) will result in

$$T_1 = D_1 D_2^{-1} \tag{20}$$

from which we get

$$\Omega = A_{11} - D_1 D_2^{-1} A_{21}$$

which implies that the dynamics of the estimator would be fixed by the system's parameters. If any eigenvalue of the matrix Ω as given above is not in the left hand plane, then the estimator does not exist. On the other hand, if Ω is a stability matrix then the state of the dynamical system can be estimated. This is most easily done by taking

$$R_1 = I \tag{21}$$

which would result in the following values for the parameters of the estimator given in Eq. (3)

$$F = \Omega \tag{22}$$

$$L = B_1 - T_1 B_2 \tag{23}$$

$$E = A_{12} - T_1 A_{22} + FT_1 \tag{24}$$

To illustrate the simplicity in the design, and the functionality of the UIO, consider the following illustrative example.

Example 1 - Consider the following time invariant linear system

$$\dot{x} = \begin{pmatrix} -1 & 1 & 0 \\ -1 & 0 & 0 \\ 0 & -1 & -1 \end{pmatrix} x + \begin{pmatrix} 1 \\ 0 \\ 1 \end{pmatrix} v \quad \text{with} \quad x(0) = \begin{pmatrix} 1 \\ -1 \\ 2 \end{pmatrix}$$

$$y = \begin{pmatrix} 0 & 1 & 0 \\ 0 & 0 & 1 \end{pmatrix}$$

The above system is open loop stable. It is desired to estimate the first state variable of this system regardless of the value of the unknown input v. Note that in this case the number of unknown inputs is less than those of outputs. Therefore, the estimator's pole can be assigned arbitrarily, and here it was assigned at -5. Using Eqs. (3-12), the dynamics of the UIO is obtained as

$$\dot{w} = -5w + [22 \quad -4]y$$

For the sake of illustration, the random signal shown in Figure 1 was applied to the system as the unknown input. Figure 2 illustrates the actual (solid curve) versus the estimated (dotted curve) trajectory for the first state variable.

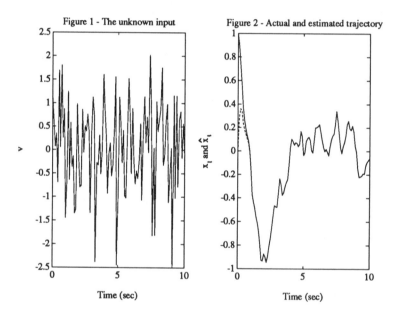

Figure 1 - The unknown input

Figure 2 - Actual and estimated trajectory

Time (sec)   Time (sec)

B. Proportional-Integral Observer (PIO)

In Section A we considered the UIO which is capable of estimating the state of the system in spite of the unknown and possibly time varying inputs. In many industrial applications the unknown inputs are constant (or changing slowly in step like fashion). In addition, in many applications (such as designing fault tolerant controllers, or controllers for cascade systems with time lags) it would be of benefit to not only estimate the state of the system, but also estimate the disturbance acting on the system. Therefore, the problem of interest here is to estimate the unknown input as well as the state of the system for constant unknown inputs.

The PIO was introduced by Shafai and Carroll [25] and was further developed in [26]. However, [25-26] consider the use of a PIO for systems driven by known inputs only. The integration path in their design is used to basically give an additional degree of freedom for designing an estimator which would be insensitive to parameter variations in the system. In this way the authors manage to recover the stability margins of an LQR problem in spite of the estimator in the loop. In this section we propose a new approach for using PIOs to estimate the state as well as the unknown input in systems that are driven by constant unknown inputs. To show how this task may be accomplished, consider a more general version of the system described in Eq. (1,2) as follow:

$$\dot{x} = Ax + Bu + Dv \tag{25}$$

$$y = Cx + Ev \tag{26}$$

where all the parameters are as defined in Eq. (1,2). The only difference here is that the unknown input is assumed to be constant, and also it is assumed that it can even contaminate the measurement through the appropriately dimensioned matrix E. All the assumptions made in the previous section are assumed here as well. Before we present the main result for design of a PIO, consider decomposing the above system in the following fashion

$$\begin{pmatrix} \dot{x}_1 \\ \dot{x}_2 \end{pmatrix} = \begin{pmatrix} A_{11} & A_{12} \\ A_{21} & A_{22} \end{pmatrix} \begin{pmatrix} x_1 \\ x_2 \end{pmatrix} + \begin{pmatrix} B_1 \\ B_2 \end{pmatrix} u + \begin{pmatrix} D_1 \\ D_2 \end{pmatrix} v \tag{27}$$

$$y = x_2 + Ev \tag{28}$$

Then the following theorem would give the desired estimator.

Theorem 4 - The state of the system in Eq. (25) as well as the unknown constant disturbance acting on the system can be estimated using a reduced order estimator

$$\dot{w} = G_1 w + G_2 u + G_3 y \tag{29}$$

where the estimator's parameters are given in the following, if and only if the pair

$$\left\{ [\mathbf{A}_{21} \quad \mathbf{D}_2 - \mathbf{A}_{22}\mathbf{E}] , \begin{bmatrix} \mathbf{A}_{11} & \mathbf{D}_1 - \mathbf{A}_{12}\mathbf{E} \\ 0 & 0 \end{bmatrix} \right\} \tag{30}$$

is completely observable. Furthermore, in such a case , by properly selecting the estimator's gain $\mathbf{M}$, its poles can be placed arbitrarily as long as the usual complex conjugacy condition hold.

$$\mathbf{G}_1 = \begin{bmatrix} \mathbf{A}_{11} - \mathbf{M}_1\mathbf{A}_{21} & (\mathbf{D}_1 - \mathbf{M}_1\mathbf{D}_2) + (\mathbf{M}_1\mathbf{A}_{22} - \mathbf{A}_{12})\mathbf{E} \\ -\mathbf{M}_2\mathbf{A}_{21} & \mathbf{M}_2(\mathbf{A}_{22}\mathbf{E} - \mathbf{D}_2) \end{bmatrix} \tag{31}$$

$$\mathbf{G}_2 = \begin{bmatrix} \mathbf{B}_1 - \mathbf{M}_1\mathbf{B}_2 \\ -\mathbf{M}_2\mathbf{B}_2 \end{bmatrix} \tag{32}$$

$$\mathbf{G}_3 = \mathbf{G}_4 + \mathbf{G}_1\mathbf{M} \tag{33}$$

$$\mathbf{G}_4 = \begin{bmatrix} \mathbf{A}_{12} - \mathbf{M}_1\mathbf{A}_{22} \\ -\mathbf{M}_2\mathbf{A}_{22} \end{bmatrix} \tag{34}$$

$$\mathbf{M} = \begin{bmatrix} \mathbf{M}_1 \\ \mathbf{M}_2 \end{bmatrix} \tag{35}$$

Proof: Using Eqs. (27,28), define

$$\mathbf{q} = \dot{\mathbf{x}}_2 - \mathbf{A}_{22}\mathbf{x}_2 - \mathbf{B}_2\mathbf{u}$$

$$= \dot{\mathbf{y}} - \mathbf{A}_{22}\mathbf{y} + \mathbf{A}_{22}\mathbf{E}\hat{\mathbf{v}} - \mathbf{B}_2\mathbf{u}$$

$$= \mathbf{A}_{21}\mathbf{x}_1 + \mathbf{D}_2\mathbf{v} \tag{36}$$

and

$$\mathbf{r} = \mathbf{A}_{12}\mathbf{x}_2 + \mathbf{B}_1\mathbf{u}$$

$$= \mathbf{A}_{12}\mathbf{y} - \mathbf{A}_{12}\mathbf{E}\hat{\mathbf{v}} + \mathbf{B}_1\mathbf{u} \tag{37}$$

where $\hat{\mathbf{x}}$ and $\hat{\mathbf{v}}$ denote the estimate of the state and disturbance, respectively. Using the above definitions Eq. (27) can be written as

$$\dot{\mathbf{x}}_1 = \mathbf{A}_{11}\mathbf{x}_1 + \mathbf{D}_1\mathbf{v} + \mathbf{r}$$

$$\mathbf{q} = \mathbf{A}_{21}\mathbf{x}_1 + \mathbf{D}_2\mathbf{v}$$

To design an estimator for the above system, let

$$z = \int (q - A_{21}\hat{x}_1 - D_2\hat{v})dt$$

or

$$\dot{z} = q - A_{21}\hat{x}_1 - D_2\hat{v} \tag{38}$$

Assuming the following estimator's dynamics which involves an integral as well as a proportional term involving the difference between q and its estimate

$$\dot{\hat{x}}_1 = A_{11}\hat{x}_1 + D_1\hat{v} + M_1\dot{z} + r \tag{39}$$

$$\dot{\hat{v}} = M_2\dot{z} \tag{40}$$

and upon substitution for q, r, and some manipulation, the following can be obtained

$$\dot{\xi} = G_1\xi + G_2u + M\dot{y} + G_4y \tag{41}$$

where G_i's are as defined in Eqs. (31-34), and

$$\xi = \begin{bmatrix} \hat{x}_1 \\ \hat{v} \end{bmatrix}$$

Eq. (41) requires the derivative of the output. To eliminate the need for differentiating the output define

$$w = \xi - My$$

Using the above in Eq. (41) will result in the estimator dynamics as given in Eq. (29). The estimates of the state and the disturbance can then be easily obtained from the following

$$\xi = w + My \tag{42}$$

$$\hat{x}_2 = y - E\hat{v} \tag{43}$$

The poles of the estimator are the eigenvalues of G_1, and it can be shown (Chen [24]) that they can be arbitrarily assigned if and only if the pair given in (30) is completely observable. Finally, the convergence of the estimates can be proved by

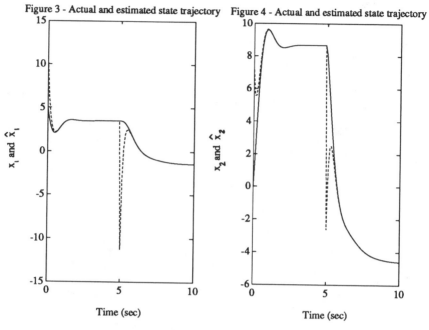

Figure 3 - Actual and estimated state trajectory

Figure 4 - Actual and estimated state trajectory

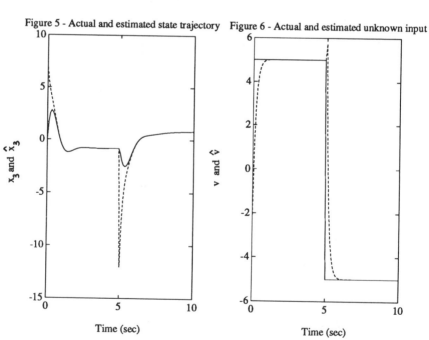

Figure 5 - Actual and estimated state trajectory

Figure 6 - Actual and estimated unknown input

constructing the estimation error and showing that the error dynamics is asymptotically stable.

Example 2 - Consider the following system

$$\dot{x} = \begin{pmatrix} -3 & 1 & -1 \\ 2 & -1 & 4 \\ 3 & -2 & -1 \end{pmatrix} x + \begin{pmatrix} 1 \\ 0 \\ 1 \end{pmatrix} u + \begin{pmatrix} 0 \\ 1 \\ 1 \end{pmatrix} v \quad \text{with} \quad x(0) = \begin{pmatrix} 5 \\ 0 \\ 0 \end{pmatrix}$$

$$y = \begin{pmatrix} 0 & 1 & 0 \\ 0 & 0 & 1 \end{pmatrix} x + \begin{pmatrix} 1 \\ 1 \end{pmatrix} v$$

The above system is open loop stable. It is desired to design a PIO with its poles at $\{-8,-6\}$ for estimating both the state variables as well as the unknown disturbance. Using Eq. (30) the estimator's gain was obtained as

$$M = \begin{pmatrix} 1.4286 & 0.7143 \\ -1.2857 & 0.8571 \end{pmatrix}$$

and the PIO dynamics was obtained to be

$$\dot{w} = \begin{pmatrix} -8 & 0 \\ 0 & -6 \end{pmatrix} w + \begin{pmatrix} 0.2857 \\ -0.8571 \end{pmatrix} u + \begin{pmatrix} -7.5714 & -11.7143 \\ 8.1429 & 0.8571 \end{pmatrix} y$$

For a unit step input and a step varying unknown input of the form given in Figure 6 (solid curve), the estimates of the observer along with the actual trajectories are shown in Figures 3 to 6. Clearly the PIO performs as desired.

III) EFFECT OF MODEL UNCERTAINTIES

The discussion of the UIO and the PIO in the previous section was based upon the fact that the dynamics of the system are known perfectly. In practice however, the dynamics of most practical systems are known with some degree of uncertainty. Also, plant parameters in many systems can vary due to components aging, corrosion, etc.

Typical Luenberger Observer or a Kalman Filter require precise knowledge of the system's model. In this section we show how an UIO can sometimes be used for estimation purposes in systems with large degree of parameter variations and uncertainties. It is assumed here that only the nominal value of A, and B matrices in Eqs. (1,2) are known. In order to characterize the uncertainties in the model of our system appropriately, consider the following definition:

Definition - The nxk_A *uncertainty indicator matrix* of any nxm matrix A is defined as $I_A(a_1, a_2, \ldots\ldots, a_{k_A})$, where k_A is the number of rows of A that contain unknown elements. The jth column of this matrix has zero entries except for the a_jth entry which has a value of one.

As an example, if A is a 4x4 matrix and there are uncertain elements in the second and third rows (therefore $k_A=2$), then $a_1 = 2$ and $a_2 = 3$.

Since it is assumed that only the nominal value of the system's matrices are known, one can write

$$A = A_o + \Delta A \tag{44}$$

where A_o is the nominal value, and

$$\Delta A = \begin{pmatrix} \Delta A_1 \\ \Delta A_2 \\ \cdot \\ \cdot \\ \cdot \\ \Delta A_n \end{pmatrix} \tag{45}$$

is the uncertainty matrix associated with A.

Now assuming that k_A rows of A have uncertain elements associated with them ($k_A < n$), Eq. (45) can then be written as

$$A = BC$$

where B is a pxr full rank matrix and C is a rxq full rank matrix.

proof: According to the Singular Value Decomposition Theorem, matrix A can be decomposed as

$$A = GDH$$

where G and H are orthogonal pxp and qxq matrices and D is a non-negative definite matrix of the following form

$$D = \begin{pmatrix} \Sigma & 0 \\ 0 & 0 \end{pmatrix}$$

and has the strictly positive singular values of A on its diagonal. Now define the rxr matrix σ as

$$\sigma^2 = \Sigma$$

Then A can be written as

$$A = G \begin{bmatrix} \sigma \\ 0 \end{bmatrix} [\sigma \quad 0] H$$

Defining

$$B = G \begin{bmatrix} \sigma \\ 0 \end{bmatrix} \qquad C = [\sigma \quad 0] H$$

Then the pxr matrix B, and the rxq matrix C are both of full ranks.

Therefore, if matrix D in Eq. (47) turn out to be rank deficient, then it can be decomposed into product of two full rank matrices $\bar{D}\bar{D}$, and thus Eq. (47) can be written as

$$\dot{x} = A_o x + B_o u + \bar{D}(\bar{D}v) \tag{50}$$

where now the term $\bar{D}v$ is taken to be a new unknown input to the system.

To summarize the results of this section, we will state the following result.

$$\Delta A = I_A(a_1, a_2, ..., a_{k_A}) \begin{pmatrix} \Delta A_{a_1} \\ \Delta A_{a_2} \\ \cdot \\ \cdot \\ \cdot \\ \Delta A_{a_{k_A}} \end{pmatrix} = I_A \Delta A_a \tag{46}$$

Where ΔA_{a_i} is the a_ith row of the matrix ΔA.

Using the above notation, an uncertain dynamical system given by

$$\dot{x} = Ax + Bu$$

can be written as

$$\dot{x} = A_o x + B_o u + Dv \tag{47}$$

where

$$D = [I_A \quad I_B] \tag{48}$$

$$v = \begin{pmatrix} \Delta A_a x \\ \Delta B_b u \end{pmatrix} \tag{49}$$

From Eq. (47), It is evident that the uncertainties in the system's dynamics can be grouped together and be treated as an unknown input to the nominal system. It should also be pointed out that a similar procedure can be employed when linearizing a nonlinear system. That is, the effect of the higher order terms can be modeled as an unknown input to the system. Once this is done, the state of the system can be estimated using an UIO given that the conditions to its existence (given in Theorem 1) are satisfied.

It should be mentioned however that the matrix D in Eq. (47) may often turn out to be singular. Since the application of UIO requires D to be nonsingular (see Section II.A), this problem can be remedied by using a decomposition scheme stated in the following.

Proposition 1 - Any pxq matrix A, whose rank is r can be decomposed as follows

Corollary 2 - Consider an uncertain system as described above where only the nominal values of the system's matrices are known. The state of this system can be estimated using an UIO as described in Eq. (3) so long that

$$k_A + k_B - k_{AB} \le p$$

where k_{AB} is the number of rows for which $a_i = b_j$.

Example 3 - Consider a LTI system described by

$$\dot{x} = \begin{pmatrix} 0 & 1 & 0 \\ -1 & 0 & 1 \\ a_{31} & a_{32} & a_{33} \end{pmatrix} x + \begin{pmatrix} 1 \\ -1 \\ 1 \end{pmatrix} u \quad \text{with} \quad x(0) = \begin{pmatrix} 5 \\ 0 \\ 0 \end{pmatrix}$$

$$y = \begin{pmatrix} 0 & 1 & 0 \\ 0 & 0 & 1 \end{pmatrix} x$$

where a_{ij} are uncertain parameters, however, their nominal values are known. Specifically, in this example the following values for the nominal and actual values of **A** were used.

$$A_o = \begin{pmatrix} 0 & 1 & 0 \\ -1 & 0 & 1 \\ -8 & -10 & -5 \end{pmatrix}, \quad A = \begin{pmatrix} 0 & 1 & 0 \\ -1 & 0 & 1 \\ -6 & -13 & -7 \end{pmatrix}$$

For the sake of illustration two estimators are to be designed for the above problem. The first is simply a Luenberger observer, and a second is based on the above discussion of robust estimator design using UIO theory. Both estimators are to have their pole at -8. The standard Luenberger estimator is given by

$$\dot{w} = -8w + [-9 \quad 3]y + 2u$$

To design the robust estimator, the uncertain system described above is rewritten in the form of Eq. (47) as

$$\dot{x} = \begin{pmatrix} 0 & 1 & 0 \\ -1 & 0 & 1 \\ -8 & -10 & -5 \end{pmatrix} x + \begin{pmatrix} 1 \\ -1 \\ 1 \end{pmatrix} u + \begin{pmatrix} 0 \\ 0 \\ 1 \end{pmatrix} v$$

where the unknown input v is given by

$$v = \Delta A_{a_3} x$$

The UIO for the above system is given by

$$\dot{w} = -8w + [65 \qquad 8]y + -7u$$

Figure 7 shows the response of the two estimators for a unit step input to the system. Clearly, the performance of the standard Luenberger estimator (dotted line) is unacceptable as its estimate never converge. The robust estimator (dashed line) provides perfect estimate of the unavailable state of the system despite considerable parameter uncertainty in the model.

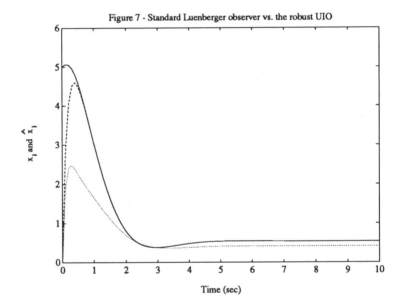

Figure 7 - Standard Luenberger observer vs. the robust UIO

IV. APPLICATIONS

In this section the UIO theory presented will be extended to suit a number of application areas. Numerical examples are presented along with the necessary theoretical developments.

A. Optimal Modal Robust Servomechanism Problem

Basically the robust servomechanism problem can be defined as that of designing a closed loop feedback system so that one or a number of outputs would track one or a number of reference inputs, in spite of the disturbances acting on the plant. The multivariable version of this problem has been addressed by a number of authors in the past (e.g. [1-8]). In all these approaches integral action is utilized in order to offset the steady state error due to the disturbances or the unmodeled dynamics of the system. Mahalanabis and Pal [4], proposed the optimal modal LQR theory to the problem. However, it is well known that the standard LQR theory can not accomplish the eigenspectrum placement. Thus [4] uses an approach for calculating the weighting matrices in the quadratic cost functional so that the optimal control law would also place the closed loop eigenspectrum of the augmented system at desired locations. The main problem with [4] is that the computation of the desired weighting matrix is extremely involved, and requires polynomial matrix multiplications and solution of a set of nonlinear algebraic equations. In addition, if the original system is of high order, this makes the calculation of the weighting matrices even more complicated.

In this section we will address the same problem as in [4]. That is, it is desired to find an optimal PI control law that would achieve tracking or regulation of the system outputs in the face of plant disturbances, minimize the performance index, and also achieve a desired set of eigenvalues for satisfactory response. The approach presented here is an extension of the author's recent work on designing modal LQRs [27,28]. Since the controller proposed by [4] as well as [27,28] requires the availability of the state of the system, the UIO theory will be slightly modified for use in the servomechanism problem.

Consider the system in Eq. (1,2), and assume that it is desired that the output **y** to track some constant reference input signal **r**. Note that the last term in Eq. (1)

arise quite naturally due to modeling errors or disturbances acting on the plant, and v is assumed to be constant (or slowly stepwise varying).

Defining

$$\dot{\alpha} = r - y \qquad (51)$$

and augmenting Eq. (1) with Eq. (51) will result in the following (n+p) dimensional system

$$\dot{\bar{x}} = \bar{A}\bar{x} + \bar{B}u + \bar{T}r + \bar{D}v \qquad (52)$$

$$\bar{y} = \bar{C}\bar{x} \qquad (53)$$

where

$$\bar{x} = \begin{bmatrix} x \\ \alpha \end{bmatrix}, \quad \bar{A} = \begin{bmatrix} A & 0 \\ -C & 0 \end{bmatrix}, \quad \bar{B} = \begin{bmatrix} B \\ 0 \end{bmatrix}, \quad \bar{D} = \begin{bmatrix} D \\ 0 \end{bmatrix}, \quad \bar{T} = \begin{bmatrix} 0 \\ I \end{bmatrix}, \quad \bar{C} = \begin{bmatrix} C & 0 \\ 0 & I \end{bmatrix}$$

It is now desired to obtain the proportional-integral (PI) optimal control law u of the form

$$u = K_1 x + K_2 \alpha \qquad (54)$$

such that the following performance measure is minimized subject to Eq. (52).

$$J = \frac{1}{2} \int_0^\infty (||\bar{x}||_Q^2 + ||u||_R^2) dt \qquad (55)$$

Additionally, in order to achieve desired response characteristics, it is required that the control law in Eq. (54) should place the closed loop eigenspectrum of Eq. (52) at preassigned locations. Given a fixed R, if Q in Eq. (55) is appropriately selected, the optimal control law that would achieve the above objectives would be given by [27,28]

$$u = -R^{-1}\bar{B}^T\bar{P}\bar{x} = K\bar{x} = [K_1 \quad K_2]\bar{x} \qquad (56)$$

where $\bar{P}$ is the positive semidefinite solution of the algebraic matrix Riccati equation (AMRE). It is well known that the control law in Eq. (56) will exit if and only if none of the 2(n+p) eigenvalues of the following Hamiltonian matrix

$$\overline{H} = \begin{bmatrix} \overline{A} & -\overline{B}R^{-1}\overline{B}^T \\ -Q & -\overline{A}^T \end{bmatrix} \tag{57}$$

are on the imaginary axis, and the pair $\{\overline{A}, \overline{B}\}$ is completely controllable [27]. Furthermore, it can be shown [5] that the above pair is completely controllable if and only if the matrix

$$\begin{bmatrix} B & A \\ 0 & -C \end{bmatrix} \tag{58}$$

has a rank (n+p).

Given the control law in Eq. (56) the closed loop system can be represented by

$$\dot{\overline{x}} = (\overline{A} - \overline{B}R^{-1}\overline{B}^T P)\overline{x} + \overline{D}v + \overline{T}r = \overline{A}_c \overline{x} + \overline{D}v + \overline{T}r \tag{59}$$

In the following we will briefly outline a sequential procedure based on [27,28] for selecting an appropriate weighting matrix Q in Eq. (55) and the resulting optimal control law in Eq. (56).

Suppose that we are at the ith stage of the sequential process. The system under consideration described by[1]

$$\dot{\overline{x}}_i = \overline{A}_i \overline{x}_i + \overline{B} u_i \tag{60}$$

is aggregated (reduced) to an lth order dynamical system, where l=1, or 2 depending on whether a real, or a complex conjugate pair of pole(s) is being placed. The reduced order system is described by

$$\dot{\hat{x}}_i = \hat{A}_i \hat{x}_i + \hat{B}_i \hat{u}_i \tag{61}$$

---

[1] The sequential process starts at stage i=1, where $\overline{A}_1 = \overline{A}$, and $\overline{A}$ is as given in Eq. (52), also since the terms containing v and r do not effect the analysis of this part, they are dropped for brevity in this discussion.

where l of the eigenvalues of $\overline{\mathbf{A}}_i$ in Eq. (60) are contained in $\hat{\mathbf{A}}_i$, that is $\Lambda(\hat{\mathbf{A}}_i) \subset \Lambda(\mathbf{A}_i)$

, where $\Lambda(.)$ is a set that contains the eigenvalues of the matrix in the argument. Note that in this subsection the variables with the ($\wedge$) will refer to the reduced order system. The transformation (aggregation) matrix that accomplishes the model reduction is given by

$$\Gamma = [\mathbf{I}_l \qquad 0]\Xi^{-1}(\overline{\mathbf{A}}_i) \tag{62}$$

where $\Xi(.)$ is the modal matrix of the matrix argument, and in this case its first l columns are the eigenvectors corresponding to the l eigenvalues of $\hat{\mathbf{A}}_i$. The matrices $\hat{\mathbf{A}}_i$, and $\hat{\mathbf{B}}_i$ are obtained by using the following transformations

$$\begin{aligned} \hat{\mathbf{A}}_i &= \Gamma \ \overline{\mathbf{A}}_i \ \Gamma^+ \\ \hat{\mathbf{B}}_i &= \Gamma \ \overline{\mathbf{B}} \end{aligned} \qquad \text{if} \quad l = 1 \tag{63}$$

$$\begin{aligned} \hat{\mathbf{A}}_i &= \Theta^{-1}\Gamma \ \overline{\mathbf{A}}_i \ \Gamma^+ \Theta \\ \hat{\mathbf{B}}_i &= \Theta^{-1}\Gamma \ \overline{\mathbf{B}} \end{aligned} \qquad \text{if} \quad l = 2 \tag{64}$$

$$\Theta = \begin{pmatrix} 0.5 & +j0.5 \\ 0.5 & -j0.5 \end{pmatrix}$$

where Γ^+ is the Moore-Penroze pseudo inverse of Γ. At this point the desired state weighting matrix ($\hat{\mathbf{Q}}_i$) in the following quadratic cost functional

$$\hat{\mathbf{J}}_i = \frac{1}{2} \int_0^\infty \left(||\hat{\mathbf{x}}_i||_{\hat{\mathbf{Q}}_i}^2 + ||\hat{\mathbf{u}}_i||_{\mathbf{R}}^2 \right) dt \tag{65}$$

is obtained using either the approach in [27] or [28] so that the optimal control law which would minimize Eq. (65) would place the real or the complex conjugate poles of the aggregated system ($\hat{\mathbf{A}}_i$) at desired locations.

Once the appropriate matrices $\hat{\mathbf{Q}}_i$, and optimal controller gain $\hat{\mathbf{K}}_i$ for the reduced order system are found, they are transformed back to the original higher dimensional space via the following transformations to obtain $\mathbf{Q}_i$, and $\mathbf{K}_i$ that would assign the same poles for the system in Eq. (60).

$$\mathbf{K}_i = \hat{\mathbf{K}}_i \Gamma$$
$$\mathbf{Q}_i = \Gamma^T \hat{\mathbf{Q}}_i \Gamma \qquad \text{if} \quad l = 1 \qquad (66)$$

$$\mathbf{K}_i = \hat{\mathbf{K}}_i \Theta^{-1} \Gamma$$
$$\mathbf{Q}_i = \Gamma^T \Theta^{-T} \hat{\mathbf{Q}}_i \Theta^{-T} \Gamma \qquad \text{if} \quad l = 2 \qquad (67)$$

Next the system dynamics will be updated by

$$\overline{\mathbf{A}}_{i+1} = \overline{\mathbf{A}}_i + \overline{\mathbf{B}} \mathbf{K}_i \qquad (68)$$

The above describes one stage of the sequential procedure. Now letting i=i+1, Eq. (60) will be aggregated to a first or second order system for placing another real or complex conjugate pair of pole(s). The sequential process will continue in this manner until all or a number of the dominant poles of the system are placed. At this time the overall desired weighting matrix $\mathbf{Q}$ in Eq. (55), and the optimal state feedback gain $\mathbf{K}$ in Eq. (56), that would achieve the pole placement will be calculated from

$$\mathbf{Q} = \sum_i \mathbf{Q}_i \qquad (69)$$

$$\mathbf{K} = \sum_i \mathbf{K}_i \qquad (70)$$

This completes the robust servomechanism design procedure. Using the above feedback gain, and weighting matrix, the control law in Eq. (56) will minimize the cost functional in Eq. (55). Also, the output $\mathbf{y}$ in Eq. (2) will now track the reference input $\mathbf{r}$ in spite of the unknown disturbances $\mathbf{v}$ acting on the plant.

The robust servomechanism discussed above is based on the assumption of availability of all state variables. Without the external disturbances present in Eq. (1) a standard Luenberger observer can be designed for estimating the state of the system. However, with the presence of the external disturbances, we could employ either the UIO or the PIO for the problem considered in the above. However, a slight modification to the structure of the estimator given in Eq. (3) or Eq. (29) is necessary. For the robust servomechanism problem the UIO dynamics in Eq. (3) should be modified to

$$\dot{w} = Fw + Ey + Lu + Sr \qquad (71)$$

where essentially everything remains as defined before with matrix S defined as

$$S = (\overline{T}_1 - M\overline{T}_2) \qquad (72)$$

where

$$\overline{T}_i = \overline{T}_i - D_i D_3^{-1} \overline{T}_3 \qquad for \qquad i = 1, 2 \qquad (73)$$

Example 4 - Here we will apply the optimal robust servomechanism described for controlling the position of a brushless dc motor with unknown load torque. Due to the ease of control and their versatility, dc motors are employed in variety of industrial applications such as industrial manipulators. A number of authors have suggested various sophisticated control strategies for controlling them. The model reference adaptive controller (MRAC) of [29,30], and optimal model following of [31] are just a few examples of such attempts.

The system's block diagram is given in Figure 8. Taking the state variables as assigned in Figure 8, and using the numerical values of the parameters as given in [31], the following state space representation of the system is obtained.

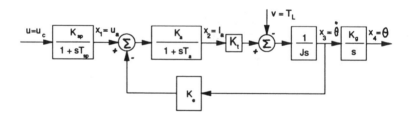

Figure 8 - Block diagram of the brushless D.C. motor

U_a	Input voltage of the transistor inverter	K_{sp}	Gain of the supplying transistor inverter
T_{sp}	Time constant of the supplying transistor inverter	U_a	Armature voltage
K_a	Gain of the armature circuit	T_a	Time constant of the armature circuit
I_a	Armature current	K_t	Torque constant
T_L	Load torque	J	Total inertia
θ	Angular position	$\dot{\theta}$	Angular velocity
K_g	Gear ratio	K_e	Back emf constant

$$\dot{x} = \begin{bmatrix} -6666.667 & 0 & 0 & 0 \\ 83.333 & -133.33 & -50.0 & 0 \\ 0 & 60.0 & 0 & 0 \\ 0 & 0 & 1 & 0 \end{bmatrix} x + \begin{bmatrix} 133333.33 \\ 0 \\ 0 \\ 0 \end{bmatrix} u + \begin{bmatrix} 0 \\ 0 \\ -100 \\ 0 \end{bmatrix} v$$

$$y = \begin{bmatrix} 0 & 0 & 1 & 0 \\ 0 & 0 & 0 & 1 \end{bmatrix} x$$

The open loop poles of this system are located at $\{0,-104.67,-28.61,-6666.67\}$. It is desired to design a robust servomechanism so that the motor's shaft position would track a reference position input. The augmented system Eq. (52) was obtained as

$$\dot{\bar{x}} = \begin{bmatrix} -6666.667 & 0 & 0 & 0 & 0 \\ 83.333 & -133.33 & -50.0 & 0 & 0 \\ 0 & 60.0 & 0 & 0 & 0 \\ 0 & 0 & 1 & 0 & 0 \\ 0 & 0 & 0 & -1 & 0 \end{bmatrix} \bar{x} + \begin{bmatrix} 133333.33 \\ 0 \\ 0 \\ 0 \\ 0 \end{bmatrix} u + \begin{bmatrix} 0 \\ 0 \\ 0 \\ 0 \\ 1 \end{bmatrix} r + \begin{bmatrix} 0 \\ 0 \\ -100 \\ 0 \\ 0 \end{bmatrix} v$$

$$\bar{y} = \begin{bmatrix} 0 & 0 & 1 & 0 & 0 \\ 0 & 0 & 0 & 1 & 0 \\ 0 & 0 & 0 & 0 & 1 \end{bmatrix} \bar{x}$$

It is desired to find the suitable weighting matrix in Eq. (55) so that the optimal PI control law obtained from Eq. (56), would place the closed loop eigenspectrum of the system at $\{-6666.67,-30,-20,-5\pm j5\}$. Notice that the nondominant open loop pole located at -6666.67 is to be retained since the effect of this mode will not be significant in the response of the system. The sequential procedure was used and for a control weight of $R=0.01$, the final state weighting matrix was obtained. The optimal proportional integral feedback gain was computed from Eq. (70) and is given by

$$K = [5.5e-4 \quad 4.3286e-2 \quad 1.8157e-2 \quad -8.5045e-2 \quad 3.0e-1]$$

Since the armature voltage, and the current are not available for measurement, it is necessary to estimate them. A second order UIO was designed for this purpose by using the results outlined in Section II.A. It turned out that for this system the pair $\{\bar{A}_{11}, \bar{A}_{21}\}$ is not observable. Therefore, the closed loop eigenspectrum of the

UIO could not be freely assigned. However, an asymptotically stable UIO with fixed eigenspectrum does exist for this system.

For an unknown load torque of $T_L=5$ Nm, and zero initial condition on the state of the system (however, for the sake of illustration the initial condition on the initial estimate of the first two state variables was taken to be 60, and 10 respectively), the response of the system for a desired reference input as shown in Figure 9, is given in Figures 9-11.

B. Decentralized Estimation

In this subsection, the UIO theory will be applied to decentralized estimation problem in large scale dynamical systems. The problem of decentralized estimation has been addressed by a number of researchers, notably [32-38]. Here, it is shown that for a class of interconnection patterns, whether linear or nonlinear, the UIO may be applied in order to design totally decentralized estimators for large scale systems. This is done by treating the interaction terms as unknown inputs when designing the local observers. In this way there will be no need for communication among the local observers as was the case in the aforementioned works. However, the restrictions imposed on the interconnection patterns by this approach is more severe and thus the previous works are applicable over a wider range of interconnection patterns.

Suppose that the system in Eq. (1,2) is a large scale linear time invariant dynamical system. Assume that the system is broken into r subsystems, and the dynamics of the ith subsystem has been written in the decentralized form as

$$\dot{\mathbf{x}}_i = \mathbf{A}_i\mathbf{x}_i + \mathbf{B}_i\mathbf{u}_i + \sum_{\substack{j \neq i \\ j=1}}^{r} \mathbf{A}_{ij}\mathbf{x}_j \tag{74}$$

$$\mathbf{y}_i = \mathbf{C}_i\mathbf{x}_i \tag{75}$$

where $\mathbf{x}_i \in \mathfrak{R}^{n_i}$, $\mathbf{u}_i \in \mathfrak{R}^{q_i}$, $\mathbf{y}_i \in \mathfrak{R}^{p_i}$, are the state, control, and output vectors of the ith subsystem.

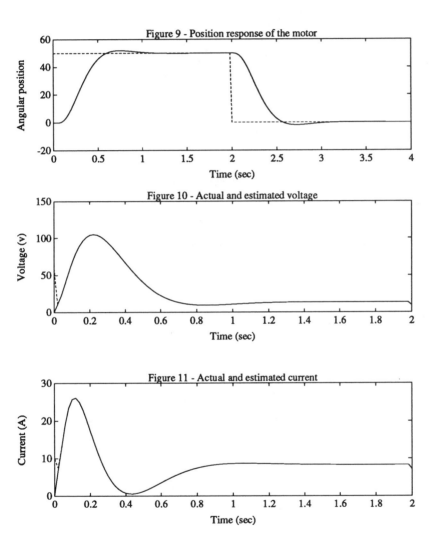

Figure 9 - Position response of the motor

Figure 10 - Actual and estimated voltage

Figure 11 - Actual and estimated current

In addition matrices A_i, A_{ij}, B_i, C_i are all constant matrices of appropriate dimension. It is clear that the ith subsystem interacts with the others through the terms containing A_{ij}.

Assume that the triplet (A_i, B_i, C_i) is a minimum realization, and the rank of $C_i = p_i$ for all i=1,2,...,r. In addition, since we are dealing with the estimation problem, it is assumed that there exists a decentralized feedback law that can stabilize the system.

Assuming that not all of the local state variables are measurable for control purposes, the problem of interest is then to design r local observers to estimate the unmeasurable components of x_i, i=1,2,...,r.

Using a similar decomposition procedure as in Section III, we can rewrite Eqs. (74,75) as

$$\dot{x}_i = A_i x_i + B_i u_i + \overline{D}_i v_i \qquad (76)$$

$$y_i = C_i x_i = [0 \qquad I]x_i \qquad (77)$$

where

$$v_i = \tilde{D}_i \overline{v}_i$$

$$D_i = \overline{D}_i \tilde{D}_i$$

$$D_i = [A_{i1}, A_{i2}, ..., A_{i(i-1)}, A_{i(i+1)}, ..., A_{ir}] \qquad with \quad rank(D_i) = m_i$$

$$\overline{v}_i^T = \begin{pmatrix} x_1 & x_2 & \cdot & \cdot & x_{i-1} & x_{i+1} & \cdot & \cdot & x_r \end{pmatrix}$$

where $\overline{D}_i \in \Re^{n_i \times m_i}$ with $\rho(\overline{D}_i) = m_i$, and $v_i \in \Re^{m_i}$.

Simply all that needs to be done from here on is to design individual UIOs (assuming they exist) for each of the subsystems.

C. Sensor Failure Detection & Identification Problem

The UIO theory has proved useful in studying the problem of instrument failure detection, isolation and accommodation (FDIA) using the notion of analytical redundancy. Detailed survey of analytical redundancy techniques could be found in [39-43].

It is well known that the FDIA is most difficult if the system under consideration is uncertain. In this subsection, we will present a sensor fault detection and identification approach using a single UIO for systems with uncertainties. To illustrate the approach, consider an uncertain dynamical system whose sensors are subject to failure. Using the approach outlined in Section III, the dynamics of such a system can be written as

$$\dot{x} = A_o x + B_o u + D v \tag{78}$$

$$y = Cx + Ef \tag{79}$$

where the vector f is an unmeasurable vector of sensor failures.

In order to employ the UIO theory, the system in Eq. (78,79) must be transformed in the proper form given in Eq. (1,2). This can be done using the following result.

Proposition 2 - Given any vector function $f \in \Re^r$ and a stable rxr matrix A_f, there will always exist an input vector $\mu \in \Re^r$ such that

$$\dot{f} = A_f f + \mu \tag{80}$$

Augmenting Eq. (78) with (80) will result in the following (n+r)th order dynamical system

$$\begin{pmatrix} \dot{x} \\ \dot{f} \end{pmatrix} = \begin{pmatrix} A_o & 0 \\ 0 & A_f \end{pmatrix} \begin{pmatrix} x \\ f \end{pmatrix} + \begin{pmatrix} B_o \\ 0 \end{pmatrix} u + \begin{pmatrix} D & 0 \\ 0 & I \end{pmatrix} \begin{pmatrix} v \\ \mu \end{pmatrix} \tag{81}$$

$$y = [C \quad E] \begin{pmatrix} x \\ f \end{pmatrix} \tag{82}$$

Notice that the above system is now in the form described by Eqs. (1,2).

Now the sensor failure detection, and identification can be easily accomplished by designing a UIO for the above system. If the necessary conditions for the existence of the UIO are satisfied, the estimate of the state of the system in Eqs. (81,82) can be obtained by using the UIO theory. Therefore, monitoring the state estimates (i.e. $\hat{x}$, and $\hat{f}$), would provide an immediate means for the detection of sensor failures. That is, $\hat{f}$ would provide us with the estimate of the every component of the failure vector f as given in Eq. (79). The failure detection logic here is very simple, under no sensor failures all the components of f should be zero or close to it. A nonzero component of the $\hat{f}$ would be indicative of a presence of a sensor failure. It is also extremely simple to detect which sensor has failed by checking which component of the $\hat{f}$ is nonzero. Similarly, presence of two nonzero entries in the $\hat{f}$ vector would indicate multiple sensor failures, and so on.

Example 5 - In the following control and FDI problem for the vertical takeoff and landing (VTOL) aircraft is considered. The linearized dynamics of the VTOL aircraft in the vertical plane was obtained in [44], and is given by

$$\mathbf{x} = \begin{bmatrix} -0.0366 & 0.0271 & 0.0188 & -0.4555 \\ 0.0482 & -1.01 & 0.0024 & -4.0208 \\ 0.1002 & 0.3681 & -0.707 & 1.420 \\ 0.0 & 0.0 & 1.0 & 0.0 \end{bmatrix} \mathbf{x} + \begin{bmatrix} 0.4422 & 0.1761 \\ 3.5446 & -7.5922 \\ -5.52 & 4.49 \\ 0.0 & 0.0 \end{bmatrix} \mathbf{u}$$

with

$$\mathbf{x} = \begin{pmatrix} horizontal\ velocity\ (knots) \\ vertical\ velocity\ (knots) \\ pitch\ rate\ (degrees/\sec) \\ pitch\ angle\ (degrees) \end{pmatrix} \quad and \quad \mathbf{u} = \begin{pmatrix} collective\ pitch\ control \\ longitudal\ cyclic\ pitch\ control \end{pmatrix}$$

where the collective control is used for the vertical motion, and the other control input is used to control the horizontal velocity of the aircraft. In this example the output was taken as

$$y = \begin{bmatrix} 1 & 0 & 0 & 0 \\ 0 & 1 & 0 & 0 \\ 0 & 0 & 1 & 0 \\ 0 & 1 & 1 & 1 \end{bmatrix} x$$

The above dynamics hold for a typical loading and flight condition of the VTOL at the air speed of 135 knots. As the airspeed changes the dynamical equation of the model will changes. The most significant of these changes will occur at the a_{32}, a_{34}, and b_{21} elements of the A and B matrices. In the rest of this example it is assumed that at different flight conditions the above parameters may change from their nominal values, and all other parameters remain constant.

The open loop poles of the nominal system are located at $\{-2.0727, 0.2758 \pm j0.2576, -0.2325\}$. Clearly, the system is unstable and needs to be stabilized. The closed loop poles of the system were placed at $\{-4, -2, -2.4, \text{ and } -3\}$.

To illustrate the approach a soft failure in sensor three under large variations (as given below) in the plant parameters was considered.

$$\Delta A = \begin{bmatrix} 0 & 0 & 0 & 0 \\ 0 & 0 & 0 & 0 \\ 0 & 0.5 & 0 & 2.0 \\ 0 & 0 & 0 & 0 \end{bmatrix} \qquad \Delta B = \begin{bmatrix} 0 & 0 \\ 2.0 & 0 \\ 0 & 0 \\ 0 & 0 \end{bmatrix}$$

The sensor failure is modeled as a soft bias fault (f=0.2) occurring at t≥2 secs.

For a unit step command inputs, Figure 12, shows the plot of the detection function (i.e. $\hat{f}$, shown with the dotted line) along with the actual bias failure occurring at t=2 seconds. It can be seen that monitoring $\hat{f}$ will provide us with the capability to detect the failure almost immediately, and in addition it will give us the actual shape of the particular fault. It should be mentioned that here it is assumed that no failure can take place during the first few seconds of the observer transient.

As further illustration of the approach, it was assumed that all four sensors are subject to failure while the aircraft is flying under nominal conditions. The actual time and shape of the failures are given by

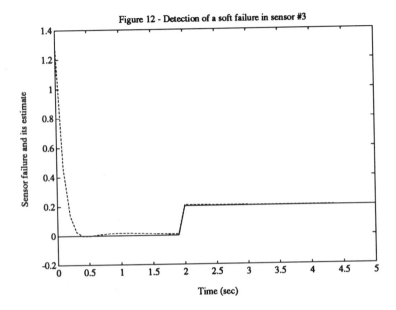

Figure 12 - Detection of a soft failure in sensor #3

$$\mathbf{f} = [0.2 \quad 0.4 \quad \sin(t-2) \quad \cos(t-2)]^T, \quad \text{for} \quad t \geq 2 \quad secs$$

Figures 13-16 show that the FDI scheme can detect and identify simultaneous failure of all the sensors.

It should be noted that unlike many other FDIA techniques, the above approach can actually identify the shape of the failure. This is important since this information can be used in accommodating for the failure of the sensors. Also, it should be pointed out that actuator failure detection may be possible using the UIO theory in a similar fashion [45].

V. CONCLUSIONS

In this article two approach for estimating the state of linear time invariant systems subject to unknown disturbances were presented. The UIO is more general in that the unknown disturbance can be time varying and the estimation of the state would still be possible if the conditions for existence of the UIO are satisfied.

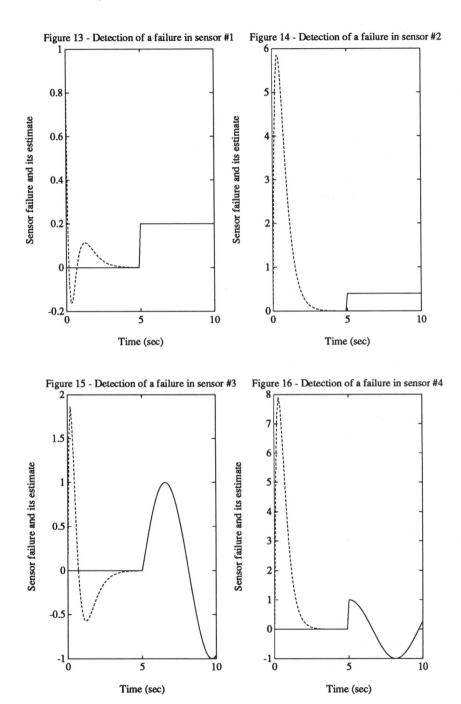

Figure 13 - Detection of a failure in sensor #1

Figure 14 - Detection of a failure in sensor #2

Figure 15 - Detection of a failure in sensor #3

Figure 16 - Detection of a failure in sensor #4

On the other hand, the PIO can only deal with constant or slowly stepwise varying disturbances, but it would provide the estimate of the disturbance as well as the state of the system if the conditions for its existence are satisfied. In addition, the PIO can perform these tasks even if the plant as well as the measurements are corrupted with the disturbance. It was also illustrated that the UIO can be used as a robust estimator for estimating the state of uncertain systems. This was done by rewriting the dynamics of an uncertain system as a nominal part, and another part consisting of the uncertainties. Then the uncertain part was considered as the unknown input to the system.

Various applications of the estimators was illustrated. The main point that should be made in this regard is that necessary conditions for existence of the estimators must be satisfied in all these cases, otherwise the approach will fail. However, in those cases where the estimators exist, it can be seen that the theory can be used effectively.

ACKNOWLEDGEMENT

The financial support from the Natural Sciences and Engineering Research Council (NSERC) of Canada, and the Center for Systems Science (CSS) at Simon Fraser University are gratefully acknowledged.

References

[1]   C.D. Johnson, *Further Study of the Linear Regulator with Disturbances-The Case of Vector Disturbances Satisfying a Linear Differential Equation*, IEEE Trans. Aut. Cont., Vol. AC-15, pp. 222-228, 1970.

[2]   C.D. Johnson, *Accommodation of External Disturbances in Linear Regulator and Servomechanism Problems*, IEEE Trans. Aut. Cont., Vol. AC-16, pp. 635-644, 1971.

[3]   B. Porter, and A. Bradshaw, *Design of Linear Multivariable Continuous-Time Tracking Systems Incorporating Error-Actuated Dynamic Controllers*, Int. J. Syst. Sci., Vol. 9, pp. 627-637, 1978.

[4]   A.K. Mahalanabis, and J.K. Pal, *Optimal Regulator Design of Linear Multivariable Systems with Prescribed Pole Locations in the Presence of System Disturbances*, IEE Proceedings-Part D, Vol. 132, pp. 231-236, 1985.

[5]   E.J. Davison, and H.W. Smith, *Pole Assignment in Linear Time-Invariant Multivartiable Systems with Constant Disturbances*, Automatica, Vol. 7, pp. 489-498, 1971.

[6]   H.W. Smith and E.J. Davison, *Design of Industrial Regulators: Integral Feedback and Feedforward Control*, Proceedings of IEE, Vol. 119, pp. 1210-1216, 1972.

[7]   E.J. Davison and A. Goldenberg, *Robust Control of a General Servomechanism Problem: The Servo Compensator*, Automatica, Vol. 11, pp. 461-471, 1975.

[8]   E.J. Davison and I.J. Ferguson, *The Design of Controllers for Multivariable Robust Servomechanism Problem Using Parameter Optimization Methods*, IEEE Trans. Aut. Cont., Vol. AC-26, pp. 93-110, 1981.

[9]   G. Basile, and G. Marro, *On the Observability of Linear, Time-Invariant Systems with Unknown Inputs*, J. Optim. Thy. & Appl. (JOTA), Vol. 3, pp. 411-415, 1969.

[10]   K. Cevik, and I.C. Goknar, *Observable, Unobservable Subspaces and Observability of an Unknown Input System*, in Frequency Domain and State Space Methods for Linear Dynamical Systems, C.I. Byrnes, and A. Lindquist (Editors), pp. 451-459, Elsevier Science Publishers, 1986.

[11] C.D. Johnson, *On Observers for Linear Systems with Unknown and Inaccessible Inputs*, Int. J. Cont., Vol. 21, pp. 825-831, 1975.

[12] J.S. Meditch, and G.H. Hostetter, *Observers for Systems with Unknown and Inaccessible Inputs*, Int. J. Cont., Vol. 19, pp. 473-480, 1974.

[13] J. O'Reilly, *Minimal-Order Observers for Linear Multivariable Systems with Unmeasurable Disturbances*, Int. J. Cont., Vol. 28, pp. 743-751, 1978.

[14] S.H. Wang, E.J. Davison, and P. Dorato, *Observing the States of Systems with Unmeasurable Disturbances*, IEEE Trans. on Aut. Cont., Vol. AC-20, pp. 716-717, 1975.

[15] K.K. Sundareswaran, P.J. McLane, and M.M. Bayoumi, *Observers for Linear Systems with Arbitrary Plant Disturbances*, IEEE Trans. Aut. Cont., Vol. AC-22, pp. 870-871, 1977.

[16] N. Kobayashi, and T. Nakamizo, *An Observer Design for Linear Systems with Unknown Inputs*, Int. J. Cont., Vol. 35, pp. 605-619, 1982.

[17] P. Kudva, N. Viswanadham, and A. Ramakrishna, *Observers for Linear Systems with Unknown Inputs*, IEEE Trans. on Aut. Cont., Vol. AC-25, pp.113-115, 1980.

[18] R.J. Miller, and R. Mukundan, *On Designing Reduced Order Observers for Linear Time Invariant Systems Subject to Unknown Inputs*, Int. J. Cont., Vol. 35, pp. 183-188, 1982.

[19] J.E. Kurek, *The State Vector Reconstruction for Linear Systems with Unknown Inputs*, IEEE Trans. on Aut. Cont., Vol. AC-28, pp. 1120-1122, 1983.

[20] F. Yang, and R.W. Wilde, *Observers for Linear Systems with Unknown Inputs*, IEEE Trans. on Aut. Cont., Vol. 33, pp. 677-681, 1988.

[21] Y. Park, and J.L. Stein, *Closed Loop State and Input Observer for Systems with Unknown Inputs*, Int. J. Cont., Vol. 48, pp. 1121-1136, 1988.

[22] Y. Guan, and M. Saif, *A Novel Approach to the Design of Unknown Input Observers*, IEEE Trans. Aut. Cont., Vol. 36, pp. 632-635, 1991.

[23] N. Viswanadham, and R. Srichander, *Fault Detection Using Unknown Input Observers*, Control-Theory and Advanced Technology (C-TAT), Vol. 3, pp. 91-101, 1987.

[24] Chen, C.T., **Linear System Theory and Design,** HRW Publishing, NY, 1984.

[25] Shafai, B., and Carroll, R.L., *Design of Proportional-Integral Observer for Linear Time-Varying Multivariable Systems*, **Proceedings of the 24th IEEE CDC**, pp. 597-599, December 1985.

[26] Beale, S.R., and Shafai, B., *Robust Control Systems Design with a Proportional Integral Observer*, **Proceedings of the 27th IEEE CDC**, pp. 554-557, December 1988, also **Int. J. Control**, Vol. 50, pp. 97-111, 1989.

[27] M. Saif, *Optimal Linear Regulator Pole Placement by Weight Selection*, **Int. J. of Control**, Vol. 50, pp. 399-414, 1989.

[28] M. Saif, *Optimal Modal Controller Design by Entire Eigenstructure Assignment*, **IEE Proceedings-Part D.**, Vol. 136, pp. 341-344, 1989.

[29] B. Courtion, and I.D. Landau, *High Speed Adaptation System for Controlled Electrical Drives*, **Automatica**, Vol. 11, pp. 119-127, 1975.

[30] T. Egami, H. Morita, and T. Tsuchiya, *Efficiency Optimized Model Reference Adaptive Control System for a dc Motor*, **IEEE Trans. on Ind. Elec.**, Vol. 37, pp. 28-33, 1990.

[31] P.M. Pelczewski, and U.H. Kunz, *The Optimal Control of a Constrained Drive System with Brushless dc Motor*, **IEEE Trans. on Ind. Elec.**, Vol. 37, pp. 342-348, 1990.

[32] Siljak, D.D., and Vukcevic, M.B., *Decentralization, Stabilization, and Estimation of Large-Scale Linear Systems*, **IEEE Transactions on Automatic Control**, Vol. AC-21, 363-366, 1976.

[33] Siljak, D.D., and Vukcevic, *On Decentralized Estimation*, **Int. J. Cont.**, Vol. 27, 113-131, 1978.

[34] Siljak, D.D., **Large-Scale Dynamic Systems: Stability and Structure,** Amsterdam: North Holland, 1978.

[35] Sundareshan, M.K., *Decentralized Observation in Large Scale Systems*, **IEEE Trans. on Syst. Man and Cybernetics**, Vol. SMC-7, 863-867, 1977.

[36] Sundareshan, M.K., and Huang, P.C.K, *On the Design of Decentralized Observation Schemes for Large Scale Systems*, **IEEE Trans. on Aut. Cont.**, Vol. AC-29, 274-276, 1984.

[37]   Sundareshan, M.K., and Elbanna, R.M., *Design of Decentralized Obser-vation Schemes for Large Scale Syustems: Some New Results*, Automatica,Vol. 26, 789-796, 1990.

[38]   Viswanadham, N., and Ramakrishna, A., *Decentralized Estimation and Control for Interconnected Systems*, Large Scale Systems, *Vol.3*, 255-266, 1982.

[39]   Frank, P.M., *Fault Diagnosis in Dynamic Systems Via State Estimation-A Survey*, System Fault Diagnosis, Reliability and Related Knowledge Based Approaches, S. Tzafestas et.al. (Eds), Vol. I, pp. 35-98, D. Reidel Publishing Co., 1987.

[40]   Frank, P.M., *Fault Diagnosis in Dynamic Systems Using Analytical and Knowledge-Based Redundancy-A Survey and Some New Results*, Automatica, Vol. 26, pp. 459-474, 1990.

[41]   Isermann, R., *Process Fault Detection Based on Modeling and Estimation Methods-A Survey*, Automatica, Vol. 20, pp. 387-404, 1984.

[42]   Willsky, A.S., *A Survey of Design Methods for Failure Detection*, Automatica, Vol. 12, pp. 601-611, 1976.

[43]   Merrill, W.C., *Sensor Failure Detection for Jet Engines Using Analytical Redundancy*, AIAA Journal of Guidance, Control and Dynamics, Vol. 8, pp. 673-682, 1985, also in Control and Dynamic Systems-Advances in Theory and Applications, *Vol. 33: Advances in Aerospace Systems Dynamics and Control Systems*, C.T. Leondes (Editor), pp. 1-34, Academic Press, 1990.

[44]   Narendra, K., and Tripathi, S.S., *Identification of Optimization of Aircraft Dynamics*, AIAA J. of Aircraft, Vol. 10, pp. 193-199, 1973.

[45]   Saif, M., and Guan, Y., *Robust Fault Detection in Systems with Uncertainties*, Proceedings of the 5th IEEE Int. Symp. on Intelligent Cont., pp. 570-575, September 1990.

Sliding Control Design in Robust Nonlinear Control Systems

Liang-Wey Chang

Department of Mechanical Engineering
Naval Postgraduate School
Monterey, CA 93943

I. Introduction

Control design of modern complex systems is challenging due to the system nonlinear effects and the demands of high performance. In addition to the fundamental requirements concerning stability and accuracy, the modern nonlinear systems tolerate hostile environment and face many unknown factors. Thus, there exists evitable system uncertainty which includes modeling errors, unmodeled dynamics, and environmental disturbances. The robustness to the uncertainty becomes an essential issue in designing a control system.

Variable-Structure Control (VSC) has been known for its robustness to system uncertainty and its simplicity of physical implementation (Itkis, 1976; DeCarlo et. al., 1988). The structure of control systems is intentionally altered as its state crosses certain surfaces in the phase space in accordance with a prescribed control law. The VSC is a time-domain control law, and is inherently effective for nonlinear time-varying plants. Examples or variations of the VSC include the uncertain system control (Corless and Leitmann, 1981; Barmish and Leitmann, 1982; Chen, 1986) and the sliding (mode) control

(Utkin, 1977). In this article, the sliding control techniques will be introduced.

Consider a scalar nth-order nonlinear time-varying dynamical system

$$x^{(n)} = bu + f + d \tag{I.1}$$

where $x^{(n)}$ is the nth-order time-derivative of an output variable of interest x. u is a control input and d is a disturbance. b and f are the functions of output variables and time. A nominal mathematical model of the system can be obtained and written as

$$x^{(n)} = \hat{b}u + \hat{f} \tag{I.2}$$

$\hat{b}$ and $\hat{f}$ are the nominal values of b and f, which are obtained through experiment or theoretical evaluations. The confinement of the uncertainty is required for the VSC, and the form of the confinement could have many forms. Consider a type of confinement (Slotine, 1984) as

$$\left| f - \hat{f} \right| < F \text{ and } \beta^{-1} \le \frac{\hat{b}}{b} \le \beta \tag{I.3}$$

where $F = (f_{max} - f_{min})/2$ and $\beta = (b_{max}/b_{min})^{1/2}$. It implies that the maximum and minimum values of system parameters and their nominal values are related as $\hat{f} = (f_{max} + f_{min})/2$ and $\hat{b} = (b_{max}b_{min})^{1/2}$. In addition, it is assumed that the disturbance d is also bounded, i.e., $|d| < D$.

The sliding control provides a systematic way of designing the VSC, which drives the system state onto a user-designed surface (the sliding surface) and maintains the state on the surface until a desired state is reached. The Lyapunov stability criterion is used to derive a sliding condition such that the sliding surface can be reached within a finite time. A Lyapunov function V is picked as $V = \frac{1}{2}s \cdot s$. By applying the Lyapunov stability criterion, $\dot{V} \le 0$, a sliding condition is obtained as

$$s \cdot \dot{s} \le 0 \tag{I.4}$$

which represents a stability criterion for a damper-spring system. Equation (I.4) can be rewritten in a general form as

$$s \cdot \dot{s} \leq -\eta \, |s| \qquad (I.5)$$

where η is a positive real number to adjust the approaching speed. Note that Equation (I.5) is a first-order sliding condition originally used in sliding control.

When the state is on the sliding surface, the system is in the sliding mode, and the closed-loop control system is equivalent to a system of lower order. Sliding surfaces determine the stability of the sliding mode and they could be expressed in terms of state variables (Young et. al., 1977; Dorling and Zinober, 1986) or tracking errors (Slotine and Sastry, 1983; Fernandez and Hedrick, 1987). The latter expression results in a simpler design of sliding surfaces than the former. A stable sliding mode can be readily obtained using the tracking-error expression as long as a stable linear differential operator on the tracking errors are selected. A sliding surface $s = 0$ expressed in tracking errors is now considered, where

$$s = \begin{cases} \prod\limits_{i=1}^{n-1} \left(\dfrac{d}{dt} + p_i \right) e & \text{for } n > 1 \\ e & \text{for } n = 1 \end{cases} \qquad (I.6)$$

e is the tracking error that is the difference between the output measurement and the desired output, i.e, $e = x - x_d$, and the p's are assigned to locate the poles of the error dynamics. The polynomial form of Equation (I.6) is often convenient for designing a robust control, i.e.,

$$s = e^{(n-1)} + c_{n-1} e^{(n-2)} + c_{n-2} e^{(n-3)} + \ldots + c_1 e \qquad (I.7)$$

As a result, an equivalent control is needed, which makes the error dynamics asymptotically stable toward zero in the sliding mode. An additional switched control is required to maintain the state on the sliding surface. The sliding control law was defined as (Slotine and Sastry, 1983)

$$u = \hat{b}^{-1} \left(\hat{u} - k \, sign(s) \right) \qquad (I.8)$$

where

$$\hat{u} = -\hat{f} + x_d^{(n)} - \varepsilon$$

$$\varepsilon = c_{n-1}e^{(n-1)} + c_{n-2}e^{(n-2)} + \ldots + c_2\ddot{e} + c_1\dot{e}$$

$$k = \beta(F + D + \eta) + (\beta - 1)|\hat{u}|$$

and $sign(s) = \begin{cases} 1 & \text{when } s \geq 0 \\ -1 & \text{when } s < 0 \end{cases}$

Note that the equivalent control $\hat{b}^{-1}\hat{u}$ was designed according to the desired motion of the sliding mode. The gain k associated with the switched control $-\hat{b}^{-1}k\,sign(s)$ was determined by the Lyapunov stability criterion. The control usually results in a fast motion to bring the state onto the sliding surface, where a slower motion proceeds until a desired state is reached.

Differentiating Equation (I.7) and combining with Equations (I.1) and (I.8), the s dynamics is obtained as

$$\dot{s} = (f - \hat{f}) + d - (1 - b\hat{b}^{-1})\hat{u} - b\hat{b}^{-1}k\,sign(s) \qquad (I.9)$$

The s dynamics, a buffer element, is the heart of the control system, which initially filters the uncertainty and provides system robustness. A resulting high-frequency output from the switched s dynamics is further filtered by a low-pass filter of error dynamics and a perfect tracking is obtained. Figure 1 describes detailed synthesis in a block diagram.

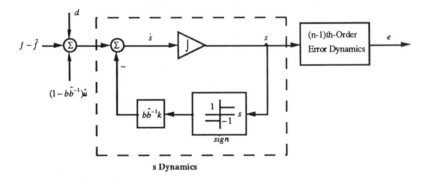

s Dynamics

Figure 1 Control System Synthesis of the Switched Sliding Control

Figures 2 shows a schematic diagram of the sliding control system for implementation purposes. The implementation diagram includes a block of an inverse error dynamics that utilizes the system error e and its derivatives to compute the s value. The diagram also includes a Lyapunov control utilizing the Lyapunov stability criterion to obtain desired s dynamics such that $s = 0$, a sliding surface, may be reached.

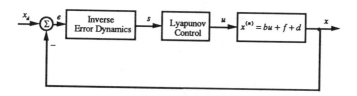

Figure 2 A Schematic Diagram of a Sliding Control System

However, the high-frequency switching results in chattering phenomenon of the control input that is not desirable for mechanical devices because of noise generation and wear. Continuous Control for nonlinear time-varying dynamical systems (Slotine and Sastry, 1983) was developed to eliminate the high-frequency switching at the expense of the perfect tracking. A boundary layer of s with thickness ϕ (a positive real number) was introduced to smooth the control. Outside the boundary layer, $|s| \geq \phi$, the sliding condition (Equation (I.5)) is kept. Inside the boundary layer, $|s| < \phi$, the switched control law is modified to impose smoothing to the s dynamics. The control law was given as

$$u = \hat{b}^{-1}\left(\hat{u} - k\, sat(\frac{s}{\phi})\right)$$

(I.10)

where the saturation function sat was defined as

$$sat(\frac{s}{\phi}) = \begin{cases} sign(\frac{s}{\phi}) & \text{when } |s| \geq \phi \\ \dfrac{s}{\phi} & \text{when } |s| < \phi \end{cases}$$

The s dynamics is thus smoothed within the boundary layer, i.e.,

$$\dot{s} + \frac{b\hat{b}^{-1}k}{\phi}s = (f - \hat{f}) + d - (1 - b\hat{b}^{-1})\hat{u} \qquad (I.11)$$

The bandwidth is determined by the gain k and ϕ. Since the gain had been selected for stability, the only design parameter for the bandwidth is the thickness of the boundary layer. The unmodeled dynamics are filtered when the bandwidth is lower than the unmodeled frequencies. Yet, there is no more parameters which could control the relative stability of the s dynamics and affect the speed and accuracy of the system. It is concluded that the high-frequency switched s dynamics is transformed into a first-order low-pass filter inside the boundary layer. However, the control scheme results in a trade-off between the robustness and the tracking accuracy.

With additional integral control, continuous control furthers the improvement and achieves a robust control without causing chattering and steady-state error. As the integral control action destabilizes the control system, the system response tends to be oscillatory. The continuous sliding control has too limited capability in shaping the system dynamics to meet the no-overshoot (no-oscillation) requirement, and the system response in turn is sluggish (longer rise time and settling time). The trade-off between the robustness and the tracking accuracy was further refined (Slotine, 1984), and the control scheme has been applied to various dynamical systems (Slotine, 1985; Yoerger and Slotine, 1985).

This article will present another sliding control using second-order sliding condition, namely, versatile sliding control, which provides better tuning capability and forms a robust-control system with better performance and more operational flexibility (Chang, 1990; Chang, 1991a). The technique will first be described in SISO (single-input-single-output) systems in section II and Section III will then present the extension of the technique to MIMO (multiple-input-multiple output) systems. The application of the control to disk files can be found in the reference (Chang, 1991b).

II. SISO Versatile Sliding Control

The idea of using a second-order sliding condition is to raise the order of the s dynamics and increase the activity of s dynamics, which results in a better control design in robustness and the accuracy. The control applies a high-pass (or band-pass) filter to the error dynamics and a low-pass filter to s dynamics.

A. SECOND-ORDER SLIDING CONDITION

The second-order sliding condition is derived from a Lyapunov function, that is

$$V = \frac{1}{2}\dot{s} \cdot \dot{s} + \frac{1}{2}s \cdot \omega_n^{2} \cdot s \qquad \text{(II.A1)}$$

To have a stable s dynamics, $\dot{V} \leq 0$, i.e.,

$$\dot{s}\left(\ddot{s} + \omega_n^{2}s\right) \leq 0 \qquad \text{(II.A2)}$$

Equation (II.A2) represents a stability criterion for a mass-spring-damper system, where the mass-spring system has positive damping. The spring constant ω_n^{2} will later be adjusted which allows more flexibility for tuning. The spring constant reflects the stiffness of the s dynamics. The sliding condition assures the attractiveness of the s dynamics toward the sliding surfaces. Equation (II.A2) can be rewritten to be

$$\dot{s}\left(\ddot{s} + \omega_n^{2}s\right) \leq -\eta|\dot{s}| \qquad \text{(II.A3)}$$

The second-order sliding condition gives the s dynamics more room to be modulated and also allows to design effective error dynamics for the system. It will be shown later that the s motion will not stay at a nonzero position when $\dot{V} = 0$.

B. SLIDING SURFACE AND ERROR DYNAMICS

By adding a zero, a stable error dynamics is defined to be

$$\Sigma = \left(\frac{d}{dt} + z_0\right)s = \prod_{i=1}^{n}\left(\frac{d}{dt} + p_i\right)\int_0^t e\,dt \tag{II.B1}$$

and the sliding surface is defined as $\Sigma = 0$. The integral control in the equation provides an assurance of zero steady-state errors. Equation (II.B1) is a high-pass filter (or band-pass filter) where the break frequency is determined by the selections of the poles and zeros. The polynomial version of the above equation is

$$\dot{s} + z_0 s = e^{(n-1)} + c_{n-1}e^{(n-2)} + c_{n-2}e^{(n-3)} + \ldots + c_0\int_0^t e\,dt \tag{II.B2}$$

The above formula could have been more sophisticated if more integral terms of s and e had been included. For example, the error dynamics could have been defined as

$$\dot{s} + z_0 s + z_1\int_0^t s\,dt + \ldots + z_p\int_0^t\ldots\int_0^t s(dt)^p$$
$$= e^{(n-1)} + c_{n-1}e^{(n-2)} + c_{n-2}e^{(n-3)} + \ldots + c_0\int_0^t e\,dt + \ldots + c_{1-m}\int_0^t\ldots\int_0^t e(dt)^m \tag{II.B3}$$

where p and m are integers. For simplicity, only Equation (II.B2) will be applied in this article. Equation (II.B2) indicates that s and its derivative are inputs to the error dynamics (a high-pass filter), and the error e becomes a filtered version of s. It will be shown that s is influenced by the control input u and therefore the s dynamics can be modulated by the control input. By differentiating Equation (II.B2),

$$\ddot{s} + z_0\dot{s} = e^{(n)} + c_{n-1}e^{(n-1)} + c_{n-2}e^{(n-2)} + \ldots + c_1\dot{e} + c_0 e \tag{II.B4}$$

The control input u can influence the s dynamics through the following relation:

$$\ddot{s} + z_0\dot{s} = bu + f + d - x_d^{(n)} + \varepsilon \tag{II.B5}$$

where $\varepsilon = c_{n-1}e^{(n-1)} + c_{n-2}e^{(n-2)} + \ldots + c_1\dot{e} + c_0 e$

The ε can be computed with feedback signal x and its derivatives measured. The role of the control u is to control the s dynamics such that the sliding surface can be reached in a finite time. Once the s dynamics reaches the sliding surface, zero steady-state tracking error can be achieved on the sliding surface with stable error dynamics.

C. CONTROL ALGORITHM AND SYSTEM STABILITY

The continuous control law is designed as

$$u = \hat{b}^{-1}\left(\hat{u} - k\ sat(\frac{\dot{s}}{\phi})\right) \qquad \text{(II. C1)}$$

where

$$\hat{u} = -\hat{f} + x_d^{(n)} - \varepsilon + z_0\dot{s} - \omega_n^2\int_0^t \dot{s}\ dt$$

and

$$\dot{s} = e^{(n-1)} + c_{n-1}e^{(n-2)} + c_{n-2}e^{(n-3)} + ... + c_0\int_0^t e\,dt - z_0\int_0^t \dot{s}\ dt$$

where ϕ is the thickness of the boundary layer of the approaching speed $\dot{s}$.

To ensure the system stability, the sliding condition must be satisfied outside the boundary layer. By substituting Equation (II.B5) into the sliding condition, an inequality is obtained as

$$\dot{s}\cdot\left(bu + f + d - x_d^{(n)} + \varepsilon - z_0\dot{s} + \omega_n^2 s\right) \leq -\eta\,|\dot{s}| \qquad \text{(II.C2)}$$

It implies that

$$s\cdot\left((f - \hat{f}) + d - (1 - b\hat{b}^{-1})\hat{u} - b\hat{b}^{-1}k\ sign(s)\right) \leq -\eta\,|s| \qquad \text{(II.C3)}$$

k is found to be

$$k \geq \beta(F + D + \eta) + (\beta - 1)|\hat{u}| \qquad \text{(II.C4)}$$

As long as the gain k is chosen according to the above equation, the Lyapunov stability is guaranteed outside the boundary layer. For control design purposes, the minimum value of k is selected since the least control effort is desired. From Equation (II.C4), it is found that the gain k is a measure of the uncertainties. For systems without uncertainties, the computed value of the gain k is η.

From the control law and Equation (II.B5), the s dynamics outside the boundary layer can be expressed as

$$\ddot{s} + b\hat{b}^{-1}k\,sign(\dot{s}) + \omega_n^2 s = (f - \hat{f}) + d - (1 - b\hat{b}^{-1})\hat{u} \qquad \text{(II.C5)}$$

This proves that the s motion moves toward zero position and never stops at a nonzero position. Even though the approaching speed $\dot{s} = 0$ when $\dot{V} = 0$, the approaching acceleration $\ddot{s}$ is never zero. The right-hand-side terms in Equation (II.C5) represents excitations to the s dynamics. The excitations consist of modeling errors, disturbances, desired trajectory dynamics, and error feedback. The s dynamics can also be viewed as a filter of the excitations; in other words, the value of s becomes a filtered version of the excitations.

Similarly, one will find that the smooth s dynamics within the boundary layer is governed by

$$\ddot{s} + \left[(1 - b\hat{b}^{-1})z_0 + b\hat{b}^{-1}\frac{k}{\phi}\right]\dot{s} + b\hat{b}^{-1}\omega_n^2 s = (f - \hat{f}) + d - (1 - b\hat{b}^{-1})(-\hat{f} + x_d^{(n)} - \varepsilon)$$

$$\text{(II.C6)}$$

The second-order s dynamics is more active than its first-order counterpart. The system performance can be improve since it critically depends on the dynamic behavior within the boundary layer, which are determined by the parameters ϕ, z_0, and ω_n^2. For instance, a wider ϕ gives lesser damping, and the s dynamics is more vibrant and more easily excited. In other words, the s dynamics is more sensitive to its excitation with a greater value of ϕ. To be able to control the s dynamics effectively, the tuning mechanisms of the parameters will be developed.

D. RELATIVE STABILITY AND TUNING MECHANISMS

The relative stability of the s dynamics and its parametric description provide the indicators to the development of tuning mechanisms. By letting the bandwidth of the s dynamics be ω_s and the damping ratio of that be ς_s, a set of inequalities are written as

$$\omega_l^2 \equiv \frac{1}{\beta}\omega_n^2 \leq b\hat{b}^{-1}\omega_n^2 \equiv \omega_s^2 \leq \beta\omega_n^2 \equiv \omega_u^2 \qquad \text{(II.D1)}$$

and

$$(1-\beta)z_0 + \frac{1}{\beta}\frac{k}{\phi} \equiv 2\varsigma_l\omega_l$$

$$\leq (1-b\hat{b}^{-1})z_0 + b\hat{b}^{-1}\frac{k}{\phi} \equiv 2\varsigma_s\omega_s \qquad \text{(II.D2)}$$

$$\leq (1-\frac{1}{\beta})z_0 + \beta\frac{k}{\phi} \equiv 2\varsigma_u\omega_u$$

Given the spring constant ω_n^2 and the boundary thickness ϕ, the lower bounds and the upper bounds of ω_s and ς_s can be obtained as follows,

$$\omega_l \leq \omega_s \leq \omega_u \qquad \text{(II.D3)}$$

and

$$\varsigma_l \leq \varsigma_s \leq \varsigma_u \qquad \text{(II.D4)}$$

where

$$\omega_l = \frac{1}{\sqrt{\beta}}\omega_n \qquad \text{(II.D5)}$$

$$\omega_u = \sqrt{\beta}\,\omega_n \qquad \text{(II.D6)}$$

$$\varsigma_l = \frac{(1-\beta)z_0 + \frac{1}{\beta}\frac{k}{\phi}}{2\omega_l} \qquad \text{(II.D7)}$$

$$\text{and} \quad \varsigma_u = \frac{(1-\frac{1}{\beta})z_0 + \beta\frac{k}{\phi}}{2\omega_u} \qquad \text{(II.D8)}$$

A time-varying ϕ can be derived from Equations (II.D7) (or Equation (II.D8)) to control the s dynamics and to improve the tracking performance. A non-oscillatory s response is desirable and $\varsigma_l > 1$ guarantees under-damped

s dynamics. By assigning ς_l, the boundary layer thickness can be computed as

$$\phi = \frac{\dfrac{k}{\beta}}{2\varsigma_l\omega_l - (1-\beta)z_0} \tag{II.D9}$$

and, with the computed ϕ, the corresponding upper bound of the damping ratio can be obtained from Equation (II.D8).

The range of the bandwidth for the s dynamics is given by Equations (II.D3, II.D5-6), where the spring constant ω_n^2 is the only control parameter for the bandwidth. To provide more tuning capability, two other control parameters, both the lower bound of the damping ratio and the placed zero, would affect ϕ, that would further shape the s dynamics. With this great versatility, the trade-off between tracking accuracy and robustness to the system disturbances no longer exists.

III. MIMO Versatile Sliding Control

The MIMO plant description is similar to the SISO's except it is in a vector form. Consider a nth-order time-varying nonlinear dynamical system

$$\mathbf{x}^{(n)} = \mathbf{Bu} + \mathbf{f} + \mathbf{d} \tag{III.1}$$

where

$$\mathbf{x}^{(n)} = [x_1^{(n)} \quad x_2^{(n)} \quad \cdots \quad x_i^{(n)} \quad \cdots \quad x_m^{(n)}]^T$$
$$\mathbf{u} = [u_1 \quad u_2 \quad \cdots \quad u_i \quad \cdots \quad u_m]^T$$
$$\mathbf{f} = [f_1 \quad f_2 \quad \cdots \quad f_i \quad \cdots \quad f_m]^T$$
$$\mathbf{d} = [d_1 \quad d_2 \quad \cdots \quad d_i \quad \cdots \quad d_m]^T$$

The system has m inputs and m outputs. An associated nominal model is given as

$$\mathbf{x}^{(n)} = \hat{\mathbf{B}}\mathbf{u} + \hat{\mathbf{f}} \tag{III.2}$$

The discrepancies between the model and the physical system are specified by $\Delta\mathbf{B}$, $\Delta\mathbf{f}$, and $\mathbf{d}$, where

$$\Delta\mathbf{B} = \mathbf{B} - \hat{\mathbf{B}} \quad \text{and} \quad \Delta\mathbf{f} = \mathbf{f} - \hat{\mathbf{f}} \tag{III.3}$$

A general confinement is defined in matrix norms as

$$\|\Delta \mathbf{B}\| \le \beta \quad \text{and} \quad \|\Delta \mathbf{f}\| \le \alpha \tag{III.4}$$

where $\|\bullet\|$ denotes a norm of $\bullet$ which is a vector or a matrix. Note that the norm of $\mathbf{A}$, for example, is defined as

$$\|\mathbf{A}\| = \sqrt{\left\{ eig(\mathbf{A}^T \mathbf{A}) \right\}_{\max}} \tag{III.5}$$

where $\left\{ eig(\bullet) \right\}_{\max}$ stands for the maximum eigenvalue (i.e., the spectral radius) of $\bullet$. In the matrix-theory community, this matrix norm is called the spectral norm and the corresponding vector norm is called the Euclidean vector norm (Horn and Johnson, 1988). The confining parameters β and α can be found as

$$\beta = \left\| \frac{\mathbf{B}_{\max} - \mathbf{B}_{\min}}{2} \right\| \quad \text{and} \quad \alpha = \left\| \frac{\mathbf{f}_{\max} - \mathbf{f}_{\min}}{2} \right\| \tag{III.6}$$

where the components of $\bullet_{\max}$ (or $\bullet_{\min}$) are the maximum (or minimum) values of the corresponding components of $\bullet$. It implies that the system parameters and their nominal values are related as $\hat{\mathbf{B}} = \dfrac{\mathbf{B}_{\max} + \mathbf{B}_{\min}}{2}$ and $\hat{\mathbf{f}} = \dfrac{\mathbf{f}_{\max} + \mathbf{f}_{\min}}{2}$. In addition, the disturbance $\mathbf{d}$ is also bounded, i.e.,

$$\|\mathbf{d}\| \le \gamma \tag{III.7}$$

Note that α, β, and γ are non-negative real numbers.

A. SLIDING SURFACE AND SLIDING CONDITION

The MIMO sliding control begins with placing an additional zero $\mathbf{z}_0$ (an $m \times 1$ vector) in the error dynamics. Let $\mathbf{e}$ be the tracking error, i.e, $\mathbf{e} = \mathbf{x} - \mathbf{x}_d$, and $\mathbf{S} = 0$ be a set of sliding surfaces ($\mathbf{S}$ is an $m \times 1$ vector). A stable error dynamics is defined to relate to $\mathbf{s}$ (an $m \times 1$ vector) as

$$\mathbf{S} = \left(\frac{\mathbf{d}}{\mathbf{dt}} + \mathbf{z}_0 \right) * \mathbf{s} = \prod_{i=1}^{n} \left(\frac{\mathbf{d}}{\mathbf{dt}} + \mathbf{p}_i \right) * \int_0^t \mathbf{e} \, dt \tag{III.A1}$$

where the element-by-element vector multiplications are denoted by $*$ or Π. Its polynomial form is written as

$$\dot{s} + z_0 * s = e^{(n-1)} + c_{n-1} * e^{(n-2)} + c_{n-2} * e^{(n-3)} + \ldots + c_0 * \int_0^t e \, dt$$

(III.A2)

The $m \times 1$ coefficient vectors c can be obtained from the pole placement of the error dynamics. Therefore, the relation between the control input and the s dynamics is given as

$$\ddot{s} + z_0 * \dot{s} = Bu + f + d - x_d^{(n)} + e_p \qquad \text{(III.A3)}$$

where $\qquad e_p = c_{n-1} * e^{(n-1)} + c_{n-2} * e^{(n-2)} + \ldots + c_1 * \dot{e} + c_0 * e$

The sliding condition will be derived from a Lyapunov function V, which is defined as

$$V = \frac{1}{2} \dot{s}^T \dot{s} + \frac{1}{2} s^T \left(\omega_n^2 * s \right)$$

where ω_n^2 is a vector with positive elements, i.e.,

$$\omega_n^2 = \begin{bmatrix} \omega_1^2 & \omega_2^2 & \cdots & \omega_i^2 & \cdots & \omega_m^2 \end{bmatrix}^T$$

By applying the Lyapunov stability criterion, $\dot{V} \leq 0$, a sliding condition is written as

$$\dot{s}^T \left(\ddot{s} + \omega_n^2 * s \right) \leq 0 \qquad \text{(III.A4)}$$

The above equation represents a stability criterion for a set of uncoupled mass-spring-damper systems. Note that ω_n^2 is a spring constant vector that adds the versatility to the control and the vector reflects the stiffness of the s dynamics. To be able to adjust the approaching speed $\dot{s}$, Equation (III.A4) is rewritten as

$$\dot{s}^T \left(\ddot{s} + \omega_n^2 * s \right) \leq -\eta \|\dot{s}\| \qquad \text{(III.A5)}$$

where η is a non-negative real number.

B. CONTROL LAW AND SYSTEM STABILITY

The control law is design using the predictor and corrector concept as

$$u = \hat{B}^{-1}\left(\hat{u} - k\,\text{sat}(\frac{\dot{s}}{\phi})\right) \qquad\qquad \text{(III.B1)}$$

where

$$\hat{u} = -\hat{f} + x_d^{(n)} - e_p + z_0 * \dot{s} - \omega_n^2 * \int_0^t \dot{s}\,dt$$

and

$$\dot{s} = e^{(n-1)} + c_{n-1} * e^{(n-2)} + c_{n-2} * e^{(n-3)} + \ldots + c_0 * \int_0^t e\,dt - z_0 * \int_0^t \dot{s}\,dt$$

The sat function is defined as

$$\text{sat}(\frac{\dot{s}}{\phi}) = [sat(\frac{\dot{s}_1}{\phi}) \quad sat(\frac{\dot{s}_2}{\phi}) \quad \cdots \quad sat(\frac{\dot{s}_i}{\phi}) \quad \cdots \quad sat(\frac{\dot{s}_m}{\phi})]^T$$

and $\qquad sat(\frac{\dot{s}_i}{\phi}) = \begin{cases} sign(\frac{\dot{s}_i}{\phi}) & \text{when } |\dot{s}_i| \geq \phi \\[2mm] \dfrac{\dot{s}_i}{\phi} & \text{when } |\dot{s}_i| < \phi \end{cases} \qquad \text{for } i = 1,2,\ldots,m$

$\hat{B}^{-1}\hat{u}$ is a predictor and $-\hat{B}^{-1}k\,\text{sat}(\frac{\dot{s}}{\phi})$ is a corrector. Outside the boundary layer, the Lyapunov stability is guaranteed as long as the gain k is selected such that the sliding condition is satisfied. To obtain the value of the gain k, the control input outside the boundary layer is used, i.e.,

$$u = \hat{B}^{-1}[\hat{u} - k\,\text{sign}(\dot{s})] \qquad\qquad \text{(III.B2)}$$

The gain k is a non-negative real number and is determined by using matrix-norm techniques. By substituting the Equation (III.A3) into the sliding condition of Equation (III.A5), an inequality is written as

$$\dot{\mathbf{s}}^T\left(\mathbf{Bu} + \mathbf{f} + \mathbf{d} - \mathbf{x_d}^{(n)} + \mathbf{e_p} - \mathbf{z_0} * \dot{\mathbf{s}} + \boldsymbol{\omega_n}^2 * \mathbf{s}\right) \le -\eta \|\dot{\mathbf{s}}\| \tag{III.B3}$$

The form of the control input (Equation(III.B2)) is used to relate the gain k with the system uncertainties. Therefore,

$$\dot{\mathbf{s}}^T\left\{\mathbf{B}\hat{\mathbf{B}}^{-1}[\hat{\mathbf{u}} - k\,\text{sign}(\dot{\mathbf{s}})] + \mathbf{f} + \mathbf{d} - \mathbf{x_d}^{(n)} + \mathbf{e_p} - \mathbf{z_0} * \dot{\mathbf{s}} + \boldsymbol{\omega_n}^2 * \mathbf{s}\right\} \le -\eta \|\dot{\mathbf{s}}\| \tag{III.B4}$$

Since $\hat{\mathbf{u}} = -\hat{\mathbf{f}} + \mathbf{x_d}^{(n)} - \mathbf{e_p} + \mathbf{z_0} * \dot{\mathbf{s}} - \boldsymbol{\omega_n}^2 * \int_0^t \dot{\mathbf{s}} dt$, the above equation is rearranged and rewritten as

$$\dot{\mathbf{s}}^T\left\{\left(\mathbf{f} - \hat{\mathbf{f}}\right) + \mathbf{d} - \left(\mathbf{I} - \mathbf{B}\hat{\mathbf{B}}^{-1}\right)\hat{\mathbf{u}} - k\mathbf{B}\hat{\mathbf{B}}^{-1}\,\text{sign}(\dot{\mathbf{s}})\right\} \le -\eta \|\dot{\mathbf{s}}\| \tag{III.B5}$$

The uncertainties described in Equation (III.3) are applied to the above equation and the sliding condition is given as

$$\dot{\mathbf{s}}^T\left\{\Delta\mathbf{f} + \mathbf{d} + \Delta\mathbf{B}\hat{\mathbf{B}}^{-1}\hat{\mathbf{u}} - k\left(\mathbf{I} + \Delta\mathbf{B}\hat{\mathbf{B}}^{-1}\right)\text{sign}(\dot{\mathbf{s}})\right\} \le -\eta \|\dot{\mathbf{s}}\| \tag{III.B6}$$

This indicates a condition for the gain k to guarantee the system stability. To quantify the gain value, the analytic properties of the matrix-norm are applied and

$$\dot{\mathbf{s}}^T\left\{\Delta\mathbf{f} + \mathbf{d} + \Delta\mathbf{B}\hat{\mathbf{B}}^{-1}\hat{\mathbf{u}} - k\left(\mathbf{I} + \Delta\mathbf{B}\hat{\mathbf{B}}^{-1}\right)\text{sign}(\dot{\mathbf{s}})\right\}$$
$$\le \|\dot{\mathbf{s}}\| \cdot \left\{\|\Delta\mathbf{f}\| + \|\mathbf{d}\| + \|\Delta\mathbf{B}\| \cdot \left\|\hat{\mathbf{B}}^{-1}\hat{\mathbf{u}}\right\| - k\left(1 - \|\Delta\mathbf{B}\| \cdot \left\|\hat{\mathbf{B}}^{-1}\text{sign}(\dot{\mathbf{s}})\right\|\right)\right\} \tag{III.B7}$$

Note that

$$-k\dot{\mathbf{s}}^T\text{sign}(\dot{\mathbf{s}}) \le -k\|\dot{\mathbf{s}}\| \tag{III.B8}$$

and $\quad -k\dot{\mathbf{s}}^T\Delta\mathbf{B}\hat{\mathbf{B}}^{-1}\text{sign}(\dot{\mathbf{s}}) \le k\|\dot{\mathbf{s}}\| \cdot \|\Delta\mathbf{B}\| \cdot \left\|\hat{\mathbf{B}}^{-1}\text{sign}(\dot{\mathbf{s}})\right\| \tag{III.B9}$

Therefore,

$$\|\dot{\mathbf{s}}\| \cdot \left\{\|\Delta\mathbf{f}\| + \|\mathbf{d}\| + \|\Delta\mathbf{B}\| \cdot \left\|\hat{\mathbf{B}}^{-1}\hat{\mathbf{u}}\right\| - k\left(1 - \|\Delta\mathbf{B}\| \cdot \left\|\hat{\mathbf{B}}^{-1}\text{sign}(\dot{\mathbf{s}})\right\|\right)\right\} \le -\eta \|\dot{\mathbf{s}}\| \tag{III.B10}$$

By the uncertainty confinements (Equation(III.4) and Equation (III.7)), Equation (III.B10) is simplified as

$$\|\dot{s}\| \cdot \left\{ \alpha + \gamma + \beta \left\| \hat{B}^{-1} \hat{u} \right\| - k \left(1 - \beta \left\| \hat{B}^{-1} \text{sign}(\dot{s}) \right\| \right) \right\} \le -\eta \|\dot{s}\| \qquad \text{(III.B11)}$$

Finally, the gain k is found as

$$k \ge \frac{\alpha + \gamma + \beta \left\| \hat{B}^{-1} \hat{u} \right\| + \eta}{1 - \beta \left\| \hat{B}^{-1} \text{sign}(\dot{s}) \right\|} \qquad \text{(III.B12)}$$

Since the gain k is a non-negative real number, a sufficient condition is given as

$$\beta \left\| \hat{B}^{-1} \text{sign}(\dot{s}) \right\| < 1 \qquad \text{(III.B13)}$$

The Lyapunov stability is guaranteed as long as the gain k is chosen according to Equation (III.B12).

The **s** dynamics outside the boundary layer can then be given as

$$\ddot{s} + \left(I - B\hat{B}^{-1} \right)\left(z_0 * \dot{s} \right) + k B\hat{B}^{-1} \text{sign}(\dot{s}) + B\hat{B}^{-1}\left(\omega_n^2 * s \right)$$
$$= (f - \hat{f}) + d - \left(I - B\hat{B}^{-1} \right)\left(-\hat{f} + x_d^{(n)} - e_p \right) \qquad \text{(III.B13)}$$

Within the boundary layer, the **s** dynamics is smoothed, i.e.,

$$\ddot{s} + \left(I - B\hat{B}^{-1} \right)\left(z_0 * \dot{s} \right) + \frac{k}{\phi} B\hat{B}^{-1}\dot{s} + B\hat{B}^{-1}\left(\omega_n^2 * s \right)$$
$$= (f - \hat{f}) + d - \left(I - B\hat{B}^{-1} \right)\left(-\hat{f} + x_d^{(n)} - e_p \right) \qquad \text{(III.B14)}$$

This represents a set of second-order low-pass filters. Unlike the first-order **s** dynamics derived from a first-order sliding condition, the second-order **s** dynamics has more than one controlling parameters which include not only the boundary-layer thickness ϕ but also the placed zero z_0 and the spring constant ω_n^2. The undamped frequency spectrum, which reflects the bandwidth of the **s** dynamics, can be adjusted by the spring constant alone. The damping of the **s** dynamics is then determined by ϕ and z_0.

C. TUNING MECHANISMS

From the SISO sliding control, it was found that the value of ϕ is crucial to the system response and the s dynamics can be deactivated by a smaller ϕ, where ϕ is automatically tuned by selecting a lower bound of damping ratio and a placed zero. Due to the complexity of the MIMO dynamics, a simplified tuning procedure is required. The spring constant is used to limit the bandwidth of the s dynamics and the relative stability (i.e., tracking accuracy) is controlled by ϕ and z_0. In addition, with the tuning versatility, constant ϕ and z_0 should be able to provide sufficient tuning flexibility to control the relative stability.

To limit the bandwidth of the s dynamics, a maximum bandwidth λ_{max} is assigned to the s dynamics within the boundary layer (Equation (III.B14)). Let λ be the spectral radius of $\mathbf{B}\hat{\mathbf{B}}^{-1}\Omega^2$, i.e.,

$$\lambda = \left\{ eig\left(\mathbf{B}\hat{\mathbf{B}}^{-1}\Omega^2\right) \right\}_{max}$$

where Ω^2 is a diagonal matrix form of $\omega_n{}^2$, i.e.,

$$\Omega^2 = \begin{bmatrix} \omega_1{}^2 & & & & & \\ & \omega_2{}^2 & & & & \\ & & \ddots & & & \\ & & & \omega_i{}^2 & & \\ & & & & \ddots & \\ & & & & & \omega_m{}^2 \end{bmatrix}$$

Let the maximum element in Ω^2 be $\left(\omega^2\right)_{max}$ which is to be designed such that $\lambda \le \lambda_{max}$. From the analytic properties of matrix norm, it can be shown that

$$\left\{ eig\left(\mathbf{B}\hat{\mathbf{B}}^{-1}\Omega^2\right) \right\}_{max} \le \left\| \left(\omega^2\right)_{max} \mathbf{B}\hat{\mathbf{B}}^{-1} \right\| \tag{III.C1}$$

Therefore,

$$\lambda \le \left(\omega^2\right)_{max} \left\| \mathbf{B}\hat{\mathbf{B}}^{-1} \right\| \tag{III.C2}$$

Since

$$\left\|\mathbf{B}\hat{\mathbf{B}}^{-1}\right\| \le 1 + \left\|\Delta\mathbf{B}\right\| \cdot \left\|\hat{\mathbf{B}}^{-1}\right\| \quad \text{and} \quad \left\|\Delta\mathbf{B}\right\| \le \beta$$

The spectral radius is confined as

$$\lambda \le \left(\omega^2\right)_{\text{max}} \left(1 + \beta\left\|\hat{\mathbf{B}}^{-1}\right\|\right) \tag{III.C3}$$

Accordingly, let

$$\lambda_{\text{max}} = \left(\omega^2\right)_{\text{max}} \left(1 + \beta\left\|\hat{\mathbf{B}}^{-1}\right\|\right)$$

and $\left(\omega^2\right)_{\text{max}}$ can be designed as

$$\left(\omega^2\right)_{\text{max}} = \frac{\lambda_{\text{max}}}{\left(1 + \beta\left\|\hat{\mathbf{B}}^{-1}\right\|\right)} \tag{III.C4}$$

In the above equation, the maximum bandwidth λ_{max} determines the selection of the maximum value for the elements in Ω^2. Equation (III.C4) also assures that the bandwidth of the s dynamics never exceeds λ_{max} and the unmodeled dynamics can be filtered by selecting a proper value of λ_{max}.

IV.   Conclusion

The sliding control algorithm is effective in dealing with filtering uncertainty for time-varying nonlinear dynamical systems. The sliding control design involves two filters, i.e., s dynamics and error dynamics. The s dynamics serves as a buffer filter to filter unwanted high-frequency signals that include unmodeled dynamics and disturbances. The error dynamics is designed as a second filter filtering s and obtaining the system error e as an output. The implementation is simple involving an inverse error dynamics and a Lyapunov control. A saturation control will counteract the chattering and the remaining steady-state system error e can be eliminated with an integral feature in the error dynamics. The control system bandwidth can be tuned by the adjustable parameters within the two dynamics.

Second-order sliding condition has been applied to the sliding control to raise the order of the s dynamics and to increase the activity of the s dynamics and the versatility of the tuning mechanisms. Within the boundary layer region, the second-order s dynamics are more active than their first-order counterparts. The system dynamic response primarily depends on the s dynamics in the boundary layer region. Therefore, the activity and the tuning of the s dynamics are critically important. Since more tuning parameters are introduced while using the second-order sliding condition, the versatility of the tuning capability has been increased.

The versatile sliding control has been developed, which utilizes a low-pass filter for a second-order s dynamics and a high-pass (or band-pass) filter for an error dynamics in series. The control places zeros within the error dynamics and applies the second-order sliding condition. The second-order s dynamics sharply rejects unwanted high-frequency uncertainty signals. The bandwidth of the s dynamics can be designed low such that the excitation to the error dynamics will be minimum. The trade-off between tracking accuracy and the robustness to the uncertainties no longer exists. A general tuning procedure needs further research for choosing the design parameters, i.e., the thickness of the boundary layer, placed zeros and the spring constant in the sliding condition, to fine tune the s dynamics and in turn the error dynamics.

V.   References

Barmish, B.R., and Leitmann, G., 1982, " On Ultimate Boundedness Control of Uncertain Systems in the Absence of Matching Conditions, " IEEE *Trans. Automatic Control*, Vol. AC-27, No. 1, pp. 153-158.

Chang, L.W., 1990, "A MIMO Sliding Control with a Second-Order Sliding Condition," 90-WA/DSC-5, ASME *Winter Annual Meeting*, Dallas, TX.

Chang, L.W., 1991a, "A Versatile Sliding Control with a Second-Order Sliding Condition," Proceedings of the 1991 *American Control Conference*, Vol. 1, Boston, MA, pp. 56-57.

Chang, L.W., 1991b, "A Robust Motion Control of Actuators in Disk Files," Proceedings of the 1991 *American Control Conference*, Vol. 1, Boston, MA, pp. 49-55.

Chen, Y.H., 1986, "On the Deterministic Performance of Uncertain Dynamical Systems," *Int. Journal of Control*, Vol. 43, No. 5, pp. 1557-1579.

Corless, M.J., and Leitmann, G., 1981, "Continuous State Feedback Guaranteeing Uniform Ultimate Boundedness for Uncertain Dynamical Systems," IEEE *Trans. Automatic Control*, Vol. AC-26, No. 5, pp. 1139-1144.

DeCarlo, R.A., et. al., 1988, "Variable Structure Control of Nonlinear Multivariable Systems: A Tutorial," *Proceedings of the IEEE*, Vol. 76, No. 3, pp. 212-232.

Dorling, C.M. and Zinober, A.S.I., 1986, "Two Approaches to Hyperplane Design in Multivariable Variable Structure Control Systems," *Int. Journal of Control*, Vol. 44, No. 1, pp. 65-82.

Fernandez, B., and Hedrick, J.K., 1987, "Control of Multivariable Non-Linear Systems by the Sliding Mode Method," *Int. Journal of Control*, Vol. 46, No.3., pp. 1019-1040.

Horn, R.A., and Johnson, C.R., 1988, *Matrix Analysis*, Cambridge University Press.

Itkis, U., 1976, *Control Systems of Variable Structure*, Halsted Press.

Slotine, J.J., and Sastry, S.S., 1983, "Tracking Control of Non-Linear Systems Using Sliding Surfaces, with Application to Robot Manipulators," *Int. Journal of Control*, Vol. 38, No. 2, pp. 465-492.

Slotine, J.J., 1984, "Sliding Controller Design for Non-Linear Systems," *Int. Journal of Control*, Vol. 40, No. 2, pp. 421-434.

Slotine, J.J., 1985, "The Robust Control of Robot Manipulators," *The International Journal of Robotics Research*, Vol. 4, No. 2, pp. 49-64.

Utkin, V.I., 1977, "Survey Paper- Variable Structure Systems with Sliding Modes," IEEE *Trans. Automatic Control*, Vol. AC-22, No. 2, pp. 212-222.

Yoerger, D., and Slotine, J.J., 1985, "Robust Trajectory Control of Underwater Vehicles," IEEE *Journal of Oceanic Engineering*, Vol. OE-10, No. 4, pp. 462-470.

Young, K.K.D., Kokotovic, P.V., and Utkin, V.I., 1977, "A Singular Perturbation analysis of High-Gain Feedback Systems," IEEE *Trans. Automatic Control*, Vol. AC-22, No. 6, pp. 931-938.

Techniques in Robust Broadband Beamforming

Meng H. Er

School of Electrical & Electronic Engineering
Nanyang Technological University
Nanyang Avenue, Singapore 2263

Antonio Cantoni

Department of Electrical & Electronic Engineering
University of Western Australia
Nedlands 6009, W.A., Australia

I. INTRODUCTION

The use of an array of sensors has long been an attractive solution to signal detection and estimation in harsh environments. An array offers a means of exploiting in a flexible way the spatial characteristics of signal and noise thereby overcoming the directivity and beamwidth limitations imposed by a single sensor [1]. Arrays have been used in many signal acquisition applications such as sonar [2-7], radar [8-11], communications [12-16] and seismology [17-20].

There are two broad classes of application of antenna array

CONTROL AND DYNAMIC SYSTEMS, VOL. 53

processors. One is spatial spectrum estimation in which the objective is to determine the direction and power of sources incident on an array. The other class is to estimate a signal arriving from a specified direction but in the presence of noise and other directional sources. In this latter case the objective is to determine a set of weighting coefficients which generate an output time series y(t) that has a better signal to noise and signal to interference ratio than that of a single element. Although the spectrum estimation problem may appear to be different from the signal estimation problem, it may be treated as a signal estimation problem by hypothesising the existence of a signal in a number of discrete directions and posing the optimum signal estimation problem for each direction. Thus, in this approach to spatial spectrum estimation, the array processor is used to estimate the power in a selected number of look directions, sufficient to provide the necessary coverage and sampling interval. Other techniques for high resolution spatial spectrum estimation have been investigated, although many apply only to narrowband scenarios or are frequency domain based [21].

In general, beamforming may be performed either on a data independent basis, or as a statistically optimum process [22]. Data independent beamformers are designed such that the beamformer response approximates some desired response independent of the array data statistics. Synthesis techniques for determining weights which result in a desired pattern response for a given array have been available for more than 40 years [23]. The great majority of this work has focused on methods which result in a pattern having prescribed mainlobe width and reduced sidelobe levels [24].

Statistically optimum beamformers, on the other hand, have their weights chosen so as to optimise the beamformer response in some prescribed sense on the basis of measured array data statistics. The aim is to optimise the beamformer response so that the output contains minimal contributions due to noise and to interference sources arriving from directions other than that of the desired signal or signals. Since the spatial and temporal characteristics of the source scenario may be nonstationary in nature, adaptive processing techniques have been devised for the realization of optimum processors [25].

In order for the beamformer to reject noises and interferences without rejecting signal, it is necessary to make some assumptions about the signal characteristics. In actual operating conditions, the assumptions of plane wave signals, perfectly matched sensor channels, perfect presteering, perfect knowledge of sensor locations and perfect match between assumed look direction and actual direction of arrival of the desired signal do not hold. Under these non ideal conditions, the performance of the zero order linearly constrained minimum power processor degrades rapidly [26]. The development of robust array processors is concerned with the formulation of antenna array processor optimization that results in performance that is relatively insensitive to the deviations from the ideal noted above. The use of multiple linear constraints [26-30] and non-linear norm-type constraints [29, 31-34] has been described in the literature as means for handling beam-steer mismatch in narrowband systems. The adverse effects due to sensor amplitude and phase errors have been handled in [29, 35-37] by the introduction of norm constraints on the weight vector. The use of artificial noise injection to achieve robustness has been considered in [38-39].

For broadband systems, very little work of this nature has been reported in the literature. Most techniques for broadband systems are based on linearly constrained optimization which are applicable only to presteered broadband arrays. Some preliminary work on the use of a non-linear norm-type constraint to design exactly presteered broadband processor robust against phase errors has been reported in [40]. A constrained adaptive beamformer which is tolerant of array gain and phase errors has been proposed in [41]. Approaches based on the matching of the optimum filter matrix of a multichannel processor to any signal covariance structure even under perturbed conditions, defined in a statistical sense, have been considered in [25, 42].

In view of the paucity of work on robust broadband antenna array processors, we have limited the scope of this chapter to techniques relevant to broadband array processing.

In the first instance, we address the problem of the degradation due to mismatch between the assumed look direction and the actual direction of incidence of the desired signal or the beam-steer error problem as it is often called. The technique we present is relevant only to presteered processors and the basis for the approach has been described in [43]. We show that to a large extent the beam-steer error problem can be handled by introducing linear constraints on the array processor weights that ensure a broadening of the beam in the steered direction. The approach we present in this chapter differs from the original work in [43] and enables more powerful results to be established.

The ability to handle other non ideal operating conditions requires a more radical departure from simply augmenting the traditional presteered processor designs with additional constraints. A technique is presented that enables the design of robust broadband antenna array processors tolerant to many of the imperfections previously noted. A significant characteristic of the technique is the ability to handle a wide range of presteering conditions, including no presteering. The basis for the technique has been described in [44-45]. In this chapter we include additional material and examples of the application of the technique. Although the technique was originally developed for broadband systems, many of the ideas can be applied to narrowband systems [34, 46]. It is beyond the scope of this chapter to cover this application of the technique.

The organization of the chapter is as follows. In Section II, we introduce the signal model, the linear broadband array processor structure and the notation that is used throughout the chapter. We introduce in Section III the concept of minimum power linearly constrained broadband processors. In Section IV, we describe a technique for enhancing the tolerance of presteered minimum power processors to beam-steer errors. In Section V, we present the formulation of an optimum broadband processor with robustness capabilities against a wide variety of errors and imperfections. We also give examples of the application of the new robust processor. In Section VI, we introduce an alternate formulation for the new robust processor that is based on a partitioned processor structure. The alternate approach leads to a robust processor that can be implemented more easily in an adaptive form and offers a reduced computational load.

II. BROADBAND PROCESSOR AND SIGNAL MODEL-LING

In this section we introduce mathematical models for the signals and processing structures that arise in our study of robust broadband antenna array processors. We highlight the various simplification and assumptions that are necessary to make the problems considered tractable. Much of the notation and terminology that is used throughout the work is also introduced in this section. A glossary of symbols is provided in Appendix A.

A.   Broadband Processor and Signals

Fig. 1 illustrates the structure for broadband antenna array processing that is most often considered in the literature. The processor consists of an array of L sensors distributed in space, J tapped delay line sections with altogether LJ adjustable coefficients and in general L presteering delay elements with delays $(T_1,T_2,...,T_L)$. In a subsequent section we consider the design of processor that can handle a variety of presteering conditions which we term, exact, coarse and no presteering. Briefly, in the case of exact presteering the delays are adjusted such that the samples of a wavefront incident on the array as produced by the sensors are all time aligned for a given look direction. This and the other presteering concepts will be made precise in sequel.

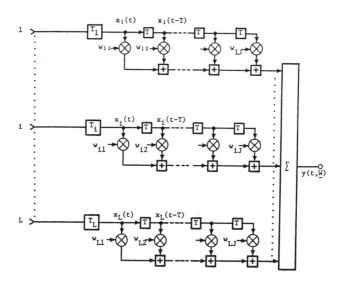

Fig. 1 Structure of a broadband time domain antenna array processor.

The locations of the array sensors in spaced are specified with respect to a chosen co-ordinate system as shown in Fig. 2. The position of the ith sensor is given by the position vector $\underline{r}_i$. The phase centre of the array, denoted by $\underline{r}_0$ is given by

$$\underline{r}_0 = \frac{1}{L}\sum_{i=1}^{L}\underline{r}_i \tag{1}$$

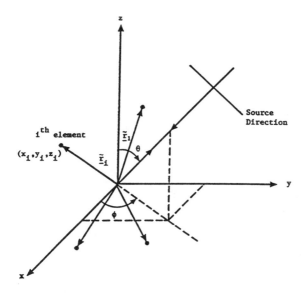

Fig. 2 Reference co-ordinates of an arbitrary array.

Throughout the work we consider a scenario in which signals propagate through a linear, isotropic and non dispersive medium with a known speed υ. Furthermore, we consider only the case in which the antenna array is in the far field of the signal sources so that the signals incident on the array can be considered plane waves. More precisely, the function $\psi : R^3xR \to C$ is a plane wave incident from direction (θ,ϕ) if

$$\psi(\underline{r},t) = s\big(t+\tau_i(\theta,\phi)\big) \qquad (2)$$

where

$$s(t) = \psi(\underline{0},t) \qquad (3)$$

$$\tau_i(\theta,\phi) = \frac{1}{\upsilon}\underline{r}_i^T\underline{u}(\theta,\phi) \qquad (4)$$

$\underline{u}(\theta,\phi)$ is the unit vector in direction (θ,ϕ)

$$\underline{u}(\theta,\phi) \;=\; \begin{bmatrix} \sin\theta \; \cos\phi \\ \sin\theta \; \sin\phi \\ \cos\theta \end{bmatrix} \tag{5}$$

The wavefront containing $\underline{r}$ is a plane defined by

$$\underline{r}+\left[\underline{u}(\theta,\phi)\right]^{\perp} \;=\; \left\{\underline{r}+\underline{x} \in R^3 : \underline{x}^T\underline{u}(\theta,\phi) \;=\; 0\right\} \tag{6}$$

It is clear from Eq. (3) that we have taken the origin as the common reference point and that we have not represented explicitly the bulk transmission delay from the source to the array. This is justified since, as will become clear soon, for plane waves only the differential propagation delay among the elements is important.

We will assume that the output of the ith array sensor can be modelled as

$$\tilde{x}_i(t) \;=\; n_{W,i}(t)+n_{I,i}(t)+\sum_{j=1}^{N}s_j\!\left(t+\tau_i\!\left(\theta_j,\phi_j\right)\right),\; i \in Z_L^+ \tag{7}$$

The component $n_{W,i}(t)$ is used to model noise that is uncorrelated from sensor to sensor. The component $n_{I,i}(t)$ represents isotropic noise incident on the array [47]. Finally, signal model includes a contribution from N discrete, directional sources represented as plane waves incident on the array. The direction of the discrete sources is defined by $\left\{(\theta_j,\phi_j),\; j \in Z_N^+\right\}$.

We will further assume that $n_{W,i}(t)$, $n_{I,i}(t)$, $\left\{s_j(t),\; j \in Z_N^+\right\}$ can be modelled as zero mean stationary stochastic processes with the following second order properties:

$$E\left[s_k(t+\tau)s_\ell(t)\right] \;=\; \rho_k(\tau)\delta_{k,\ell} \tag{8}$$

$$E\left[s_k(t+\tau)n_{W,\ell}(t)\right] \;=\; 0 \tag{9}$$

$$E\left[s_k(t+\tau)n_{I,\ell}(t)\right] \;=\; 0 \tag{10}$$

$$E\left[n_{W,k}(t+\tau)n_{I,\ell}(t)\right] \;=\; 0 \tag{11}$$

where E[•] denotes expected value. These indicate that various stochastic processes are mutually uncorrelated. Furthermore, the discrete directional sources are characterized by their correlation function $\rho_k(\tau)$ that is related to the source spectrum through the Fourier transform.

We will find it convenient to at times deal with the signals post presteering. These signals are related to the sensor signals according to

$$x_i(t) = \tilde{x}_i(t - T_i), \quad i \in Z_L^+ \tag{12}$$

The state of the multidimensional tapped delay line filter is represented by the vector $\underline{X}(t) \in R^{LJ}$, where

$$[\underline{X}(t)]_{(j-1)L+i} = x_i(t - (j-1)T), \quad j \in Z_J^+, \ i \in Z_L^+ \tag{13}$$

Similarly, the tapped delay line filter coefficients are represented by the vector $\underline{W} \in R^{LJ}$, where

$$[\underline{W}]_{(j-1)L+i} = w_{i,j}, \quad j \in Z_J^+, \ i \in Z_L^+ \tag{14}$$

The output, $y(t)$, of the array processor can now be expressed as

$$y(t) = \underline{W}^T \underline{X}(t) \tag{15}$$

Under the assumptions previously made, the mean square output, or more loosely the mean power output, is given by

$$p(\underline{W}) = E\left[|y(t)|^2\right] = \underline{W}^T R \underline{W} \tag{16}$$

where $R \in M_{LJ \times LJ}(R)$ is the nonnegative definite array correlation matrix defined by

$$R = E\left[\underline{X}(t)\underline{X}^T(t)\right] \tag{17}$$

B. Frequency Response and Presteering

The response of the array processor to a monochromatic sinusoidal plane wave incident from direction (θ,ϕ) is given by

$$y(t) = \text{Real}\left[\exp(j2\pi ft)\underline{W}^T\underline{d}(f)\otimes D(f)\underline{S}(f,\theta,\phi)\right] \qquad (18)$$

where $\underline{d}(f)\in C^J$, $\underline{S}(f,\theta,\phi)\in C^L$ and $D(f)\in M_{L\times L}(C)$ are defined by

$$[\underline{d}(f)]_k = \exp(-j2\pi f(k-1)T), \quad k\in Z_J^+ \qquad (19)$$

$$[\underline{S}(f,\theta,\phi)]_\ell = \exp(j2\pi f\tau_\ell(\theta,\phi)), \quad \ell\in Z_L^+ \qquad (20)$$

$$[D(f)]_{k,\ell} = \exp(-j2\pi fT_k)\delta_{k,\ell}, \quad k,\ell\in Z_L^+ \qquad (21)$$

In view of Eq. (18), we define $H(f,\theta,\phi)$ given by

$$H(f,\theta,\phi) = \underline{W}^T\underline{d}(f)\otimes D(f)\underline{S}(f,\theta,\phi) \qquad (22)$$

to be the frequency response of the processor to a monochromatic plane wave incident from direction (θ,ϕ). The power response is given by

$$\rho(f,\theta,\phi) = |H(f,\theta,\phi)|^2 \qquad (23)$$

We say that the array processor is presteered in the look direction (θ_0,ϕ_0) if for some $T_0>0$

$$T_\ell = T_0+\tau_\ell(\theta_0,\phi_0) \geq 0, \quad \ell\in Z_L^+ \qquad (24)$$

It is clear from Eq. (24) that

$$T_0 \geq -\min_{\ell\in Z_L^+}\left(\tau_\ell(\theta_0,\phi_0)\right) \qquad (25)$$

Consider a single plane wave incident on the array from direction (θ_0,ϕ_0), then from Eq. (7), Eq. (12) and Eq. (24) it follows that

$$x_\ell(t) = s\left(t+\tau_\ell(\theta_0,\phi_0)-T_0-\tau_\ell(\theta_0,\phi_0)\right), \quad \ell\in Z_L^+ \qquad (26)$$

This implies that $x_\ell(t)$ is independent of ℓ

$$x_\ell(t) = s(t-T_0) \qquad (27)$$

The effect of presteering the array is now clear. The signals at the output of the steering delays are identical for a planewave incident from the presteered look direction.

The frequency response of a broadband processor presteered in a look direction (θ_0,ϕ_0) can now be shown to be given by

$$H(f,\theta_0,\phi_0) \;=\; \exp\!\left(-j2\pi fT_0\right)\underline{h}^T\underline{d}(f) \tag{28}$$

where

$$\underline{h} \;=\; C_0\underline{W}, \;\; \underline{h}\in R^J \tag{29}$$

$$C_0 \;=\; I_J\otimes\underline{1}_L^T, \;\; C_0\in M_{JxLJ}(R) \tag{30}$$

In view of the definition of $\underline{d}(f)$ in Eq. (19), Eq. (28) can be interpreted as the frequency response of a one dimensional tapped delay line filter with tap coefficients given by the elements of the vector $\underline{h}$. Furthermore, it follows from Eq. (28) and Eq. (29) that any frequency response specified by $\underline{h}$ can be achieved in the look direction by constraining the array processor weight vector $\underline{W}$ to satisfy Eq. (29). Frequently, Eq. (29) is referred to as the zero order constraint.

A flat frequency response in the presteered look direction can be achieved by setting all but one of the elements of $\underline{h}$ to zero.

III. MINIMUM POWER LINEARLY CONSTRAINED BROADBAND PROCESSORS AND SENSITIVITY

One widely used weight selection criterion is linearly constrained power minimization, in which the weights are linearly constrained to achieve some characteristic of the array processor, e.g. desired frequency response in the look direction, while minimising the mean output power [48]. Crucial to this approach is the assumption that the sources incident on the array from directions other than the look direction signal are uncorrelated with the look direction signal as assumed in our signal scenario model. In this case, power minimization subject to a fixed gain for the look direction results in the minimization of the effect of interferences and noise.

In view of Eq. (16) and Eqs. (28)-(30), the simplest presteered broadband array processor optimization that has been posed, [48], is

$$\underset{\underline{W}}{\text{minimise}} \quad \underline{W}^T R \underline{W} \tag{31}$$

$$\text{subject to} \quad C_0 \underline{W} = \underline{h} \tag{32}$$

The solution to this problem, assuming R is strictly positive definite, is given by

$$\hat{\underline{W}} = R^{-1} C_0^T \left(C_0 R^{-1} C_0^T \right)^{-1} \underline{h} \tag{33}$$

and

$$p(\hat{\underline{W}}) = \underline{h}^T \left(C_0 R^{-1} C_0^T \right)^{-1} \underline{h} \tag{34}$$

It is possible to express the optimum weight vector and optimum power in terms of the noise plus interference correlation matrix and hence relate the linearly constrained minimum processor to other weight selection criteria. Let $\underline{X}(t)$ be decomposed into the look direction signal component $\underline{X}_s(t)$ and noise plus directional interference component $\underline{X}_n(t)$:

$$\underline{X}(t) = \underline{X}_s(t) + \underline{X}_n(t) \tag{35}$$

It can be shown using Eq. (27) and Eq. (30) that

$$\underline{X}_s(t) = C_0^T \underline{s}(t - T_0) \tag{36}$$

where $\underline{s}(t) \in R^J$ is given by

$$[\underline{s}(t)]_j = s(t - (j-1)T), \quad j \in Z_J^+ \tag{37}$$

Now we have that

$$R = R_n + C_0^T S C_0 \tag{38}$$

where

$$R_n = E\left[\underline{X}_n(t) \underline{X}_n^T(t) \right] \tag{39}$$

$$S = E\left[\underline{s}(t - T_0) \underline{s}^T(t - T_0) \right] \tag{40}$$

Furthermore, using the matrix inversion lemma

$$(A + BC)^{-1} = A^{-1} - A^{-1} B \left(I + C A^{-1} B \right)^{-1} C A^{-1}, \tag{41}$$

it can be shown that

$$\hat{\underline{W}} = R_n^{-1} C_0^T \left(C_0 R_n^{-1} C_0^T \right)^{-1} \underline{h} \qquad (42)$$

$$p(\hat{\underline{W}}) = \underline{h}^T \left(C_0 R_n^{-1} C_0^T \right)^{-1} \underline{h} \qquad (43)$$

The weight vector defined by Eq. (42) yields an array processor output that can be shown to be the maximum likelihood estimate of a stationary stochastic signal in Gaussian noise if the angle of arrival is known [20]. Hence, the term "maximum likelihood distortionless estimator" is often used when referring to the zero order linearly constrained minimum power array processor.

IV. MAXIMALLY FLAT PRESTEERED BROADBAND PROCESSORS

In this section, we consider an approach for designing presteered broadband antenna array procesors that are robust against mismatch between the chosen look direction and the actual direction of arrival of the signal that we wish to estimate. If the desired signal does not arrive exactly in the presteered look direction, it will be rejected to some degree since it appears to be an interference. This suppression is particularly severe when the element level signal to noise ratio is of the order of 0 dB or higher and when the array processor has many degrees of freedom as represented by the unconstrained dimensionality of the weight vector. This problem has been studied extensively for narrowband processors by Cox [42]. The use of additional linear constraints to achieve broadening of the beam in the look direction for narrowband processors has been described in [30]. The additional linear constraints were derived either by fixing the response of the array processor at additional points in (θ, ϕ) space in the vicinity of the look direction or by employing constraints that ensure that the partial derivatives of the power response $\rho(f, \theta, \phi)$ up to a specified order were zero when evaluated in the look direction (θ_0,ϕ_0). The latter approach results in what might be termed a maximally flat spatial response.

In the case of broadband processors, the majority of the reported work was based on the use of multiple linear constraints that fixed the gain of the array processor at a number of points in a region around (f_0,θ_0,ϕ_0) bounded by specified limits $f_0\pm\Delta f$, $\theta_0\pm\Delta\theta$ and $\phi_0\pm\Delta\phi$.

A.   Maximally Flat Spatial Response

A broadband presteered array processor has a kth order maximally flat spatial response in the look direction (θ_0,ϕ_0) if

$$\left.\frac{\partial^m\rho(f,\theta,\phi)}{\partial\theta^{m-n}\partial\phi^n}\right|_{(\theta_0,\phi_0)} = 0 \qquad (44)$$

$$\forall\ \ f\geq 0,\ m\in Z_k^+,\ n\in Z_m^+\cup\{0\}$$

In practice, only the first and second order cases are of interest, since beyond these, the ability of the array processor to reject interferences outside the broadened acceptance angle is severely degraded. Therefore, we consider only the first and second order maximally flat spatial response cases.

For convenience of presentation of results, we first introduce the matrices $J_n\in M_{JxJ}(R)$, $\Lambda(\theta,\phi)$, $A(\theta,\phi)$ and $B(\theta,\phi)\in M_{LxL}(R)$ defined by

$$[J_n]_{k,\ell} = \delta_{k+n,\ell},\ \ n\in Z,\ k,\ell\in Z_J^+ \qquad (45)$$

$$[\Lambda(\theta,\phi)]_{k,\ell} = \tau_\ell(\theta,\phi)\delta_{k,\ell},\ \ k,\ell\in Z_L^+ \qquad (46)$$

$$A(\theta,\phi) = \Lambda(\theta,\phi)\underline{1}_L\underline{1}_L^T \qquad (47)$$

$$B(\theta,\phi) = A(\theta,\phi)-A^T(\theta,\phi) \qquad (48)$$

We can show using Eqs. (19)-(22) that the first and second order partial derivatives of the spatial power response can be expressed as follows in terms of the above matrices:

$$\frac{\partial \rho}{\partial \theta}(f,\theta_0,\phi_0) = -4\pi f \sum_{n=1}^{J-1} \sin(2\pi f n T)\underline{W}^T(J_n - J_{-n}) \otimes \frac{\partial A}{\partial \theta}(\theta_0,\phi_0)\underline{W} \quad (49)$$

$$\frac{\partial \rho}{\partial \phi}(f,\theta_0,\phi_0) = -4\pi f \sum_{n=1}^{J-1} \sin(2\pi f n T)\underline{W}^T(J_n - J_{-n}) \otimes \frac{\partial A}{\partial \phi}(\theta_0,\phi_0)\underline{W} \quad (50)$$

$$\frac{\partial^2 \rho}{\partial \theta^2}(f,\theta_0,\phi_0) = -4\pi f \sum_{n=1}^{J-1} \sin(2\pi f n T)\underline{W}^T(J_n - J_{-n}) \otimes \frac{\partial^2 A}{\partial \theta^2}(\theta_0,\phi_0)\underline{W}$$
$$+8\pi^2 f^2 \sum_{n=0}^{J-1} \cos(2\pi f n T)\underline{W}^T(J_n + J_{-n}) \otimes \frac{\partial B}{\partial \theta}(\theta_0,\phi_0)\frac{\partial \Lambda}{\partial \theta}(\theta_0,\phi_0)\underline{W} \quad (51)$$

$$\frac{\partial^2 \rho}{\partial \phi^2}(f,\theta_0,\phi_0) = -4\pi f \sum_{n=1}^{J-1} \sin(2\pi f n T)\underline{W}^T(J_n - J_{-n}) \otimes \frac{\partial^2 A}{\partial \phi^2}(\theta_0,\phi_0)\underline{W}$$
$$+8\pi^2 f^2 \sum_{n=0}^{J-1} \cos(2\pi f n T)\underline{W}^T(J_n + J_{-n}) \otimes \frac{\partial B}{\partial \phi}(\theta_0,\phi_0)\frac{\partial \Lambda}{\partial \phi}(\theta_0,\phi_0)\underline{W} \quad (52)$$

$$\frac{\partial^2 \rho}{\partial \theta \partial \phi}(f,\theta_0,\phi_0) = -4\pi \sum_{n=1}^{J-1} \sin(2\pi f n T)\underline{W}^T(J_n - J_{-n}) \otimes \frac{\partial^2 A}{\partial \theta \partial \phi}(\theta_0,\phi_0)\underline{W}$$
$$+8\pi^2 f^2 \sum_{n=0}^{J-1} \cos(2\pi f n T)\underline{W}^T(J_n + J_{-n}) \otimes \frac{\partial B}{\partial \theta}(\theta_0,\phi_0)\frac{\partial \Lambda}{\partial \phi}(\theta_0,\phi_0)\underline{W} \quad (53)$$

B. Necessary and Sufficient Conditions for a Maximally Flat Spatial Response

To derive the necessary and sufficient conditions for a maximally flat spatial response from Eqs. (49)-(53), we make use of the following results. If

$$c(f) = \sum_{n=0}^{J} a_n \cos(2\pi f n T) \quad (54)$$

$$s(f) = \sum_{n=1}^{J} b_n \sin(2\pi f n T) \quad (55)$$

then

$$fc(f) + s(f) = 0, \ \forall \ f \geq 0 \Leftrightarrow a_0 = a_n = b_n = 0, \ \forall \ n \in Z_J^+ \quad (56)$$

We can now state the main result of the section as follows:

$$\frac{\partial \rho}{\partial \theta}(f,\theta_0,\phi_0) \;=\; 0, \;\; \forall \;\; f \geq 0 \Leftrightarrow \underline{W}^T \tilde{Q}_{\theta,n} \underline{W} = 0, \;\; n \in Z_{J-1}^+ \tag{57}$$

$$\frac{\partial \rho}{\partial \phi}(f,\theta_0,\phi_0) \;=\; 0, \;\; \forall \;\; f \geq 0 \Leftrightarrow \underline{W}^T \tilde{Q}_{\phi,n} \underline{W} = 0, \;\; n \in Z_{J-1}^+ \tag{58}$$

$$\left.\begin{array}{l} \dfrac{\partial^2 \rho}{\partial \theta^2}(f,\theta_0,\phi_0) \;=\; 0 \\[6pt] \forall \; f \geq 0 \end{array}\right\} \Leftrightarrow \begin{cases} \underline{W}^T \tilde{Q}_{\theta,\theta,1,n} \underline{W} = 0, \;\; n \in Z_{J-1}^+ \\[4pt] \underline{W}^T \tilde{Q}_{\theta,\theta,2,n} \underline{W} = 0, \;\; n \in Z_{J-1}^+ \bigcup\{0\} \end{cases} \tag{59}$$

$$\left.\begin{array}{l} \dfrac{\partial^2 \rho}{\partial \phi^2}(f,\theta_0,\phi_0) \;=\; 0 \\[6pt] \forall \; f \geq 0 \end{array}\right\} \Leftrightarrow \begin{cases} \underline{W}^T \tilde{Q}_{\phi,\phi,1,n} \underline{W} = 0, \;\; n \in Z_{J-1}^+ \\[4pt] \underline{W}^T \tilde{Q}_{\phi,\phi,2,n} \underline{W} = 0, \;\; n \in Z_{J-1}^+ \bigcup\{0\} \end{cases} \tag{60}$$

$$\left.\begin{array}{l} \dfrac{\partial^2 \rho}{\partial \theta \partial \phi}(f,\theta_0,\phi_0) \;=\; 0 \\[6pt] \forall \; f \geq 0 \end{array}\right\} \Leftrightarrow \begin{cases} \underline{W}^T \tilde{Q}_{\theta,\phi,1,n} \underline{W} = 0, \;\; n \in Z_{J-1}^+ \\[4pt] \underline{W}^T \tilde{Q}_{\theta,\phi,2,n} \underline{W} = 0, \;\; n \in Z_{J-1}^+ \bigcup\{0\} \end{cases} \tag{61}$$

where the $\tilde{Q}$ matrices $\in M_{LJxLJ}(R)$ are defined by

$$\tilde{Q}_{\theta,n} \;=\; (J_n - J_{-n}) \otimes \frac{\partial A}{\partial \theta}(\theta_0,\phi_0) \tag{62}$$

$$\tilde{Q}_{\phi,n} \;=\; (J_n - J_{-n}) \otimes \frac{\partial A}{\partial \phi}(\theta_0,\phi_0) \tag{63}$$

$$\tilde{Q}_{\theta,\theta,1,n} \;=\; (J_n - J_{-n}) \otimes \frac{\partial^2 A}{\partial \theta^2}(\theta_0,\phi_0) \tag{64}$$

$$\tilde{Q}_{\theta,\theta,2,n} \;=\; (J_n + J_{-n}) \otimes \frac{\partial B}{\partial \theta}(\theta_0,\phi_0) \frac{\partial \Lambda}{\partial \theta}(\theta_0,\phi_0) \tag{65}$$

$$\tilde{Q}_{\phi,\phi,1,n} \;=\; (J_n - J_{-n}) \otimes \frac{\partial^2 A}{\partial \phi^2}(\theta_0,\phi_0) \tag{66}$$

$$\tilde{Q}_{\phi,\phi,2,n} \;=\; (J_n + J_{-n}) \otimes \frac{\partial B}{\partial \phi}(\theta_0,\phi_0) \frac{\partial \Lambda}{\partial \phi}(\theta_0,\phi_0) \tag{67}$$

$$\tilde{Q}_{\theta,\phi,1,n} \;=\; (J_n - J_{-n}) \otimes \frac{\partial^2 A}{\partial \theta \partial \phi}(\theta_0,\phi_0) \tag{68}$$

$$\tilde{Q}_{\theta,\phi,2,n} = (J_n + J_{-n}) \otimes \frac{\partial B}{\partial \theta}(\theta_0,\phi_0) \frac{\partial \Lambda}{\partial \phi}(\theta_0,\phi_0) \qquad (69)$$

The above equations and relations state that the necessary and sufficient conditions for a maximally flat spatial response are in general nonlinear quadratic constraints on the weight vector $\underline{W}$ and involve the $\tilde{Q}$ matrices defined by Eqs. (62)-(69). Note that the matrices are neither negative definite nor positive definite. If they were, there would be no possible solution except $\underline{W} = 0$. In general, the $\tilde{Q}$ matrices are not symmetric. However, since they appear in a quadratic form, it is necessary to consider only the symmetric part of the matrices. The symmetric part of a matrix A is defined to be the symmetric matrix B defined by

$$B = \frac{A + A^T}{2} \qquad (70)$$

which has the property that for $\forall\, \underline{W}$

$$\underline{W}^T B \underline{W} = \underline{W}^T A \underline{W} \qquad (71)$$

C.   Zero Order Plus Maximal Flatness Constraints

The zero order look direction constraint defined by Eq. (29) is required to ensure that the array processor has the desired frequency response in the look direction specified by $\underline{h}$. In view of this basic requirement, we examine the combination of the zero order constraint and the constraints derived in the previous section for achieving a maximally flat spatial response in the look direction.

1.   First Order Maximally Flat Spatial Response

In this section, we consider the optimization problem relevant to a minimum power array processor subject to the zero order look direction constraint plus the necessary and sufficient conditions for a first order maximally flat response defined by Eq. (57) and Eq. (58).

For convenience of presentation, we define the following subsets of R^{LJ}:

$$G_0 = \left\{ \underline{W} \mid C_0 \underline{W} = \underline{h} \right\} \tag{72}$$

$$G_{\theta,1} = \left\{ \underline{W} \mid \underline{W}^T \tilde{Q}_{\theta,n} \underline{W} = 0, \; n \in Z_{J-1}^+ \right\} \tag{73}$$

$$G_{\phi,1} = \left\{ \underline{W} \mid \underline{W}^T \tilde{Q}_{\phi,n} \underline{W} = 0, \; n \in Z_{J-1}^+ \right\} \tag{74}$$

$$L_{\theta,1} = \left\{ \underline{W} \mid \underline{h}^T (J_n - J_{-n}) \otimes \underline{1}_L^T \frac{\partial \Lambda}{\partial \theta}(\theta_0,\phi_0)\underline{W} = 0, \; n \in Z_{J-1}^+ \right\} \tag{75}$$

$$L_{\phi,1} = \left\{ \underline{W} \mid \underline{h}^T (J_n - J_{-n}) \otimes \underline{1}_L^T \frac{\partial \Lambda}{\partial \phi}(\theta_0,\phi_0)\underline{W} = 0, \; n \in Z_{J-1}^+ \right\} \tag{76}$$

The array processor optimization problem to achieve a maximally flat spatial response and specified frequency response in the look direction while rejecting noise and interferences can be stated as

$$\underset{\underline{W}}{\text{minimise}} \quad \underline{W}^T R \underline{W} \tag{77}$$

$$\text{subject to} \quad \underline{W} \in S_{0,1} \tag{78}$$

where

$$S_{0,1} \triangleq G_0 \cap G_{\theta,1} \cap G_{\phi,1} \tag{79}$$

This problem is similar to the minimum power linearly constrained problem defined by Eq. (31) and Eq. (32) except that $\underline{W}$ is constrained to lie in the intersection of G_0 with two other sets $G_{\theta,1}$ and $G_{\phi,1}$. As stated in Eqs. (77)-(79) the problem is quite difficult to solve because the additional constraints as defined by Eq. (73) and Eq. (74) are nonlinear in $\underline{W}$.

Fortunately, we can show the following important result

$$S_{0,1} \equiv \tilde{S}_{0,1} \triangleq G_0 \cap L_{\theta,1} \cap L_{\phi,1} \tag{80}$$

The importance of this result lies in the fact that the sets on the right hand side of Eq. (80) are defined by linear constraints on $\underline{W}$. We briefly illustrate the steps required to show that

$$G_0 \cap G_{\theta,1} \equiv G_0 \cap L_{\theta,1} \tag{81}$$

since it is then straight forward to show the complete result. We make use of the following Kronecker product results [49].

$$(A \otimes B)(C \otimes D) = AC \otimes BD \tag{82}$$

$$(A \otimes B)^T = A^T \otimes B^T \tag{83}$$

From Eq. (73), Eq. (62) and Eq. (47), it follows that

$$G_{\theta,1} = \left\{ \underline{W} \ \middle| \ \underline{W}^T (J_n - J_{-n}) \otimes \frac{\partial \Lambda}{\partial \theta}(\theta_0, \phi_0) \underline{1}_L \underline{1}_L^T \underline{W} = 0, \ n \in Z_{J-1}^+ \right\} \tag{84}$$

Consider now the quadratic in Eq. (84)

$$\underline{W}^T (J_n - J_{-n}) \otimes \frac{\partial \Lambda}{\partial \theta}(\theta_0, \phi_0) \underline{1}_L \underline{1}_L^T \underline{W}$$

$$\Leftrightarrow \underline{W}^T \left((J_n - J_{-n}) I_J \right) \otimes \frac{\partial \Lambda}{\partial \theta}(\theta_0, \phi_0) \underline{1}_L \underline{1}_L^T \underline{W}$$

$$\Leftrightarrow \underline{W}^T \left((J_n - J_{-n}) \otimes \frac{\partial \Lambda}{\partial \theta}(\theta_0, \phi_0) \underline{1}_L \right) (I_J \otimes \underline{1}_L^T) \underline{W}$$

$$\Leftrightarrow \underline{W}^T (I_J \otimes \underline{1}_L^T)^T \left((J_n - J_{-n}) \otimes \frac{\partial \Lambda}{\partial \theta}(\theta_0, \phi_0) \underline{1}_L \right)^T \underline{W}$$

$$\Leftrightarrow \underline{W}^T (I_J^T \otimes \underline{1}_L)(J_n - J_{-n})^T \otimes \underline{1}_L^T \frac{\partial \Lambda}{\partial \theta}(\theta_0, \phi_0) \underline{W}$$

$$\Leftrightarrow -\underline{W}^T C_0^T (J_n - J_{-n})^T \otimes \underline{1}_L^T \frac{\partial \Lambda}{\partial \theta}(\theta_0, \phi_0) \underline{W}$$

We now use

$$G_0 = \left\{ \underline{W} \ \middle| \ C_0 \underline{W} = \underline{h} \right\}$$

to complete the proof that

$$G_0 \cap G_{\theta,1} \equiv G_0 \cap L_{\theta,1}$$

The minimum power processor with maximally flat spatial response and specified frequency response is given by the solution to the linearly constrained convex problem defined by

$$\underset{\underline{W}}{\text{minimise}} \quad \underline{W}^T R \underline{W} \tag{85}$$

$$\text{subject to} \quad C_1 \underline{W} = \underline{h}_1 \tag{86}$$

where $C_1 \in M_{(3J-2) \times LJ}(R)$ is given by

$$C_1^T = \left[C_0^T, C_{\theta,1,1}^T, \ldots, C_{\theta,1,n}^T, C_{\phi,1,1}^T, \ldots, C_{\phi,1,n}^T, \right] \tag{87}$$

$$C_{\theta,1,n} = \underline{h}^T (J_n - J_{-n}) \otimes \underline{1}_L^T \frac{\partial \Lambda}{\partial \theta}(\theta_0, \phi_0), \quad n \in Z_{J-1}^+ \tag{88}$$

$$C_{\phi,1,n} = \underline{h}^T (J_n - J_{-n}) \otimes \underline{1}_L^T \frac{\partial \Lambda}{\partial \phi}(\theta_0, \phi_0), \quad n \in Z_{J-1}^+ \tag{89}$$

and $\underline{h}_1 \in R^{(3J-2)}$ is given by

$$\underline{h}_1^T = \left[\underline{h}^T, \underline{0}^T, \underline{0}^T \right] \tag{90}$$

The linear constraints may not all be linearly independent, in which case a linearly independent set needs to be derived. Assuming that R is invertible, the solution to Eqs. (85)-(86) is given by

$$\underline{\hat{W}} = R^{-1} C_1^T \left(C_1 R^{-1} C_1^T \right)^{-1} \underline{h}_1 \tag{91}$$

and

$$p(\underline{\hat{W}}) = \underline{h}_1^T \left(C_1 R^{-1} C_1^T \right)^{-1} \underline{h}_1 \tag{92}$$

The solution defined by Eq. (91) is independent of the choice of the co-ordinate systems since the constraints for a maximally flat spatial response are derived from necessary and sufficient conditions. Furthermore, constraints are valid for arbitrary $\underline{h}$. In [43], sufficient linear constraints were derived for an arbitrary $\underline{h}$ case, including a flat frequency response. However, the linear constraints derived in [43] were found to be dependent on the choice of origin [50-52].

2.   Second Order Maximally Flat Spatial Response

In this section, we briefly consider the second order maximally flat spatial response case. It is clear from Eqs. (59)-(61) that the necessary and sufficient conditions are defined by quadratic constraints on $\underline{W}$. While it is possible to derive equivalent necessary and sufficient linear constraints for the quadratics involving $\tilde{Q}_{\theta,\theta,1,n}$, $\tilde{Q}_{\theta,\phi,1,n}$ and $\tilde{Q}_{\phi,\phi,1,n}$ given the zero order constraint, it has not been possible to effect the same simplification for the remaining quadratic constraints. Hence, the minimum power antenna array processor optimization problem involves both linear and nonlinear constraints. Unfortunately, the nonlinear constraints do not appear to have any nice properties and their study remains an open research problem. We present some numerical results to illustrate the effectiveness of second order constraints.

It is possible to derive linear constraints on $\underline{W}$ that are sufficient conditions for a maximally flat second order spatial response [43]. However, these constraints are dependent on the choice of co-ordinate system origin. The sufficient constraints have been used to determine the initial value for the nonlinear programming problem that results from the necessary and sufficient conditions.

3.   Numerical Results

The performance achievable with the first and second order derivative constraints in a spatial spectrum estimation application has been computed for a number of different cases. The results are illustrated in Figs. 3-5. A uniformly spaced linear array with ten elements and a processor with seven tapped delay line sections was used, corresponding to $L = 10$ and $J = 7$. The inter-element spacing of the array is usually specified in terms of a dimensionless quantity obtained by dividing the inter-element distance by the wavelength of the highest frequency of interest. In our examples, the inter-element spacing was 0.4.

It is common to specify the spectrum of signals relative to the Nyquist sampling frequency (1/T). In our examples, a single broadband source with a flat spectrum over [0.125, 0.25] was incident on the array from the broadside direction. The white noise level on each element was set at -30 dB and the spherically isotropic noise level was set at 0 dB. The linear array axis was located in the xy plane and aligned to the x axis of the co-ordinate system and the phase centre of the array was altered to illustrate different effects.

Figs. 3-5 show the optimal power estimate of the array processor as the look direction is varied over the angle θ. The following notation is used to describe the different processors:

ZOL: Zero Order Linear constraints only

FOLNS: First Order Linear Necessary and Sufficient constraints plus zero order constraints

FOLS: First Order Linear Sufficient constraints, [43], plus zero order constraints

SONS: Second Order Necessary and Sufficient constraints plus zero order constraints

SOLS: Second Order Linear Sufficient constraints, [43], plus zero order constraints

Note for both SONS and SOLS the first order necessary and sufficient constraints are used.

Fig. 3 shows that when only the zero order constraint is used, a rapid attenuation of the signal occurs should the optimum processor look direction be mismatched with the actual angle of arrival, as might occur when a small number of scan directions are used. Fig. 3 also shows that the first order maximally flat processor is quite effective in broadening the acceptance angle and would enable a smaller number of scan directions to be used. It is also clear that there is only a small loss in performance against the isotropic noise as the processor is pointed away from the source. The array phase centre was located at the origin of the co-ordinate system.

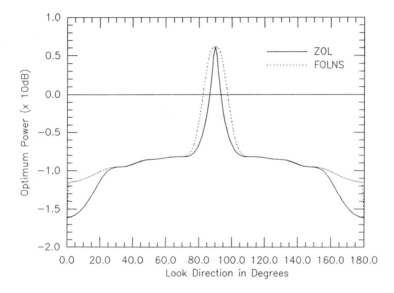

Fig. 3 Optimum power estimate plots for the ZOL and FOLNS processors.

Fig. 4 is included to illustrate the effect of the dependence on the choice of co-ordinate system origin on the first order sufficient conditions initially given in [43]. The array phase centre was located at $(x = 1, y = 2)$, where the units are measured in terms of the wavelength of the upper frequency limit of the band of interests. It is clear that the sufficient conditions behave very badly as the look direction approaches the endfire direction of the array. We also note that the necessary and sufficient conditions yield the same performance as shown in Fig. 3, since they are independent of the choice of origin.

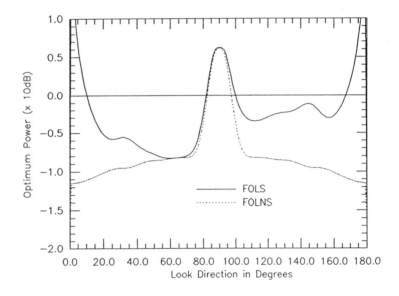

Fig. 4 Optimum power estimate plots for the FOLS and FOLNS processors.

Fig. 5 shows the performance for second order constraints when the array phase centre is located at the origin of the co-ordinate system. Both the linear second order sufficient conditions and the nonlinear second order necessary and sufficient conditions yield the same result. However, we are not certain that the results for the nonlinear constraints correspond to the global optimum. The nonlinear programming problem was solved using a NAG library routine and the solution for the sufficient conditions was used as the initial value. Nevertheless, it is clear that the second order constraints are very effective in achieving a wider acceptance angle. A comparison of Fig. 5 and Fig. 3 shows that there is a further loss in performance against isotropic noise when the processor is pointed away from the source.

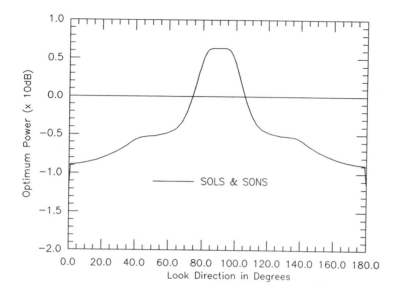

Fig. 5 Optimum power estimate plots for SONS and SOLS processors.

V. QUADRATIC CONSTRAINT APPROACH TO THE DESIGN OF ROBUST BROADBAND PROCESSORS

The maximally flat spatial response approach described in the previous section can be effective in handling direction of arrival mismatch problems. However, it cannot handle many other errors and imperfections that arise. Furthermore, the approach is limited to exactly presteered antenna array processors. The need for presteering cannot be ignored in the formulation of an approach to the design of robust broadband antenna array processors since presteering itself produces difficulties.

There can be a significant computational cost in the implementation of adaptive versions of optimum processors that involve presteering since the correlation matrix is a function of look direction.

Another difficulty with the presteered processors is that in a digital implementation, it requires that the output of each sensor be sampled in time prior to beamforming. In this case, the only time delays available are integer multiples of the sampling interval [53-55]. For a given quantization there may be a discrete set of "look directions" which can be achieved exactly. These exact beams are referred to as synchronous beams [54-55]. The number of synchronous beam pointing directions increases with the input sampling rate. Digital interpolation of the sampled data [56-59] can be used to generate the required delayed signals so that the input sampling rate needs only to be consistent with the Nyquist rate. However, this represents an additional computational burden and the interpolation is not necessarily optimal.

To eliminate the need for the broadband time steering delays and at the same time achieve broadband capability, various types of linear and non-linear constraints have been proposed in the literature [60-61]. However, for broadband signals occupying a band as broad as one octave, these approaches achieve only limited success.

This section presents a quadratic constraint approach for designing robust broadband processors. The approach is based on the idea of minimizing the mean-square-deviation between a desired look direction response and the response of the processor over a frequency band of interest. With this approach, three types of presteering can be handled : no presteering, coarse presteering and exact presteering. The new design approach enables various types of errors and mismatches between signal model and actual scenario to be incorporated in the problem formulation, thereby producing a robust processor.

A. Response Constraint

For a given look direction (θ,ϕ), the mean-square-deviation between a desired response, $A(f,\theta,\phi)$ and the response of the processor over a frequency band of interest $[f_\ell, f_u]$ is given by

$$e_1^2 = \frac{1}{\sigma} \int_{f_\ell}^{f_u} \left| A(f,\theta,\phi) - H(f,\theta,\phi) \right|^2 df \qquad (93)$$

where $A(f,\theta,\phi)$ is the desired look direction complex frequency response and σ is a normalisation scalar given by

$$\sigma = \int_{f_\ell}^{f_u} A^*(f,\theta,\phi) A(f,\theta,\phi) df \qquad (94)$$

Substituting Eq. (22) into Eq. (94) and after some manipulation, one obtains

$$e_1^2 = \underline{W}^T Q_1 \underline{W} - 2\underline{P}_1^T \underline{W} + 1 \qquad (95)$$

where $Q_1 \in M_{LJ \times LJ}(R)$ is the nonnegative definite matrix defined by

$$[Q_1]_{k,\ell} = \psi\left[(\tau_i - \tau_j) + (T_i - T_j) + (n-m)T \right]$$
$$k = i + (m-1)L, \quad i,j \in Z_L^+$$
$$\ell = j + (n-1)L, \quad m,n \in Z_J^+ \qquad (96)$$

where

$$\psi[\tilde{\tau}] = \frac{1}{\sigma} \left[f_u \operatorname{sinc}(2\pi f_u \tilde{\tau}) - f_\ell \operatorname{sinc}(2\pi f_\ell \tilde{\tau}) \right] \qquad (97)$$

$$\operatorname{sinc}(2\pi f \tilde{\tau}) \triangleq \frac{\sin(2\pi f \tilde{\tau})}{2\pi f \tilde{\tau}} \qquad (98)$$

and $\underline{P}_1 \in R^{LJ}$ is defined by

$$[\underline{P}_1]_{(j-1)L+i} = \operatorname{Real}\left\{ \frac{1}{\sigma} \int_{f_\ell}^{f_u} A^*(f,\theta,\phi) \exp\left[j2\pi f(\tau_i - T_i - (j-1)T) \right] df \right\}$$
$$j \in Z_J^+, \; i \in Z_L^+ \qquad (99)$$

where * denotes complex conjugate.

Note that Eq. (95) can be factorized as

$$e_1^2 = \left(\underline{W}_1^0 - \underline{W} \right)^T Q_1 \left(\underline{W}_1^0 - \underline{W} \right) + \alpha_1 \qquad (100)$$

where α_1 is a scalar given by

$$\alpha_1 = 1 - \underline{W}_1^{0^T} Q_1 \underline{W}_1^0 \tag{101}$$

and $\underline{W}_1^0 \in R^{LJ}$ satisfies

$$Q_1 \underline{W}_1^0 = \underline{P}_1 \tag{102}$$

To limit signal distortion over the frequency band of interest, Eq. (100) can be constrained to be less than or equal to a small quantity, that is

$$\left(\underline{W}_1^0 - \underline{W}\right)^T Q_1 \left(\underline{W}_1^0 - \underline{W}\right) + \alpha_1 \leq \xi \tag{103}$$

Let

$$\epsilon_1 \triangleq \xi - \alpha_1 \tag{104}$$

then Eq. (103) can be expressed as

$$\left(\underline{W}_1^0 - \underline{W}\right)^T Q_1 \left(\underline{W}_1^0 - \underline{W}\right) \leq \epsilon_1 \tag{105}$$

Note that the $\underline{W}_1^0$ vector given by Eq. (102) corresponds to the vector which minimizes the mean-square-deviation function given by Eq. (95) and the scalar α_1 corresponds to the minimum mean-square-deviation. Also, note that there is a lower bound on the value of ξ which can be specified. Since Q_1 is positive semidefinite, it is clear from Eq. (105) that

$$\epsilon_1 \geq 0 \tag{106}$$

From Eq. (101), Eq. (104) and Eq. (106), it follows that for a given array and a given number of taps, the computable lower bound on ξ is given by

$$\xi \geq 1 - \underline{W}_1^{0^T} Q_1 \underline{W}_1^0 \tag{107}$$

If the processor is designed to have a flat frequency response over a frequency band of interest, $A(f,\theta,\phi)$ can be chosen to be

$$A(f,\theta,\phi) = \exp(j2\pi f\tau) \tag{108}$$

where τ is a time delay parameter which can be optimized.

The three types of presteering which can be handled with the above approach are now considered. Let (θ_0, ϕ_0) be a specific look direction.

1.   No Presteering

In this case, $D(f) = I$. Physically, this means that each sensor data is not time shifted as in the presteered processors. This case is an important application of the new processor.

2.   Coarse Presteering as Limited by Sampling Time T

In coarse presteering, the $\{T_i\}$ are chosen so as to approximate in some sense the exact time delays necessary to achieve steering, subject to the discretization introduced by the sampling of the signals at interval specified by T. In this case, $\{T_i\}$ is a set of time delays given by

$$T_i = m_i T, \ i \in Z_L^+ \tag{109}$$

where $\{m_i\}$ is a set of integer chosen such that

$$m_0 = \left\lfloor \frac{T_0}{T} \right\rfloor \tag{110}$$

and

$$m_i = \left\lfloor \frac{\tau_i(\theta_0, \phi_0) + T_0}{T} \right\rfloor, \ i \in Z_L^+ \tag{111}$$

or

$$m_0 = \left\lceil \frac{T_0}{T} \right\rceil \tag{112}$$

and

$$m_i = \left\lceil \frac{\tau_i(\theta_0, \phi_0) + T_0}{T} \right\rceil, \ i \in Z_L^+ \tag{113}$$

where $\lfloor \alpha \rfloor$ denotes the smallest integer greater than or equal to α and $\lceil \alpha \rceil$ denotes the greatest integer smaller than or equal to α.

Note that either Eq. (111) or Eq. (113) will ensure that

$$\left| \tau_i(\theta_0,\phi_0) + T_0 - T_i \right| = \left| \tau_i(\theta_0,\phi_0) + T_0 - m_i T \right| < T, \quad i \in Z_L^+ \tag{114}$$

and thus

$$\underline{S}^T(f,\theta_0,\phi_0)D(f) = \alpha_0(f)\underline{\tilde{S}}^T(f,\theta_0,\phi_0) \tag{115}$$

where

$$\alpha_0(f) = \exp(-j2\pi f m_0 T) \quad \text{for some} \quad m_0 \tag{116}$$

and

$$\left[\underline{\tilde{S}}(f,\theta_0,\phi_0)\right]_k = \exp(j2\pi f \tilde{\tau}_k), \quad k \in Z_L^+ \tag{117}$$

where

$$0 \le \left| \tilde{\tau}_i \right| < T, \quad i \in Z_L^+ \tag{118}$$

Other methods of quantizing the steering delays are possible. For example, the integers $\{m_i\}$ can be chosen such that

$$\left| \tau_i(\theta_0,\phi_0) + T_0 - T_i \right| = \left| \tau_i(\theta_0,\phi_0) + T_0 - m_i T \right| \le T/2 \tag{119}$$

The method of quantization affects the theory to be presented only in as much as it specifies a set of $\{T_i\}$. The performance of the optimized processor will depend on the type of quantization used. This case of coarse presteering is also an important application of the new approach.

3.   Exact Presteering

The broadband delays are in this case adjusted so that

$$\underline{S}^T(f,\theta_0,\phi_0)D(f) = \alpha(f)\underline{1}_L^T \tag{120}$$

where

$$\alpha(f) = \exp(-j2\pi f T_0), \quad \text{for some} \quad T_0 \tag{121}$$

The exact steering delays are given by

$$T_i = \tau_i(\theta_0,\phi_0)+T_0, \; i \in Z_L^+ \tag{122}$$

Note that the three steering situations can be treated simultaneously since each case merely determines a specific set of $\{T_i\}$.

B.   Generalized Response Deviation Constraint

Apart from the broadband capability as formulated in Section V, A, the basic approach can also be used to design broadband processors with robustness capabilities against various types of errors and mismatches between signal model and the actual scenario.

A generalized response deviation constraint of the form given by Eq. (105) can be formulated as follows:

The weighted mean-square-deviation of the processor response over variation in parameters p_1 to p_m is defined as

$$e_\gamma^2 = \frac{1}{\beta} \int_{p_m^0-\Delta p_m/2}^{p_m^0+\Delta p_m/2} \cdots \int_{p_1^0-\Delta p_1/2}^{p_1^0+\Delta p_1/2} \Omega(p_1,p_2,...,p_m) e_1^2(p_1,p_2,...,p_m) dp_1 ... dp_m$$

$$= \underline{W}^T Q_\gamma \underline{W} - 2\underline{P}_\gamma^T \underline{W} + 1 \tag{123}$$

where $\Omega(p_1,p_2,...,p_m)$ is a non-negative weighting function for deterministic type of parameters or a probability density function (pdf) for those parameters which are modelled as random variables, $Q_\gamma \in M_{LJ\times LJ}(R)$ is given by

$$Q_\gamma = \frac{1}{\beta} \int_{p_m^0-\Delta p_m/2}^{p_m^0+\Delta p_m/2} \cdots \int_{p_1^0-\Delta p_1/2}^{p_1^0+\Delta p_1/2} \Omega(p_1,p_2,...,p_m) Q_1(p_1,p_2,...,p_m) dp_1 ... dp_m,$$
$$\tag{124}$$

$\underline{P}_\gamma \in R^{LJ}$ is given by

$$\underline{P}_\gamma = \frac{1}{\beta} \int_{p_m^0-\Delta p_m/2}^{p_m^0+\Delta p_m/2} \cdots \int_{p_1^0-\Delta p_1/2}^{p_1^0+\Delta p_1/2} \underline{P}_1(p_1,p_2,...,p_m) dp_1 ... dp_m, \tag{125}$$

and

$$\beta = \int_{p_m^0-\Delta p_m/2}^{p_m^0+\Delta p_m/2} \cdots \int_{p_1^0-\Delta p_1/2}^{p_1^0+\Delta p_1/2} \Omega(p_1, p_2, ..., p_m) dp_1 ... dp_m \tag{126}$$

where $\{p_i, \ i \in Z_m^+\}$ are the parameters related to signal model or array geometry, and γ is an index used to indicate the appropriate integral used, depending on which parameters are included in the integral.

Note that Eq. (123) can also be factorized as

$$e_\gamma^2 = \left(\underline{W}_\gamma^0 - \underline{W}\right)^T Q_\gamma \left(\underline{W}_\gamma^0 - \underline{W}\right) + \alpha_\gamma \tag{127}$$

where

$$Q_\gamma \underline{W}_\gamma^0 = \underline{P}_\gamma \tag{128}$$

and

$$\alpha_\gamma = 1 - \underline{W}_\gamma^{0^T} Q_\gamma \underline{W}_\gamma^0 \tag{129}$$

Robustness in the design can be achieved by introducing the constraint

$$\left(\underline{W}_\gamma^0 - \underline{W}\right)^T Q_\gamma \left(\underline{W}_\gamma^0 - \underline{W}\right) \leq \epsilon_\gamma \tag{130}$$

where

$$\epsilon_\gamma \triangleq \xi - \alpha_\gamma \tag{131}$$

The new processor can be optimized in the sense of rejecting non-look direction interferences by minimizing the mean output power subject to the quadratic constraint which ensures low distortion of the look direction signal even under perturbed conditions.

1. Directional Mismatch

As one example, we address the problem of performance degradation due to mismatch between the assumed look direction and the actual direction of incident of the desired signal, that was considered in Section IV. To overcome this look direction mismatch problem, it is desirable to broaden the acceptance angle, i.e. the width of the beam in the look direction, whilst preserving the array processor's ability to reject noise and interferences outside the acceptance angle. Beam broadening in ϕ domain can be achieved by integrating Eq. (95) over a spatial region $\Delta\phi$ in the look direction ϕ_0 as follows:

$$e_2^2 = \frac{1}{\Delta\phi}\int_{\phi_0-\Delta\phi/2}^{\phi_0+\Delta\phi/2}e_1^2 d\phi$$

$$= \underline{W}^T Q_2 \underline{W} - 2\underline{P}_2^T \underline{W} + 1 \tag{132}$$

where $Q_2 \in M_{LJxLJ}(R)$ is given by

$$Q_2 = \frac{1}{\Delta\phi}\int_{\phi_0-\Delta\phi/2}^{\phi_0+\Delta\phi/2}Q_1 d\phi \tag{133}$$

and $\underline{P}_2 \in R^{LJ}$ is given by

$$\underline{P}_2 = \frac{1}{\Delta\phi}\int_{\phi_0-\Delta\phi/2}^{\phi_0+\Delta\phi/2}\underline{P}_1 d\phi \tag{134}$$

where $\Delta\phi$ specifies some angular tolerance over which the desired response is to be preserved.

Similar to the derivation of Eq. (130), a quadratic constraint based on Eq. (132) can be obtained, namely,

$$\left(\underline{W}_2^0 - \underline{W}\right)^T Q_2\left(\underline{W}_2^0 - \underline{W}\right) \leq \epsilon_2 \tag{135}$$

where

$$\epsilon_2 \overset{\Delta}{=} \xi - \alpha_2 \tag{136}$$

$$Q_2\underline{W}_2^0 = \underline{P}_2 \tag{137}$$

$$\alpha_2 = 1 - \underline{W}_2^{0^T} Q_2 \underline{W}_2^0 \tag{138}$$

2.   Deviations in Array Geometry

As another example of the application of the generalized response deviation constraint, we consider incorporating parameters for perturbations in array geometry.

In the first instance, we consider perturbations of the inter-element spacing of sensors in a linear array that nominally has uniform spacing between sensors. Robustness in design can be formulated by introducing the inter-element spacing as say p_1 in Eq. (123). The parameter is assumed to be in the range $[r_0 - \Delta r/2, \ r_0 + \Delta r/2]$. Thus we obtain the following integrals:

$$e_2^2 = \frac{1}{\Delta r} \int_{r_0-\Delta r/2}^{r_0+\Delta r/2} e_1^2 dr$$

$$= \underline{W}^T Q_2 \underline{W} - 2\underline{P}_2^T \underline{W} + 1 \tag{139}$$

where $Q_2 \in M_{LJ \times LJ}(R)$ is given by

$$Q_2 = \frac{1}{\Delta r} \int_{r_0-\Delta r/2}^{r_0+\Delta r/2} Q_1 dr \tag{140}$$

where

$$[Q_2]_{k,\ell} = \frac{1}{2\pi\sigma\Delta r h_{ij}} \left\{ Si\left[2\pi f_u\left(\left(r_0 + \frac{\Delta r}{2} \right) h_{ij} + \mathring{\delta}_{nm} \right) \right] \right.$$

$$- Si\left[2\pi f_u\left(\left(r_0 - \frac{\Delta r}{2} \right) h_{ij} + \mathring{\delta}_{nm} \right) \right]$$

$$- Si\left[2\pi f_\ell\left(\left(r_0 + \frac{\Delta r}{2} \right) h_{ij} + \mathring{\delta}_{nm} \right) \right]$$

$$\left. + Si\left[2\pi f_\ell\left(\left(r_0 - \frac{\Delta r}{2} \right) h_{ij} + \mathring{\delta}_{nm} \right) \right] \right\}$$

$$k = i+(m-1)L, \ i,j \in Z_L^+$$
$$\ell = j+(n-1)L, \ m,n \in Z_J^+ \tag{141}$$

where

$$Si(\alpha) = \int_0^\alpha \frac{\sin(u)}{u} du \tag{142}$$

$$h_{ij} = \frac{1}{\upsilon}(i-j)\sin\phi \tag{143}$$

$$\mathring{\delta}_{nm} = \left(T_j - T_i \right) + (n-m)T \tag{144}$$

If $h_{ij} = 0$, then

$$[Q_2]_{k,\ell} = \frac{1}{\sigma}\left[f_u\text{sinc}\left(2\pi f_u \overset{\circ}{\delta}_{nm}\right) - f_\ell\text{sinc}\left(2\pi f_\ell \overset{\circ}{\delta}_{nm}\right)\right] \qquad (145)$$

$\underline{P}_2 \in R^{LJ}$ is given by

$$\underline{P}_2 = \frac{1}{\Delta r}\int_{r_0-\Delta r/2}^{r_0+\Delta r/2}\underline{P}_1 dr \qquad (146)$$

For the case of a flat frequency response

$$[\underline{P}_2]_{(m-1)L+i} = \frac{1}{2\pi\Delta f\Delta r d_i}\left\{\text{Si}\left[2\pi f_u\left(\delta'_m-\left(r_0+\frac{\Delta r}{2}\right)d_i\right)\right]\right.$$

$$- \text{Si}\left[2\pi f_u\left(\delta'_m-\left(r_0-\frac{\Delta r}{2}\right)d_i\right)\right]$$

$$- \text{Si}\left[2\pi f_\ell\left(\delta'_m-\left(r_0+\frac{\Delta r}{2}\right)d_i\right)\right]$$

$$\left.+ \text{Si}\left[2\pi f_\ell\left(\delta'_m-\left(r_0-\frac{\Delta r}{2}\right)d_i\right)\right]\right\}$$

$$i \in Z_L^+, \quad m \in Z_J^+ \qquad (147)$$

where

$$\delta'_m = \tau+(m-1)T+T_i \qquad (148)$$

$$d_i = \frac{1}{\upsilon}\left(i-\frac{L+1}{2}\right)\sin\phi \qquad (149)$$

We now consider a second example of how to incorporate robustness against perturbations in array geometry. Consider the circular array shown in Fig. 6. Robustness against variation in the array parameter r can be incorporated in the design by associating r with the parameter say p_1 in Eq. (123). The resulting Q_2 matrix has the form given by Eq. (141) but with h_{ij} given by

$$h_{ij} = \begin{cases} \dfrac{5}{\upsilon}\cos(\phi-\alpha_i)-\dfrac{5}{\upsilon}\cos(\phi-\alpha_j), & 1\le i,j\le 5 \\[3ex] \dfrac{4}{\upsilon}\cos(\phi-\alpha_i)-\dfrac{4}{\upsilon}\cos(\phi-\alpha_j), & 6\le i,j\le 10 \\[3ex] \dfrac{5}{\upsilon}\cos(\phi-\alpha_i)-\dfrac{4}{\upsilon}\cos(\phi-\alpha_j), & \begin{aligned}1\le i\le 5\\6\le j\le 10\end{aligned} \\[3ex] \dfrac{4}{\upsilon}\cos(\phi-\alpha_i)-\dfrac{5}{\upsilon}\cos(\phi-\alpha_j), & \begin{aligned}6\le i\le 10\\1\le j\le 5\end{aligned} \end{cases} \qquad (150)$$

where

$$\alpha_k = \frac{2(k-1)\pi}{5}, \quad k\in Z_{10}^+ \qquad (151)$$

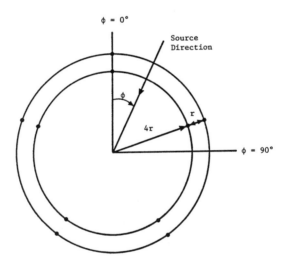

Fig. 6 Azimuth plane of a double-ring circular array.

Also, it can be shown that for the case of flat response, the kth component of $\underline{P}_2$ has the same form as Eq. (147) but with d_i given by

$$d_i = \begin{cases} \dfrac{5}{\upsilon}\cos(\phi-\alpha_i), & 1\le i\le 5 \\[3mm] \dfrac{4}{\upsilon}\cos(\phi-\alpha_i), & 6\le i\le 10 \end{cases} \tag{152}$$

3. Channel Phase Errors

As one example of the application of the generalized response deviation constraint involving probabilistic type of parameters, consider incorporating uncertainty in channel time delay. In a real system, sensor phase mismatch is a function of frequency. For simplicity, it is assumed that sensor phase differences take the form of a pure time delay, or equivalently that phase errors are proportional to f.

In this case, e_1^2 given by Eq. (95) is a function of random time delay deviations $\{\tilde{\delta}_i, \ i\in Z_L^+\}$. It is assumed that $\{\tilde{\delta}_i, \ i\in Z_L^+\}$ are independent random variables, uniformly distributed in $[-\Delta\tau/2, \Delta\tau/2]$ and are of zero mean and variance σ_δ^2. Robustness in the design can be formulated as follows:

$$e_2^2 = \frac{1}{\beta}\int_{-\Delta\tau/2}^{\Delta\tau/2}\!\!\cdots\int_{-\Delta\tau/2}^{\Delta\tau/2}\Omega\big(\tilde{\delta}_1,...,\tilde{\delta}_L\big)e_1^2\big(\tilde{\delta}_1,...,\tilde{\delta}_L\big)d\tilde{\delta}_1...d\tilde{\delta}_L \tag{153}$$

where β is a scalar given by

$$\beta = \int_{-\Delta\tau/2}^{\Delta\tau/2}\!\!\cdots\int_{-\Delta\tau/2}^{\Delta\tau/2}\Omega\big(\tilde{\delta}_1, ..., \tilde{\delta}_L\big)d\tilde{\delta}_1...d\tilde{\delta}_L \tag{154}$$

and $\Omega(\tilde{\delta}_1,...,\tilde{\delta}_L)$ is the joint pdf of $\{\tilde{\delta}_i, \ i\in Z_L^+\}$ and has the property that

$$\iint_{-\infty}^{\;\;\;\infty}\!\!\!\!\cdots\int\Omega\big(\tilde{\delta}_1, ..., \tilde{\delta}_L\big)d\tilde{\delta}_1...d\tilde{\delta}_L = 1 \tag{155}$$

Since $\{\tilde{\delta}_i, \ i\in Z_L^+\}$ are assumed to be mutually independent, it follows that

$$\Omega\big(\tilde{\delta}_1,...,\tilde{\delta}_L\big) = \prod_{i=1}^{L}\Omega_i\big(\tilde{\delta}_i\big) \tag{156}$$

where $\Omega_i(\tilde{\delta}_i)$ is the pdf of $\tilde{\delta}_i$.

Substituting Eq. (95) into Eq. (153) and after some manipulation, one obtains

$$e_2^2 = \underline{W}^T Q_2 \underline{W} - 2\underline{P}_2^T \underline{W} + 1 \qquad (157)$$

where $Q_2 \in M_{LJ \times LJ}(R)$ is given by

$$[Q_2]_{k,\ell} = \frac{1}{(\Delta\tau)^2} \int_{-\Delta\tau/2}^{\Delta\tau/2} \int_{-\Delta\tau/2}^{\Delta\tau/2} [Q_1]_{k,\ell} \, d\tilde{\delta}_i d\tilde{\delta}_j \qquad (158)$$

$$k = i+(m-1)L, \quad i,j \in Z_L^+$$
$$\ell = j+(n-1)L, \quad m,n \in Z_J^+$$

and $\underline{P}_2 \in R^{LJ}$ is given by

$$[\underline{P}_2]_{(m-1)L+i} = \frac{1}{\Delta\tau} \int_{-\Delta\tau/2}^{\Delta\tau/2} [\underline{P}_1]_{(m-1)L+i} \, d\tilde{\delta}_i, \quad i \in Z_L^+, \ m \in Z_J^+ \qquad (159)$$

In general, these integrals cannot be evaluated in a closed form.

C. Optimization of the New Processor

As already noted, the new processor weight vector can be selected to minimize the total mean output power as an indirect way of rejecting interferences incident on the array, provided a constraint to control the response in the look direction is added. A quadratic constraint of the form derived in Section V, B can be used to ensure that the optimized processor achieves approximately the desired look direction response.

The optimum weight vector is the solution to the following constrained optimization problem, namely

$$\underset{\underline{W}}{\text{minimise}} \quad \underline{W}^T R \underline{W} \qquad (160)$$

$$\text{subject to} \quad \left(\underline{W}_\gamma^0 - \underline{W}\right)^T Q_\gamma \left(\underline{W}_\gamma^0 - \underline{W}\right) \leq \epsilon_\gamma \qquad (161)$$

where R is the array correlation matrix defined by Eq. (17).

Using the standard primal-dual technique, [62], it can be shown that the optimum weight vector that solves Eqs. (160)-(161) is given by

$$\hat{\underline{W}} = \underline{W}^0_\gamma - \left(R + \hat{\lambda}Q_\gamma\right)^{-1} R\underline{W}^0_\gamma \tag{162}$$

where $\hat{\lambda}$ is the root of the transcendental equation

$$\underline{W}^{0^T}_\gamma R\left(R + \hat{\lambda}Q_\gamma\right)^{-1} Q_\gamma \left(R + \hat{\lambda}Q_\gamma\right)^{-1} R\underline{W}^0_\gamma = \epsilon_\gamma \tag{163}$$

The output of the optimum processor is given by

$$y\left(t, \hat{\underline{W}}\right) = \underline{X}^T(t)\underline{W}^0_\gamma - \underline{X}^T(t)\left(R + \hat{\lambda}Q_\gamma\right)^{-1} R\underline{W}^0_\gamma \tag{164}$$

and the optimum mean output power can be expressed as

$$p\left(\hat{\underline{W}}\right) = \hat{\lambda}^2 \underline{P}^T_\gamma \left(R + \hat{\lambda}Q_\gamma\right)^{-1} R\left(R + \hat{\lambda}Q_\gamma\right)^{-1} \underline{P}_\gamma \tag{165}$$

Recall from Eq. (108) that for the flat frequency response case, $\underline{P}_\gamma$ is a function of τ and in view of Eq. (128), $\underline{W}^0_\gamma$ is also a function of τ. We now consider a technique for determining the optimum value of τ for the flat response case. The optimum τ, denoted by $\hat{\tau}$, is found by solving the problem

$$\underset{\tau}{\text{minimize}} \quad \hat{\underline{W}}^T(\tau)R\hat{\underline{W}}(\tau) \tag{166}$$

where $\hat{\underline{W}}(\tau)$ given by Eq. (162) solves the optimization problem previously described. For clarity, the dependence on τ is explicitly indicated.

It is unlikely that $\hat{\underline{W}}^T(\tau)R\hat{\underline{W}}(\tau)$ is unimodal and therefore an exhaustive search technique is required to find the optimum τ. If τ is restricted to discrete values, for example, multiples of the sampling time T, then the optimization requires a finite search. For the case of no presteering, the maximum search interval is defined by the maximum delay across the array geometry. The computational complexity of the search can be made very reasonable by solving the original optimization problem in a particular way. It turns out that this particular method of solving the problem also has computational advantages for the case of a fixed τ.

Note that the optimum weight vector given by Eq. (162) can also be expressed as

$$\underline{\hat{W}}(\tau) = \hat{\lambda}\left(R + \hat{\lambda}Q_\gamma\right)^{-1}\underline{P}_\gamma(\tau) \tag{167}$$

where $\hat{\lambda}$ is the optimum Lagrange multiplier satisfying Eq. (163). For the case of flat response, Eq. (163) can be expressed as

$$\hat{\lambda}\underline{P}_\gamma^T(\tau)\left[\left(R + \hat{\lambda}Q_\gamma\right)^{-1} + \left(R + \hat{\lambda}Q_\gamma\right)^{-1}R\left(R + \hat{\lambda}Q_\gamma\right)^{-1}\right]\underline{P}_\gamma(\tau) = \eta \tag{168}$$

where

$$\eta = 1 - \xi \tag{169}$$

Substituting Eq. (167) into Eq. (166) and using Eq. (168), it can be shown that Eq. (166) can be expressed as

$$\underset{\tau}{\text{minimize}} \quad \hat{\lambda}\left[\eta - \underline{P}_\gamma^T(\tau)\underline{\hat{W}}(\tau)\right] \tag{170}$$

The computational complexity of determining $\hat{\lambda}$ which satisfies Eq. (168) can be very much reduced by employing matrix factorization as follows:

Since R and Q_γ are real symmetric matrices, with R positive definite, there exists a nonsingular matrix Γ [63], such that

$$R = \Gamma\Gamma^T \tag{171}$$

$$Q_\gamma = \Gamma\Lambda_\gamma\Gamma^T \tag{172}$$

where

$$\Gamma = S_1 S_2^{-1} S_3 \tag{173}$$

In Eq. (173), $S_1 \in M_{LJxLJ}(R)$ is the orthogonal matrix given by

$$R = S_1 \Sigma S_1^T \tag{174}$$

where $\Sigma \in M_{LJxLJ}(R)$ is the diagonal matrix defined by

$$[\Sigma]_{k,\ell} = \lambda_k \delta_{k,\ell} , \qquad k,\ell \in Z_{LJ}^+ \tag{175}$$

$S_2 \in M_{LJxLJ}(R)$ is the orthogonal matrix which reduces the matrix $S_2^T S_1^T Q_\gamma S_1 S_2$ to a diagonal form, that is,

$$S_2^T S_1^T Q_\gamma S_1 S_2 = S_3 \Lambda_\gamma S_3^T \tag{176}$$

where $\Lambda_\gamma \in M_{LJxLJ}(R)$ is the diagonal matrix given by

$$\left[\Lambda_\gamma\right]_{k,\ell} = \mu_k \delta_{k,\ell} , \qquad k,\ell \in Z_{LJ}^+ \tag{177}$$

Substituting Eqs. (171)-(172) into Eq. (168), one obtains

$$\underline{P}_\gamma^T(\tau)\left(\Gamma^T\right)^{-1}\left(2\hat{\lambda}I_{LJ} + \hat{\lambda}^2\Lambda_\gamma\right)\left(I_{LJ} + \hat{\lambda}\Lambda_\gamma\right)^{-2}\Gamma^{-1}\underline{P}_\gamma(\tau) = \eta \tag{178}$$

For the case of no presteering, in order to determine the optimum weights, S_1 is computed once only for a given array correlation matrix R and for all look directions. This requires approximately $(LJ)^3$ operations. S_3 and Γ^{-1} need to be computed for each look direction but are independent of τ. S_3 depends on S_1 and Q_γ. Note that Q_γ can be pre-computed for each look direction once and for all. The computational cost for S_3 and Γ^{-1} is approximately $[3(LJ)^3+2(LJ)^2]$ and $[(LJ)^3+(LJ)^2]$ operations, respectively. In Eq. (178), only $\underline{P}_\gamma(\tau)$ is a function of τ. Therefore, the optimization of τ requires approximately $n[(LJ)^2+m(LJ)]$ operations, where n is the number of searches for τ and m is the number of iterations of the root solving algorithms. It has been found that for $\xi = 0.1\%$, $m \cong 10$. Once the optimum Lagrange multiplier is obtained, the determination of $\hat{\underline{W}}$ requires approximately $(LJ)^2$ operations.

D.   Numerical Results

To demonstrate the performance characteristics of the new design approach, computer studies involving the double-ring circular array shown in Fig. 6 and a wide variety of signal and noise field configurations have been carried out.

The array dimensions are specified in terms of a dimensionless quantity which is obtained by dividing the inter-ring spacing by the wavelength of the highest frequency of interest. We refer to this quantity as the spatial sampling factor μ_u. For comparison purposes, the numerical results also include the performance of the presteered minimum power zero order constrained processor defined in Section III. These results are labelled with the acronym MPZ. A few representative results of the computer studies are now presented.

In the computer studies, one source scenario was assumed to consist of two broadband directional sources : a 0 dB source at 0° and a 6 dB source at 180°. Broadband spherically isotropic noise of 0 dB and white element noise of −30 dB were also included. All the broadband sources were assumed to cover one octave of bandwidth in the range [0.125, 0.250]. The number of taps of the tapped delay line filter used in the computer studies was fixed at five taps and the frequency band of interest was designed for one octave bandwidth [0.125, 0.250].

Fig. 7 shows the optimum power estimate plots for the new processor with no presteering and with coarse presteering as a function of bearing angle for spatial sampling factor $\mu_u = 0.25$. For the case of coarse presteering, the delay quantization of the type given by Eq. (113) is used. In the plots, an optimum beam with optimized $\hat{\tau}$ and $\xi = 0.1\%$ is scanned through a number of bearing angles ranging from 0° to 360°. It is clear from the plots that for $\mu_u = 0.25$, the new processor with no presteering can perform as well as the MPZ processor which requires exact presteering. The effect of coarse presteering on the MPZ processor is illustrated in Fig. 8. It is clear that the performance of the MPZ processor depends critically on the exact presteering.

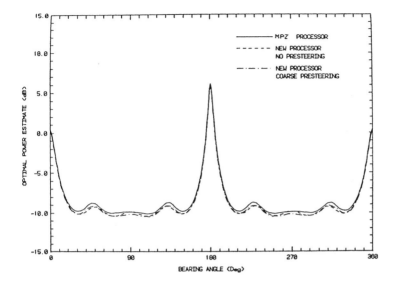

Fig. 7 Optimum power estimate plots for the new processors and the MPZ processor.

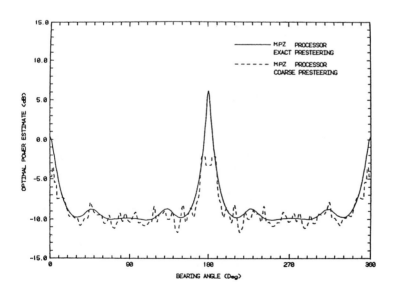

Fig. 8 Effect of coarse presteering on the MPZ processor.

Fig. 9 illustrates the frequency response of the new processor
with no presteering for the 0° look direction as well as the 180°
direction for μ_u = 0.25. In the plots, the optimum weight vector is
computed for optimum $\hat{\tau}$ and ξ = 0.1%. The amplitude response is
plotted as a function of frequency. It can be seen that the new
processor with no presteering is able to maintain very close to a flat
frequency response over the frequency band of interest [0.125, 0.250] in
the 0° direction while attenuating significantly the broadband
interference at 180°.

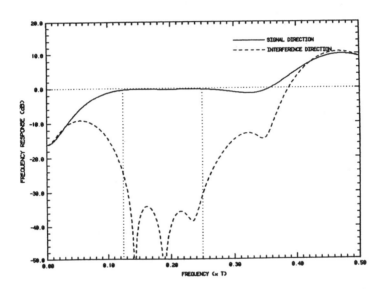

Fig. 9 Frequency response of the new processor with no presteering.

To illustrate the use of the generalized response deviation
constraint, consider the problem due to directional mismatch. Fig.
10 shows the output statistics for the new processor with no
presteering and μ_u = 0.25. In the plots, the source scenario was
assumed to consist of a 0° broadband directional source of power 6 dB.
Spherically isotropic noise of power 0 dB and white element noise of

power −30 dB were also included. An optimized beam with optimum $\hat{\tau}$ and ξ = 0.1% is scanned through a number of bearing angles ranging from 0° to 25°. It can be seen from the plots that if the arrival direction is not exactly matched to the look direction, the signal will be treated as an unwanted interference by the processor and will tend to be suppressed when the total output power is minimized.

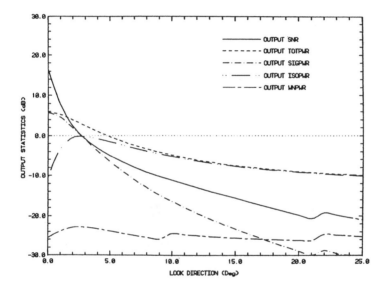

Fig. 10 Output statistics for the new processor with no presteering.

Fig. 11 illustrates the processor output statistics when a $\Delta\phi$ = 6° is incorporated in the design. A numerical integration using Simpson's rule with step size of 10 points per 1° is used to generate the Q_2 matrix and the $\underline{P}_2$ vector. The effectiveness of incorporating $\Delta\phi$ can be judged from the shape of the signal output power curve, (OUTPUT SIGPWR). This curve clearly shows the increased flatness at 0°. The loss in ability to reject interferences can be judged by examining the isotropic noise output power, (OUTPUT ISOPWR), and the white noise output power, (OUTPUT WNPWR), curves or the

output SNR curve. It is clear from Fig. 11 that the effect of
incorporating $\Delta\phi$ is to minimize the degradation due to directional
mismatch.

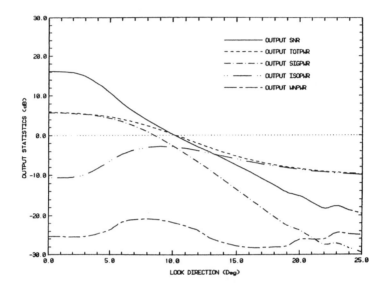

Fig. 11 Output statistics for the new processor with no presteering
and with $\Delta\phi=6°$ incorporated in the design.

As another numerical example of the use of the generalised
response deviation constraint, consider the effect of incorporating an
array geometry parameter. Fig. 12 shows the array gain of the new
processor with no presteering for the 0° direction as a function of
inter-ring spacing errors for the double-ring circular array. In the
plots, the array correlation matrix R is computed based on the array
geometry with different inter-ring spacings ranging from 0% to 25%
smaller than $r_0 = 0.25 \, \lambda_u$ which is the value assumed in the
constraint equation. The source scenario was assumed to consist of a
0 dB 180° directional broadband source, a 0 dB spherically isotropic
noise and −30 dB white noise. The power of the 0° directional source
was assumed to be 6 dB. It can be seen that the processor without Δr

incorporated is very sensitive to inter-ring spacing error. On the other hand, the processor with Δr incorporated in the design is able to retain the array gain in the presence of errors.

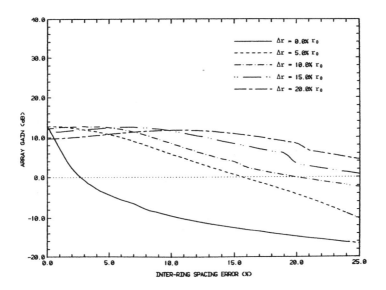

Fig. 12 Array gain of the new processor with no presteering and with Δr incorporated in the design.

VI. PARTITIONED PROCESSOR FORMULATION

An examination of the optimum weight vector defined by Eq. (164) suggests an alternate structure to that shown in Fig. 1 for realizing a robust antenna array processor. The formulation of a robust processor based on the alternate structure not only provides additional insight into the robustness problem but also leads to a significant simplification of the optimization problem that needs to be solved.

The structure suggested by Eq. (164) is the partitioned form shown in Fig. 13. This structure has been studied previously in the context of linearly constrained minimum power processors, [50]. The upper section of the processor consists of an L channel tapped delay line filter with J tapped delay line sections per channel and with weight coefficient vector $\underline{W}_\gamma^0 \in R^{LJ}$. This upper section can be considered to produce a beam output given by

$$y_0(t) = \underline{X}^T(t)\underline{W}_\gamma^0 \tag{179}$$

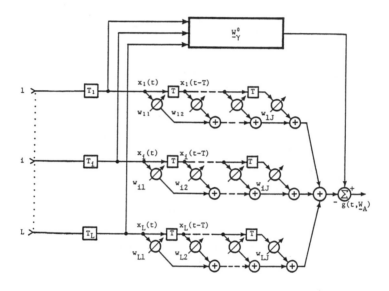

Fig. 13 Structure of the new partitioned processor.

The lower section in Fig. 13 consists of an L channel tapped delay line filter with J tapped delay line sections but with adjustable weight coefficient vector $\underline{W}_A \in R^{LJ}$. This lower section can be considered to produce a beam output given by

$$y_A(t) = \underline{X}^T(t)\underline{W}_A \tag{180}$$

The output of the partitioned processor, $g(t,\underline{W}_A)$ is given by

$$g(t,\underline{W}_A) = \underline{X}^T(t)\left[\underline{W}_\gamma^0 - \underline{W}_A\right] \tag{181}$$

We initially consider $\gamma = 1$, that is, the design of an antenna array processor with broadband capabilities for arbitrary presteering conditions. In this case, $\underline{W}_1^0$ is chosen such that the fixed upper beam has a response which is closest in a mean-square sense to a desired frequency response over some frequency band of interest $[f_\ell, f_u]$ in a specified look direction (θ_0, ϕ_0). To prevent suppression of the desired signal incident from (θ_0, ϕ_0) by the lower beam, the adjustable weights of the lower beam are constrained to satisfy

$$\underline{W}_A^T Q_1 \underline{W}_A \leq \epsilon_1 \tag{182}$$

where Q_1 is defined by Eq. (96). This constraint ensures that $y_A(t)$ contains little look direction signal power and hence cannot subtract significantly from the desired signal component of $y_0(t)$. Loosely speaking, the constraint defined by Eq. (182) forces the weights in the lower beamformer to be orthogonal in the mean-square sense to the upper beamformer in the look direction over the frequency band of interest. The degree of orthogonality is set by ϵ_1.

Robustness can be incorporated in the design of the partitioned form by introducing $\underline{W}_\gamma^0$ in the upper beam and by constraining the weight vector $\underline{W}_A$ to satisfy

$$\underline{W}_A^T Q_\gamma \underline{W}_A \leq \epsilon_\gamma \tag{183}$$

Thus, the upper beamformer weight vector $\underline{W}_\gamma^0$, defined by Eq. (128), ensures that the upper filter achieves the best mean-square approximation in the desired look direction over the frequency band of interest and over variations in the system parameters $\{p_i,\ i \in Z_m^+\}$. Furthermore, Eq. (183) ensures that the desired signal power in the lower beamformer output is limited by ϵ_γ over system parameter variations.

The total mean output power of the partitioned processor is given by

$$E\left[g^2(t, \underline{W}_A)\right] = \left(\underline{W}_\gamma^0 - \underline{W}_A\right)^T R\left(\underline{W}_\gamma^0 - \underline{W}_A\right) \tag{184}$$

Interference rejection can be achieved by minimizing the total mean output power subject to the robustness constraint defined by Eq. (183). Hence, the optimization problem associated with the partitioned processor can be expressed as:

$$\underset{\underline{W}_A}{\text{minimise}} \quad \left(\underline{W}_\gamma^0 - \underline{W}_A\right)^T R\left(\underline{W}_\gamma^0 - \underline{W}_A\right) \tag{185}$$

$$\text{subject to} \quad \underline{W}_A^T Q_\gamma \underline{W}_A \le \epsilon_\gamma \tag{186}$$

Recall that $\underline{W}_\gamma^0$ is a function of the unspecified delay τ that can be optimized.

Using the standard primal-dual technique, [62], it can be shown that the optimum weight vector that solves Eqs. (185)-(186) is given by

$$\hat{\underline{W}}_A = \left(R + \hat{\lambda} Q_\gamma\right)^{-1} R \underline{W}_\gamma^0 \tag{187}$$

where $\hat{\lambda}$ is the root of the transcendental equation

$$\underline{W}_\gamma^{0T} R\left(R + \hat{\lambda} Q_\gamma\right)^{-1} Q_\gamma \left(R + \hat{\lambda} Q_\gamma\right)^{-1} R \underline{W}_\gamma^0 = \epsilon_\gamma \tag{188}$$

It follows from Eq. (187) and Eq. (181) that

$$g\left(t, \hat{\underline{W}}_A\right) = \underline{X}^T(t)\left[\underline{W}_\gamma^0 - \left(R + \hat{\lambda} Q_\gamma\right)^{-1} R \underline{W}_\gamma^0\right] \tag{189}$$

Comparing Eq. (189) and Eq. (164), it is clear that

$$y\left(t, \hat{\underline{W}}\right) = g\left(t, \hat{\underline{W}}_A\right) \tag{190}$$

Hence, the two forms of the robust broadband array processor have identical performance when optimized according to the corresponding criterion defined by Eqs. (160)-(161) and Eqs. (185)-(186).

A.   Linearly Constrained Robust Processor

The search for a simplification of the optimization problem that needs to be solved to determine the optimum robust processor weights is motivated by the need to reduce the computational load and also to make the optimum processor amenable to implementation in a simple adaptive form with predictable convergence characteristics. One approach to reduce the computational load was described in Section V, C. The approach is particularly useful when no pre-steering is used and is relevant to an adaptive implementation based on correlation matrix estimation.

We now show how the robust antenna array processor optimization problem defined by Eqs. (185)-(186) can be approximated by a linearly constrained optimization problem. This simplfication is significant since it not only reduces the computational load for determining optimum robust processor weights but also opens the way for adaptive implementation using algorithms that have been studied extensively and have predictable convergence characteristics [25, 27, 48, 50].

Since Q_γ is a symmetric positive definite matrix, it can be factorized as

$$Q_\gamma = \Gamma_\gamma \Lambda_\gamma \Gamma_\gamma^T \tag{191}$$

where $\Gamma_\gamma \in M_{LJxLJ}(R)$ is the orthogonal matrix given by

$$\Gamma_\gamma = \left[\underline{U}_1^\gamma, \underline{U}_2^\gamma, ..., \underline{U}_{LJ}^\gamma \right] \tag{192}$$

where $\{\underline{U}_i^\gamma, \ i \in Z_{LJ}^+\}$ are the orthonormal eigenvectors of Q_γ and have the property that

$$\underline{U}_i^{\gamma T} \underline{U}_j^\gamma = \delta_{ij} \quad i,j \in Z_{LJ}^+ \tag{193}$$

and

$$\left[\Lambda_\gamma \right]_{k,\ell} = \lambda_k^\gamma \delta_{k,\ell} \quad k,\ell \in Z_{LJ}^+ \tag{194}$$

where $\{\lambda_k^\gamma, \ k \in Z_{LJ}^+\}$ are the LJ eigenvalues of Q_γ and are assumed to be ordered such that

$$\lambda_1^\gamma \geq \lambda_2^\gamma \geq ... \geq \lambda_{LJ}^\gamma \geq 0 \tag{195}$$

Consider the quadratic in Eq. (186), and use Eq. (191) then

$$\underline{W}_A^T Q_\gamma \underline{W}_A = \underline{W}_A^T \Gamma_\gamma \Lambda_\gamma \Gamma_\gamma^T \underline{W}_A \qquad (196)$$

If Q_γ has rank n_γ, a necessary and sufficient condition for

$$\underline{W}_A^T Q_\gamma \underline{W}_A = 0 \qquad (197)$$

is that

$$\underline{W}_A^T \underline{U}_k^\gamma = 0, \quad k \in Z_{n_\gamma}^+ \qquad (198)$$

Thus, in this case the quadratic constraint given by Eq. (186) with $\epsilon_\gamma = 0$ can be replaced by an equivalent set of linear constraints defined by Eq. (198).

When Q_γ has full rank, it is not possible to achieve $\epsilon_\gamma = 0$ nor is it possible to exactly replace Eq. (186) by an equivalent set of linear constraints for $\epsilon_\gamma > 0$. However, we show that it is possible to approximate the quadratic constraint by a set of linear constraints. We also note here that in practice, even if Q_γ does not have full rank, it is not possible to determine exactly the rank and hence some criterion for choosing the number of eigenvectors to be used in Eq. (198) is required.

Let n_0 linear constraints be enforced such that

$$\underline{W}_A^T \underline{U}_k^\gamma = 0, \quad k \in Z_{n_0}^+ \qquad (199)$$

It follows from Eq. (196) that

$$\underline{W}_A^T Q_\gamma \underline{W}_A = \sum_{i=n_0+1}^{LJ} \lambda_i^\gamma \left(\underline{W}_A^T \underline{U}_i^\gamma \right)^2 \qquad (200)$$

From the Cauchy-Schwarz inequality [63] and Eq. (193), we have that

$$\left(\underline{W}_A^T \underline{U}_i^\gamma \right)^2 \le \left\| \underline{W}_A \right\|_2^2 \qquad (201)$$

Thus

$$\underline{W}_A^T Q_\gamma \underline{W}_A \le \left(\sum_{i=n_0+1}^{LJ} \lambda_i^\gamma \right) \left\| \underline{W}_A \right\|_2^2 \qquad (202)$$

Equation (202) suggests that n_0 should be such that $\sum_{i=n_0+1}^{LJ} \lambda_i^\gamma$ is small. Based on this we have adopted the following normalized criterion for choosing n_0. The value of n_0 is chosen such that

$$\left(\sum_{i=1}^{n_0} \lambda_i^\gamma \bigg/ \sum_{i=1}^{LJ} \lambda_i^\gamma \right) \times 100\% \geq \varepsilon \tag{203}$$

for a given threshold ε which will usually be chosen close to 100%.

We also note that Eq. (202) suggests that it may be necessary to add a norm constraint on $\underline{W}_A$ of the form

$$\underline{W}_A^T \underline{W}_A \leq \delta \tag{204}$$

to handle cases in which $\underline{W}_A$ becomes large. If Eq. (204) is used, then Eq. (202) becomes

$$\underline{W}_A^T Q_\gamma \underline{W}_A \leq \left(\sum_{i=n_0+1}^{LJ} \lambda_i^\gamma \right) \delta \tag{205}$$

and an upper bound on $\underline{W}_A^T Q_1 \underline{W}_A$ can be guaranteed. In many examples studied we have not found it necessary to add the norm constraint defined by Eq. (204). It is worth noting that some adaptive algorithms for handling linear constraints and a simple norm constraint like Eq. (204) have been studied [37] for narrowband beamformers.

Finally, we note that in [40] it has been proposed that the norm constraint defined by Eq. (204) be used together with the zero order constraint given by Eq. (29) as a means of achieving robustness against channel errors in a presteered broadband processor. Recall that the robustness approaches we have developed so far in this work have been applicable to any form of presteering, including no presteering.

In summary, two linearly constrained robust broadband antenna array processor based on the partitioned structure have been developed.

One involves purely linear constraints and the optimum weight vector is given by the solution to the following optimization problem.

$$\underset{\underline{W}_A}{\text{minimize}} \quad \left(\underline{W}^0_\gamma - \underline{W}_A\right)^T R\left(\underline{W}^0_\gamma - \underline{W}_A\right) \tag{206}$$

$$\text{subject to } D\underline{W}_A = 0 \tag{207}$$

where $D \in M_{n_0 \times LJ}(R)$ matrix is defined by

$$D^T = \left[\underline{U}^\gamma_1, \ \underline{U}^\gamma_2, \ ..., \ \underline{U}^\gamma_{n_0}\right] \tag{208}$$

Note that D has full rank, n_0, since the eigenvectors are linearly independent.

The optimum weight vector is given by

$$\hat{\underline{W}}_A = \left[I - R^{-1}D^T\left(DR^{-1}D^T\right)^{-1}D\right]\underline{W}^0_\gamma \tag{209}$$

and the optimum mean output power is given by

$$p\left(\hat{\underline{W}}_A\right) = \underline{W}^{0^T}_\gamma D^T\left(DR^{-1}D^T\right)^{-1}D\underline{W}^0_\gamma \tag{210}$$

Recall that $\underline{W}^0_\gamma$ is a function of the unspecified delay τ that can be optimized. The other processor suggested includes a norm constraint and the optimum weight vector is the solution to

$$\underset{\underline{W}_A}{\text{minimize}} \quad \left(\underline{W}^0_\gamma - \underline{W}_A\right)^T R\left(\underline{W}^0_\gamma - \underline{W}_A\right) \tag{211}$$

$$\text{subject to } D\underline{W}_A = 0 \tag{212}$$

$$\underline{W}^T_A \underline{W}_A \leq \delta \tag{213}$$

B. Numerical Results

Fig. 14 shows the optimum power estimate plots for the new linearly constrained partitioned processor without presteering and with coarse presteering as a function of bearing angle for $\mu_u = 0.25$. The source scenario used in the plots is the same as that used in Fig. 7 . In the plots, an optimum beam with % tr $\geq$ 99.9% and optimum $\hat{\tau}$ is scanned through a number of bearing angle ranging from 0° to 360°. The optimum $\hat{\tau}$ is chosen such that the mean output power given by Eq. (210) is minimum. The maximum search interval for $\hat{\tau}$ is defined by the maximum delay across the array geometry. It is clear from the plots that the linearly constrained partitioned processor can achieve the same type of performance as the quadratically constrained processor. Note that for the case of no presteering, only a one shot computation of the array correlation matrix R is needed to determine the plot and since the constraints are linear, the computational load is significantly reduced compared to the quadratic constraint case. Furthermore, the computational load is also less than that for the exactly presteered MPZ processor since R needs to be computed once only for all look directions.

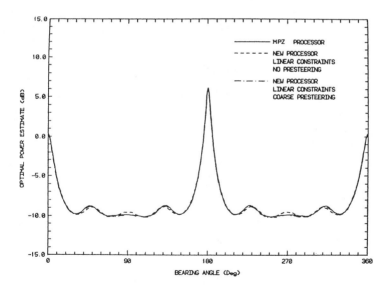

Fig. 14 Optimum power estimate plots for the new linearly constrained partitioned processors and the MPZ processor.

Fig. 15 illustrates the frequency response of the new linearly constrained partitioned processor with no presteering for the 0° look direction as well as the 180° direction for $\mu_u = 0.25$. The plots are determined for % tr ≥ 99.9% and optimum $\hat{\tau}$. It can be seen that the processor is able to maintain very close to a flat frequency response over the frequency band of interest [0.125, 0.250] in the 0° direction while attenuating significantly the broadband interference at 180°.

Figs. 16-17 show the output statistics for the new linearly constrained partitioned processor without presteering for $\Delta\phi = 0°$ and $\Delta\phi = 6°$, respectively. The source scenario used in the plots is the same as that used in Figs. 10-11. In the plots, an optimum beam with %tr ≥ 99.99% and optimum $\hat{\tau}$ is scanned through a number of bearing angles ranging from 0° to 25°. It is clear from the plots that with $\Delta\phi$ incorporated, the quadratic constraint can also be approximated by a set of linear constraints to achieve a wider acceptance angle under mismatched look direction conditions.

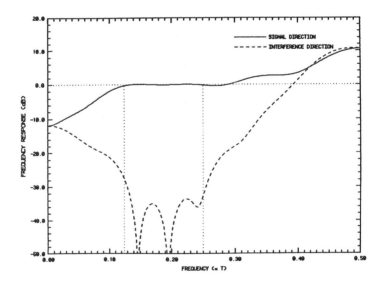

Fig. 15 Frequency response of the new linearly constrained partitioned processor.

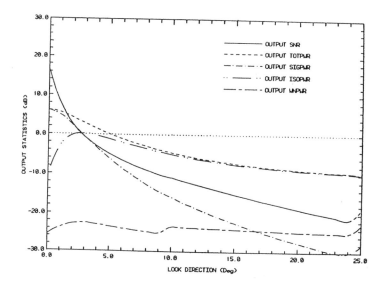

Fig. 16 Output statistics for the new linearly constrained partitioned processor with no presteering and with $\Delta\phi = 0°$.

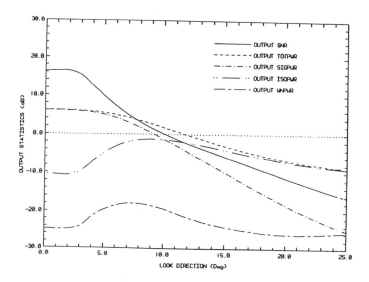

Fig. 17 Output statistics for the new linearly constrained partitioned processor with no presteering and with $\Delta\phi = 6°$.

APPENDIX A : NOTATION AND SYMBOLS

R   field of real numbers

C   field of complex numbers

Z   ring of integers

Z^+ $\{x \in Z \mid x > 0\}$

Z_a^+ $\{x \in Z \mid 0 < x \leq a\}$

j $\sqrt{-1}$

F^k set of k-tuples of elements of F

$M_{kx\ell}(F)$ set of k x ℓ matrices with elements in F

$\underline{a}^T, A^T$ transpose of vector $\underline{a}$ and matrix A respectively

$\underline{a}^H, A^H$ complex conjugate transpose of $\underline{a}$ and A respectively

A^{-1} inverse of A

$\underline{e}_{k,\ell}$ ℓ^{th} column of I^k and ℓ^{th} vector in the ordered standard basis of R^k

$\delta_{k,\ell}$ Kronecker delta

$[A]_{k,\ell}$ $(k,\ell)^{th}$ element of A

$[\underline{a}]_k$ k^{th} co-ordinate of $\underline{a}$ with respect to the standard basis

$A \otimes B$ Kronecker product of A and B

I_n n x n identity matrix

$A \Rightarrow B$ A implies B

$\forall$ for all

$\exists$ there exists, for some

$a \in A$ a is an element of A

$A \cap B$ $\{x \mid x \in A \text{ and } x \in B\}$

$A \cup B$ $\{x \mid x \in A \text{ or } x \in B\}$

$\{x \mid A(x)\}$ set of all x such that A(x) holds

$\underline{1}_k \qquad = \sum_{\ell = 1}^{k} \underline{e}_{k,\ell}$

$\underline{a}^{\perp}$ orthogonal complement of $\underline{a}$

$\|\underline{a}\|_2$ Euclidean Norm of vector $\underline{a}$

VII. REFERENCES

1.   R.A. Monzingo and T.W. Miller, *Introduction to Adaptive Arrays*, John Wiley, New York (1980).

2.   F. Bryn, "Optimum Signal Processing of Three-dimensional Arrays Operating on Gaussian Signals and Noise", *J. Acoust. Soc. Amer. 34(3)*, pp. 289-297 (1962).

3.   V. Vanderkulk, "Optimum Processing for Acoustic Arrays", *J. Brit. IRE 26(4)*, pp. 286-292 (1963).

4.   H. Mermoz, "Filtrage Adapte' et Utilisation Optimale D'une Antenne", *Proceedings of NATO Advanced Study Inst. Signal Processing Emphasis Underwater Acoust.*, Grenoble, France (1964).

5.   D. Middleton and H.I. Groginski, "Detection of Random Acoustic Signals by Receivers with Distributed Elements", *J. Acoust. Soc. Amer. 38*, pp. 727-737 (1965).

6.   S. Shor, "Adaptive Technique to Discriminate Against Coherent Noise in a Narrow-band System", *J. Acoust. Soc. Amer. 39(1)*, pp. 74-78 (1966).

7.   D.J. Edelblute, J.M. Fisk, and G.L. Kinnison, "Criteria for Optimum-Signal-Detection Theory for Arrays", *J. Acoust. Soc. Amer. 41(1)*, pp. 199-205 (1967).

8.   P.W. Howells, "Intermediate Frequency Side-lobe Canceller", *General Electric Co., Patent 3*, 202,990 (1959).

9.   S.P. Applebaum, "Adaptive Arrays", *IEEE Trans. Antennas Propagat. AP-24*, pp. 585-598 (1976).

10. H.N. Kritikos, "Optimal Signal-to-Noise Ratio for Linear Arrays by the Schwartz Inequality", *J. Franklin Inst. 276(4)*, pp. 295-304 (1963).

11. Y.T. Lo, S.W. Lee, and Q.H. Lee, "Optimization of Directivity and Signal-to-Noise Ratio of an Arbitrary Antenna Array", *Proceedings of IEEE 54*, pp. 1033-1045 (1966).

12   R.T. Compton, R.J. Huff, W.G. Swarner and A.A. Ksienki, "Adaptive Arrays for Communication Systems", *IEEE Trans. Antennas Propagat. AP-24*, pp. 599-607 (1976).

13. R.T. Compton, "An Adaptive Array in a Spread Spectrum Communications System", *Proceedings of IEEE 66*, pp. 289-298 (1978).

14. M.D. Windram, "Adaptive Antennas for UHF Broadcast Reception", *IEE Conf. Pub. 169, Part 1*, pp. 30-35 (1978).

15. D.H. Brandwood and C.J. Tarran, "Adaptive Arrays for Communications", *Proceedings of IEE, Part F 129 (3)*, pp. 223-232 (1982).

16. L.E. Brennan and I.S. Reed, "An Adaptive Array Signal Processing Algorithm for Communications", *IEEE Trans. Aerosp. Electron. Syst. AES-18 (1)*, pp. 124-130 (1982).

17. J.P. Burg, "Three-dimensional Filtering with an Array of Seismometers", *Geophysics 29(5)*, pp. 693-713 (1964).

18. P.E. Green, E.J. Kelly, Jr, and M.J. Levin, "A Comparison of Seismic Array Processing Methods", *Geophysics J. Roy. Astron. Soc. 11*, pp. 67-84 (1966).

19. J. Capon, R.J. Greenfield, and R.J. Kolker, "Multi-dimensional Maximum-Likelihood Processing of a Large Aperture Seismic Array", *Proceedings of IEEE 55*, pp. 192-211 (1967).

20. J. Capon, "High-Resolution Frequency-Wavenumber Spectrum Analysis", *Proceedings of IEEE 57*, pp. 1408-1418 (1969).

21. S.U. Pillai, *Array Signal Processing*, Springer-Verlag, New York (1989).

22. B.D. Van Veen and K.M. Buckley, "Beamforming : a Versatile Approach to Spatial Filtering", *IEEE ASSP Magazine*, pp. 4-24 (1988).

23. C.L. Dolph, "A Current Distribution for Broadside Arrays which Optimizes the Relationship between Beamwidth and Side-lobe Level", *Proceedings of IRE 34*, pp. 335-348 (1946).

24. A.W. Rudge, K. Milne, A.D. Olver, and P. Knight (Eds.), *The Handbook of Antenna Design, Vols. 1 and 2*, Peter Peregrinus Ltd., London U.K. (1983).

25. A.M. Vural, "An Overview of Adaptive Array Processing for Sonar Applications", *Proceedings of IEEE EASON 1975 Record*, pp. 34A-34M (1975).

26. A.M. Vural, "Effects of Perturbations on the Performance of Optimum/Adaptive Arrays", *IEEE Trans. Aerosp. Electron. Syst. AES-15 (1)*, pp. 76-87 (1979).

27. K. Takao, H. Fujita, and T. Niski, "An Adaptive Array Under Directional Constraint", *IEEE Trans. Antennas Propagat. AP-24 (5)*, pp. 662-669 (1976).

28. S.P. Applebaum, and D.J. Chapman, "Adaptive Arrays with Main Beam Constraints", *IEEE Trans. Antennas Propagat. AP-24 (5)*, pp. 650-662 (1976).

29. J.E. Hudson, *Adaptive Array Principles*, Peter Peregrinus, New York (1981).

30. A.K. Steele, "Comparison of Directional and Derivative Constraints for Beamformers Subject to Multiple Linear Constraints", *Proceedings of IEE, Pts. F and H 130 (1)*, pp. 41-45 (1983).

31. J.W.R. Griffiths and J.C. Hudson, "An Introduction to Adaptive Processing in a Passive Sonar System", *Aspects of Signal Processing*, Pt. 2, Dordrecht, Holland : Reidel, pp. 299-308 (1977).

32. J.N. Maksym, "A Robust Formulation of an Optimum Cross-Spectral Beamformer for Linear Array", *J. Acoust. Soc. Am. 65 (4)*, pp. 971-975 (1979).

33. K.M. Ahmed and R.J. Evans, "An Adaptive Array Processor with Robustness and Broadband Capabilities", *IEEE Trans. Antennas Propagat. AP-32 (9)*, pp. 944-950 (1984).

34. M. H. Er and A. Cantoni, "An Alternative Formulation for an Optimum Beamformer with Robustness Capability", *Proceedings of IEE, Pt. F 132*, pp. 447-460 (1985).

35. E.N. Gilbert and S.P. Morgan, "Optimum Design of Directive Antenna Arrays Subject to Random Variations", *Bell Syst. Tech. J. 34*, pp. 637-663 (1955).

36. H. Cox, R.M. Zeskind, and T. Kooij, "Practical Supergain", *IEEE Trans. Acoust., Speech, Signal Processing ASSP-34*, pp. 393-398 (1986).

37. H. Cox, R.M. Zeskind, and M.M. Owen, "Robust Adaptive Beamforming", *IEEE Trans. Acoust., Speech, Signal Processing ASSP-35 (10)*, pp. 1365-1376 (1987).

38. N.K. Jablon, "Adaptive Beamforming with the Generalized Sidelobe Canceller in the Presence of Array Imperfection", *IEEE Trans. Antennas Propagat. AP-34 (8)*, pp. 996-1012 (1986).

39. K. Takao and N. Kikuma, "Tamed Adaptive Antenna Array", *IEEE Trans. Antennas Propagat. AP-34*, pp. 388-394 (1986).

40. J.E. Hudson, "A Study of Element Space Array Processors, Part III : Norm Limited Optimum Processors", *Tech. Report EE8201*, Department of Electrical and Computer Engineering, University of Newcastle, N.S.W., 2308, Australia, January 1982.

41. J.M. McCool, "A Constrained Adaptive Beamformer Tolerant of Array Gain and Phase Errors", *Aspects of Signal Processing, Pt. 2*, Dordrecht, Holland : Reidel, pp. 517-522 (1977).

42. H. Cox, "Sensitivity Considerations in Adaptive Beamforming", *Signal Processing (Proc. NATO Advanced Study Inst. Signal Processing with Particular Reference to Underwater Acoust., Loughborough, U.K., August 1972)*, J.W.R. Griffiths, P.L. Stocklin, and C. Van Schooneveld, Eds. Academic Press, New York and London (1973).

43. M.H. Er and A. Cantoni, "Derivative Constraints for Broad-band Element Space Antenna Array Processors", *IEEE Trans. Acoust., Speech, Signal Processing ASSP-31*, pp. 1378-1393 (1983).

44. M.H. Er and A. Cantoni, "A New Approach to the Design of Broad-band Element Space Antenna Array Processors", *IEEE J. Oceanic Engineering, OE-10*, pp. 231-240 (1985).

45. M.H. Er and A. Cantoni, "A New Set of Linear Constraints for Broad-band Time Domain Element Space Processors", *IEEE Trans. Antennas Propagat. AP-34*, pp. 320-329 (1986).

46. M.H. Er and A. Cantoni, "A Unified Approach to the Design of Robust Narrow-band Antenna Array Processors", *IEEE Trans. Antennas Propagat. AP-38 (1)*, pp. 17-23 (1990).

47. B.F. Cron, B.C. Hassell and F.J. Keltonic, "Comparison of Theoretical and Experimental Values of Spatial Correlation", *J. Acoust. Soc. Am. 37(3)*, pp. 523-529 (1965).

48. O.L. Frost, III, "An Algorithm for Linearly Constrained Adaptive Array Processing", *Proceedings of IEEE 60(8)*, pp. 926-935 (1970).

49. J.W. Brewer, "Kronecker Products and Matrix Calculus in System Theory", *IEEE Trans. Circuits and Systems CAS-25(9)*, pp. 772-781 (1978).

50. K.M. Buckley and L.J. Griffiths, "An Adaptive Generalized Sidelobe Canceller with Derivative Constraints", *IEEE Trans. Antennas Propagat. AP-34 (3)*, pp. 311-319 (1986).

51. K.M. Buckey and L.J. Griffiths, "Linearly-Constrained Beamforming : A Relationship Between Phase Centre Location and Beampattern", *Proceedings of 19th Asilomar Conference on Circuits, Computer and Systems*, pp. 234-238 (1986).

52. M.H. Er and B.P. Ng, "The Dependency of Weight Vector on Array Origin for Derivative Constrained Broadband Arrays", *Journal of Electrical and Electronics Engineering, Australia 10 (3)*, pp. 202-206 (1990).

53. V.C. Anderson, "Digital Array Phasing", *J. Acoust. Soc. Am. 32 (7)*, pp. 867-870 (1960).

54. D.E. Dudgeon, "Fundamentals of Digital Array Processing", *Proceedings of IEEE 65 (6)*, pp. 898-904 (1977).

55. W.C. Knight, R.G. Pridham, and S.M. Kay, "Digital Signal Processing for Sonar", *Proceedings of IEEE 69 (11)*, pp. 1451-1507 (1981).

56. R.A. Mucci, "A Comparison of Efficient Beamforming Algorithms", *IEEE Trans. Acoust., Speech, Signal Processing ASSP-32 (3)*, pp. 548-558 (1984).

57. R.G. Pridham and R.A. Mucci, "A Novel Approach to Digital Beamforming", *J. Acoust. Soc. Am. 63 (2)*, pp. 425-434 (1978).

58. R.G. Pridham and R.A. Mucci, "Digital Interpolation Beamforming for Low-pass and Band-pass Signals", *Proceedings of IEEE 67 (6)*, pp. 904-919 (1979).

59. R.W. Schafer and L.R. Rabiner, "A Digital Signal Processing Approach to Interpolation", *Proceedings of IEEE 61 (6)*, pp. 692-707 (1973).

60. K.M. Ahmed and R.J. Evans, "Broadband Adaptive Array Processing", *Proceedings of IEE, Pt. F 130 (5)*, pp. 433-440 (1983).

61. D. Nunn, "Performance Assessments of a Time-Domain Adaptive Processor in a Broadband Environment", *Proceedings of IEE, Pts F and H 130 (1)*, pp. 139-146 (1983).

62. D.G. Luenberger, *Optimization by Vector Space Methods*, Wiley, New York (1973).

63. R. Bellman, *Introduction to Matrix Analysis*, McGraw Hill, New York (1960).

TECHNIQUES FOR THE ROBUST
CONTROL OF RIGID ROBOTS

by
C. Abdallah, D. Dawson, P. Dorato, and M. Jamshidi

C. Abdallah, P. Dorato and M. Jamshidi are with the CAD Laboratory for Systems and Robotics, Electrical and Computer Engineering Department, University of New Mexico, Albuquerque, NM 87131.

D. Dawson is with the Department of Electrical and Computer Engineering, Clemson University, Clemson, SC 29634-0915.

I. INTRODUCTION

The control of uncertain systems is usually accomplished using either an adaptive control philosophy, or a robust control philosophy. In the adaptive approach, one designs a controller which attempts to "learn" the uncertain parameters of the system and, if properly designed, will eventually be a "best" controller for the system in question. In the robust approach, the controller has a fixed-structure which yields "acceptable" performance for a class of plants which include the plant in question. In general, the adaptive approach is applicable to a wider range of uncertainties, but robust controllers are simpler to implement and no time is required to "tune" the controller to the particular plant.

In this chapter, we review different robust control designs used in controlling the motion of rigid robots. An earlier version of this study was published in [1]. A discussion of adaptive motion controllers in robotics may be found in [2] and a comprehensive survey of robust control theory is available in [3,4]. Let us point out that robust control theory may be subdivided into the unstructured uncertainties or H_∞ design and synthesis methods, the Lyapunov analysis and design methods and the structured uncertainties or Kharitonov analysis methods. The robust control of rigid robots falls within the H_∞ and the Lyapunov design methods.

The techniques discussed in this survey belong to one of five categories. The first is the linear-multivariable or feedback-linearization approach [5], where the inverse dynamics of the robot are used in order to globally linearize and decouple the robot's equations. Since one does not have access to the exact inverse dynamics, the linearization and the decoupling will not be exact. This will be manifested by uncertain feedback terms that will be handled using multivariable linear robust control techniques which include both H_∞ and Lyapunov methods. The model-based controllers such as those of [5-14] fall under this heading.

The second category contains methods that exploit the passive nature of the robot [15,16]. Rigid robots are physical systems which dissipate energy, and the techniques of this category try to maintain the passivity of the closed-loop robot/controller system, despite uncertain knowledge of the robot's parameters using Lyapunov-type methods. Although not as transparent to linear control techniques as the computed-torque approach is, passivity-based methods can nonetheless guarantee the robust stability of the closed-loop robot/controller system. The works described in [17,18] belong to this category and will be discussed in this chapter.

Next, we group methods that are for the most part Lyapunov-based nonlinear control schemes [19-29]. These include variable-structure and saturation controllers which attempt to robustly control a rigid robot. Some of these techniques may actually rely on the feedback-linearizability or the passivity of the robot dynamics and could have been studied within those approaches. We choose instead to present them in a separate section in order to illustrate some of their salient features.

Finally, we briefly survey approaches that combine robust and adaptive techniques [29-34]. These approaches are for the most part Lyapunov-based and are attempting to combine the advantages of the robust and adaptive philosophies. It should be noted that other classifications of robust controllers in robotics are possible and that this survey reflects our own philosophy rather than a universally accepted division. We will attempt throughout this survey to include examples and simulations that will illustrate the behavior of some resulting controllers.

Let the rigid robot dynamics be given in joint-space by the Lagrange-Euler equations [35] where q is an n vector of generalized coordinates representing the joints positions, and τ is the generalized n torque input vector. The matrix $D(q)$ is an $n \times n$ symmetric positive-definite inertia matrix and $h(q,\dot{q})$ is an n vector containing the Coriolis, centrifugal, and gravity terms. The Coriolis and centripetal terms $V(q,\dot{q})$ may also be separated from the gravity terms $g(q)$ as follows

$$D(q)\ddot{q} + h(q,\dot{q}) = D(q)\ddot{q} + V(q,\dot{q}) + g(q) = \tau \qquad (1)$$

In general, (1) arises as a solution to the Lagrange equations of motion for natural systems [5]. Therefore, the control of rigid robots may be thought of as a special case of the control problems of natural systems. In a way, these systems are unique since they are nonlinear, yet their physics is well understood. There are of course some effects that are not well understood such as the friction term $F(\dot{q})$ and a disturbance torque τ_d which may be included in model (1) as follows

$$D(q)\ddot{q} + V(q,\dot{q}) + F(\dot{q}) + g(q) = \tau + \tau_d \tag{2}$$

We shall first describe some of the unique properties of rigid robots which make the robust control designs more obvious. These properties have been discussed by many authors and may be found in [15].

Properties of the Inertia Matrix: As described in [16], $D(q)$ is a symmetric, positive-definite matrix. Another vital property of $D(q)$ is that it is bounded above and below. That is,

$$d_1(q)I \leq D(q) \leq d_2(q)I \tag{3}$$

where $d_1(q)$ and $d_2(q)$ are scalar constants for a revolute arm and scalar functions of q for an arm containing prismatic joints. It is easy to see then that $D^{-1}(q)$ is also bounded since

$$\frac{1}{d_2(q)}I \leq D^{-1}(q) \leq \frac{1}{d_1(q)}I \tag{4}$$

$\square$

Properties of the Coriolis and Centripetal Terms: The first observation one can make about the term $V(q,\dot{q})$ is that it is quadratic in the generalized velocity $\dot{q}$. In fact, $V(q,\dot{q})$ is bounded as follows [34]

$$\|V(q,\dot{q})\| \leq v_b(q)\|\dot{q}\|^2 \tag{5}$$

where $v_b(q)$ is a scalar constant for an all-revolute arm and a scalar function of q for arms containing prismatic joints. In addition there exists numerous factorizations of $V(q,\dot{q})$ one of which is given by

$$V(q,\dot{q}) = C(q,\dot{q})\dot{q} \tag{6}$$

This factorization is particularly important in controllers that exploit the passivity of the robot using the following property

$$\dot{M}(q) - 2C(q,\dot{q}) \quad \text{is skew-symmetric} \tag{7}$$

□

Properties of the Gravity Terms: The gravity term $g(q)$ is bounded as follows [34]

$$\|g(q)\| \leq g_b(q) \tag{8}$$

where $g_b(q)$ is a scalar constant for revolute arms and a scalar function of q for arms containing prismatic joints.

□

Properties of the Friction Terms: Since friction is a local effect, we may assume that the friction terms $F(\dot{q})$ are uncoupled among the joints so that [36]

$$F(\dot{q}) = \begin{bmatrix} f_1(\dot{q}_1) \\ f_2(\dot{q}_2) \\ . \\ . \\ f_n(\dot{q}_n) \end{bmatrix} \tag{9}$$

where each $f_i(\dot{q}_i)$ is a known scalar function. In fact, we shall assume the following forms for the viscous and static frictions:

$$F(\dot{q}) = F_v \dot{q} + F_s(\dot{q})$$
$$F_v = \text{diag}(v_i)$$
$$F_s(\dot{q}) = \text{diag}(k_i)sgn(\dot{q}) \tag{10}$$

where $sgn(\dot{q}) = [sgn(\dot{q}_1) \ sgn(\dot{q}_2) \ \cdots \ sgn(\dot{q}_n)]^T$. One can then bound the friction terms as follows

$$\|F(\dot{q})\| \leq v\|\dot{q}\| + k \tag{11}$$

□

Properties of the Disturbance Torques: The disturbance torque τ_d is by definition unknown. It is however reasonable to assume that it is bounded by a known function $d(q,\dot{q})$ [16]

$$\|\tau_d\| \le d(q,\dot{q}) \tag{12}$$

□

The above properties of each term in equation (2) are important on their own, but they also combine to give the following most useful results

Property 1: For the rigid, nonredundant robot model (2), there is an independent control input for each degree of freedom, i.e. there are n components to the input vector τ corresponding to the n components of the position vector q.

□

Property 2: The dynamic equations (2) define a passive mapping from the input torque τ and the generalized velocity $\dot{q}$. In fact, it is easy to show that [2]

$$\int_0^t \dot{q}^T(u)\tau(u)du = H(t) - H(0) \ge -H(0) \tag{13}$$

where $H(t)$ is the sum of the kinetic and potential energy of the robot. This property is intimately related to (7).

□

Property 3: The equations of motion (2) are linear in a suitably defined set of parameters [2]. In other words, one can write equation (2) as

$$Y(q,\dot{q},\ddot{q})\theta = \tau \tag{14}$$

where $Y(q,\dot{q},\ddot{q})$ is a matrix of known time functions and θ is a vector containing known functions of the robot's physical parameters such as link masses, moments of inertia etc. Note that $Y(.)$ is nxl where l is usually larger than n.

□

In this chapter, we survey methods that use the above three properties in designing controllers that will make q and $\dot{q}$ track some desired q_d and $\dot{q}_d$ when some components of $D(q)$ and $h(q,\dot{q})$ are uncertain. Note however, that some of the methods surveyed here may also be used to control the robot to follow a desired trajectory in the task or end-effector space [5,6]. This is due to the fact that properties 1-3 may be satisfied in the task space for many rigid robots.

The controllers reviewed in this chapter will guarantee the stability of the closed-loop systems. Unfortunately, the stability of the nonlinear robots is not uniquely defined [37] and some of the stability concepts may not be familiar to all readers. We will then review the concepts of L_p spaces for $1 \leq p \leq \infty$ in the following definitions [38].

Definition 1: Let $f(.) : [0, \infty) \rightarrow R$ be a uniformly continuous function. A function f is uniformly continuous if for any $\varepsilon > 0$, there is a $\delta(\varepsilon)$ such that

$$|t-t_0| < \delta(\varepsilon) \quad \text{implies} \quad |f(t)-f(t_0)| < \varepsilon$$

Then, f is said to belong to L_p if for $1 \leq p < \infty$,

$$\int_0^\infty ||f(t)||^p \, dt < \infty$$

f is said to belong to L_∞ if it is bounded i.e. if there exists a finite B such that $\sup_{0 \leq t < \infty} |f(t)| \leq B < \infty$.

$\square$

Definition 2 The p-norm of f which belongs to L_p is given by

$$||f(.)||_p = [\int_0^\infty |f(t)|^p \, dt]^{1/p} \; ; \; 1 \leq p < \infty$$

and

$$||f(.)||_\infty = \sup_{0 \leq t < \infty} |f(t)|$$

$\square$

In particular the spaces L_2 which corresponds to signals with finite energy and L_∞ which corresponds to bounded signals will be important in discussing the control schemes of this chapter.

II. LINEAR - MULTIVARIABLE APPROACH

In this section we review different designs which use linear multivariable techniques to obtain robust robot controllers. We shall assume that the friction terms $F(\dot{q})$ and the disturbance torque τ_d are unknown and thus concentrate on the model given by (1). Note that the friction terms have a stabilizing effect, while the disturbance terms will result in bounded rather than vanishing errors. In reality however, the controllers need to be robust when these effects are present.

In the early days of robot control, the idea of linearizing the nonlinear robot dynamics about their desired trajectory (using a Taylor series expansion for example) was popular, and many controllers were designed that way [39-42]. Later however, the properties of the matrix $D(q)$ in conjunction with property 1 of the previous section, led to the "global" linearization of the nonlinear robotic system. It is this later approach that is stressed in this section. It should be noted that the robot dynamics may actually be linearized in task or cartesian space, and in many other spaces as described in [5]. By defining the trajectory error vector, $e_1 = q - q_d$, $e_2 = \dot{e}_1$, and assuming that q_d, $\dot{q}_d$, and $\ddot{q}_d$ are bounded (see example 1), one is able to globally linearize the nonlinear error system, to the following

$$\dot{e} = Ae + Bv$$

where

$$A = \begin{bmatrix} 0 & I \\ 0 & 0 \end{bmatrix} ; B = \begin{bmatrix} 0 \\ I \end{bmatrix} ; e = \begin{bmatrix} e_1 \\ e_2 \end{bmatrix}$$

and

$$v = D(q)^{-1}[\tau - h(q,\dot{q})] - \ddot{q}_d. \tag{15}$$

The problem is then reduced to finding a linear control v which will achieve a desired closed-loop performance, i.e. find F, G, H, and J in

$$\dot{z} = Fz + Ge,$$

$$v = Hz + Je,$$

or,

$$v(t) = [H(sI-F)^{-1}G + J]e(t) \equiv C(s)e(t) \tag{16}$$

Note that the above notation indicates that $v(t)$ is the output of a system $C(s)$ when an input $e(t)$ is applied. The following static state-feedback controller is often used

$$F = G = H = 0, J = -K,$$

$$v = -K_1 e_1 - K_2 e_2 \equiv -Ke \tag{17}$$

leading to the nonlinear controller

$$\tau = D(q)[\ddot{q}_d + v] + h(q,\dot{q}) \tag{18}$$

which, due to the invertibility of $D(q)$ gives the following closed-loop system

$$\ddot{e}_1 + K_2\dot{e}_1 + K_1 e_1 = 0. \tag{19}$$

If K_1 and K_2 are diagonal matrices with positive elements, equation (19) reduces to n decoupled and stable second-order systems. Unfortunately, the control law (18) can not usually be implemented due to its complexity or to uncertainties present in $D(q)$ and $h(q,\dot{q})$ and to the presence of τ_d and $F(\dot{q})$. Instead, one applies τ in (20) below where $\hat{D}$ and $\hat{h}$ are estimates of D and h

$$\tau = \hat{D}[\ddot{q}_d + v] + \hat{h} \tag{20}$$

This in turn will keep some coupling between the different joints of the robot and leads to (Fig. 1)

$$\dot{e} = Ae + B(v+\eta)$$
$$\eta = E(v+\ddot{q}_d) + D^{-1}\Delta h$$
$$E = D^{-1}\hat{D} - I_n \; , \; \Delta h = \hat{h} - h. \tag{21}$$

Note first that E, Δh and therefore η are zero if $\hat{D} = D$ and $\hat{h} = h$. In general, however, the vector η is a nonlinear function of both e and v and can not be treated as an external disturbance. It represents an internal disturbance of the globally linearized error dynamics caused by modeling uncertainties, parameter variations, external disturbances, friction terms, and maybe even noisy measurements [6]. In general, we can always consider $e(0) = 0$ by choosing $q_d(0) = q(0)$ and $\dot{q}_d(0) = \dot{q}(0)$. This choice of initial conditions will be used repeatedly in many of the controllers in this chapter. Most commercial robots are in fact controlled with the controller given in (20) with the choices of $\hat{D} = I$ and $\hat{h} = 0$. The choice of $\hat{D}$ is validated by the powerful motors used to drive the robot links, and the gearing mechanisms used to torque the motor output to an acceptable level, while slowing its speed down [16]. The choice of $\hat{h}$ is validated by keeping the different motors from driving their links too fast, thus limiting the Coriolis and centripetal torques. Such commercial controllers are known as "Nonmodel-Based Controllers" and have been used since the early days of robotics. The quest for more performance is however leading researchers and manufacturers to use direct-drive robots and to attempt moving them at higher speeds with less powerful but more efficient motors [43]. This new direction is increasing the need for more robust controllers such as the ones described next.

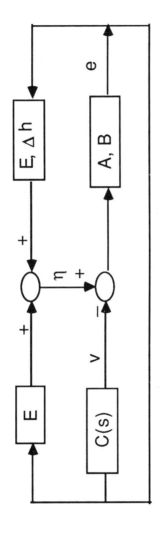

Figure 1: Block Diagram of Uncertain System

The linear multivariable approaches revolve around the design of linear controllers $C(s)$ such that the complete closed-loop system (Fig. 1) is stable in some suitable sense, e.g. uniformly ultimately bounded [28], globally asymptotically stable [12], L_p stable [6], etc. for a given class of nonlinear perturbation η. In other words, choose $C(s)$ in (16) such that the error $e(t)$ in (21) is stable in some desired sense.

The reasonable assumptions (22-24) below are often made for revolute-joint robots when using this approach [6,9]. In the following, d_1, d_2, α, β_0, β_1, and β_2 are nonnegative finite constants which depend on the size of the uncertainties.

$$\frac{1}{d_2}I_n \le ||D^{-1}|| \le \frac{1}{d_1}I_n, \tag{22}$$

$$||E|| \le \alpha \tag{23}$$

$$||\Delta h|| \le \beta_0 + \beta_1 ||e|| + \beta_2 ||e||^2 \tag{24}$$

Note that these assumptions must be modified for robots with prismatic joints. In general, the small-gain theorem [38], the passivity theorem [38], or the total stability theorem [44] are invoked to find $C(s)$. The most common of these controllers have been designed using the total-stability theorem and will be discussed first.

2.1 Static Compensators:

Static feedback compensators such as the ones given in (17) have been extensively used starting with the works of Freund [11], and Tarn et.al. [8], where

$$v = C(s)e = -Ke \tag{25}$$

such that

$$\dot{e} = Ae + B(v+\eta) = (A-BK)e + B\eta = A_c e + B\eta. \tag{26}$$

In these papers, the authors use state feedback to either place the poles sufficiently far in the left-half-plane [16], therefore guaranteeing stability in the presence of η (by the total stability theorem for example), or an extra control loop [8] to correct for the effects of η. In fact, a certain degree of robustness to the effects of η is guaranteed by choosing a stabilizing controller gain K. The question is only whether the destabilizing effect of η is overcome by the stability of A_c. Let us note from (21) and (22-24) that η is bounded as follows

$$\|\eta\| \le \alpha[\|\ddot{q}_d\| + k\|e\|] + \frac{1}{d_1}[\beta_0 + \beta_1\|e\| + \beta_2\|e\|^2] \qquad (27)$$

Example 1: In all of our examples we will use a two-link revolute-joint robot shown in Fig. 2 and whose dynamics are described by

$$\begin{bmatrix} D_{11}(q) & D_{12}(q) \\ D_{12}(q) & D_{12}(q) \end{bmatrix} \begin{bmatrix} \ddot{q}_1 \\ \ddot{q}_2 \end{bmatrix} + \begin{bmatrix} h_1(q,\dot{q}) \\ h_2(q,\dot{q}) \end{bmatrix} = \begin{bmatrix} \tau_1 \\ \tau_2 \end{bmatrix}$$

where

$$D_{11}(q) = (m_1+m_2)a_1^2 + m_2a_2^2 + 2m_2a_1a_2cos(q_2)$$

$$D_{12}(q) = m_2a_2^2 + m_2a_1a_2cos(q_2)$$

$$D_{22}(q) = m_2a_2^2$$

$$h_1(q) = -m_2a_1a_2(2\dot{q}_1\dot{q}_2+\dot{q}_2^2)sin(q_2) + (m_1+m_2)ga_1cos(q_1)$$
$$\qquad + m_2ga_2cos(q_1+q_2)$$

$$h_2(q) = m_2a_1a_2\dot{q}_1^2sin(q_2) + m_2ga_2cos(q_1+q_2)$$

The   parameters $m_1 = 1\ Kg$, $m_2 = 1\ Kg$, $a_1 = 1\ m$, $a_2 = 1\ m$,   and $g = 9.8\ m/s^2$ are given. In this case, the feedback-linearization approach yields the linear system (2) with e a 4x1 vector. The 4x2 transfer function between v and e is given by

$$G_v(s) = \frac{1}{s^2}\begin{bmatrix} I \\ sI \end{bmatrix}$$

where I is the 2x2 identity matrix. Let the desired trajectory used in all examples throughout this chapter be described by

$$q_d = \begin{bmatrix} sin(t) \\ sin(t) \end{bmatrix} ; \quad \dot{q}_d = \begin{bmatrix} cos(t) \\ cos(t) \end{bmatrix}$$

Then $\|q_d\|_\infty = 1rad$, $\|\dot{q}_d\|_\infty = 1rad/sec$, and choose the controller

$$v = -15e_1 - 30e_2$$

$$\tau = \ddot{q}_d + v$$

This controller corresponds to $\hat{D} = I$, and $\hat{h} = 0$. A simulation of the robot's trajectory is shown in Fig. 3. We also start our simulation at

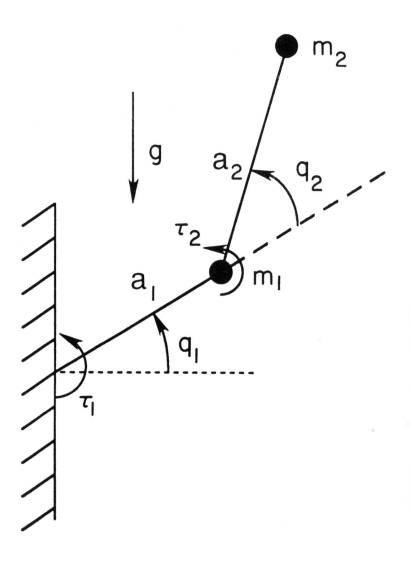

Figure 2:   2-link Planar RR Arm

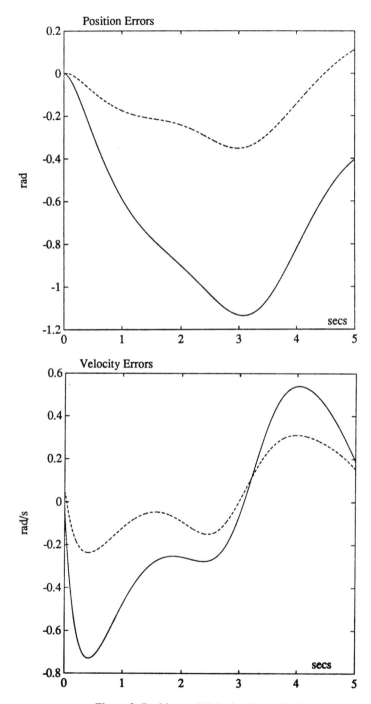

Figure 3: Position and Velocity Errors For Ex. 1

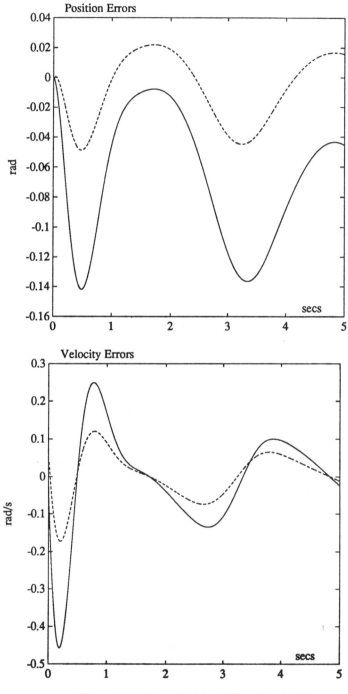

Figure 4: Position and Velocity Errors For Ex. 1

$q_1(0) = q_2(0) = 0 \; rad$, and $\dot{q}_1(0) = \dot{q}_2(0) = 1 \; rad/sec$. The effect of increasing the proportional gains is shown in Fig. 4, which corresponds to the controller

$$v = -225e_1 - 30e_2$$

$$\tau = \ddot{q}_d + v$$

□

In [12], the state-feedback controller was used to define an appropriate output Ke such that the input-output closed-loop linear systems $K(sI-A+BK)^{-1}B$ is Strictly-Positive-Real (SPR). The closed-loop stability was then assured for all η resulting from passive nonlinear uncertainties by using the passivity theorem [38]. This approach then allows η to be arbitrarily large as long as it is passive as shown in the next example.

Example 2: It can be shown that for the robot used in example 1, d_2 in (3) may be chosen larger than 6 i.e.

$$D(q) \leq 6I$$

According to the design of [12], the controller given by (17,20) will guarantee the global asymptotic stability of the closed-loop system if

$$\hat{D} = 6I \; ; \; \hat{h} = h(q)$$

$$0 < K_1 < K_2^2$$

In the simulation presented in Fig. 5, we let $K_1 = \dfrac{15}{6}I$ and $K_2 = \dfrac{30}{6}I$.

□

The most general controllers using linear multivariable techniques have been designed using Youla's parametrization and H_∞ robust control methods [3,45] and are discussed next.

2.2 Dynamic Compensators:

Spong and Vidyasagar [6] used the factorization approach [45] to design a class of linear compensators $C(s)$, parametrized by a stable transfer matrix $Q(s)$, which guarantee that the solution $e(t)$ to the linear system (8) has a bounded L_∞ norm. The authors actually assumed that the bound on Δh is linear, i.e. $\beta_2 = 0$ in (24) and found the family of all L_∞ stabilizing compensators of the nominal plant. A particular compensator may then be obtained

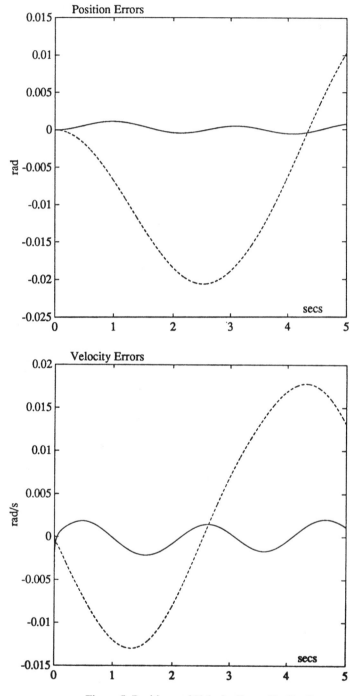

Figure 5: Position and Velocity Errors For Ex. 2

by choosing the parameter $Q(s)$ to satisfy other design criteria such as suppressing the effects of η. This design methodology is illustrated in the next example.

Example 3: Let $G_v(s)$ of example 1, be factored as

$$G_v(s) = [\tilde{D}(s)]^{-1}\tilde{N}(s) = N(s)D(s)$$

where $N(s), D(s), \tilde{N}(s), \tilde{D}(s)$ are stable rational functions. Note that $D(s)$ is not the same as $D(q)$, the robot inertia matrix. We can then find

$$N(s) = \tilde{N}(s) = \frac{1}{(s+1)^2}\begin{bmatrix} I \\ sI \end{bmatrix}$$

$$D(s) = \frac{s^2}{(s+1)^2}I$$

$$\tilde{D}(s) = \frac{1}{(s+1)^2}\begin{bmatrix} (s^2+2s)I & -2I \\ -sI & (s^2+1)I \end{bmatrix}$$

Next, we solve the Bezout identity for $X(s)$ and $Y(s)$ which are also stable rational functions

$$Y(s)D(s) + X(s)N(s) = I$$

to get

$$X(s) = \frac{1}{(s+1)^2}[(1+2s)I \quad (2+4s)I] \; ; \; Y(s) = \frac{s^2+4s+2}{(s+1)^2}I.$$

Then all stabilizing controllers are given by

$$C(s) = -[Y(s)-Q(s)\tilde{N}(s)]^{-1}[X(s)+Q(s)\tilde{D}(s)]$$

where $Q(s)$ is a stable rational function which is otherwise arbitrary. One choice is of course to let $Q(s) = 0$ which leads to the "Central Solution"

$$C(s) = \frac{2s+1}{s^2+4s+2}[I \quad 2I]$$

One can then choose $Q(s)$ to satisfy the required performance. In particular, the following choice of Q is presented in [6]

$$Q(s) = [2I \quad \frac{4k+(k+2)s}{s+k}I] \; ; \; k = 1,2,3, \cdots$$

which leads to the following controller

$$C(s) = [C_1(s) \quad C_2(s)]$$

where

$$C_1(s) = \frac{-[2s^3+(k+4)s^2+(2k+1)s+k]}{s^2(s+2)}I$$

$$C_2(s) = \frac{-[ks^3+4(k+1)s^2+5ks+2k]}{s^2(s+2)}I$$

As k increases, the disturbance rejection property of the controller is enhanced at the expense of higher gains as seen from the expression of $C(s)$. A simulation of this controller for $k = 10$ is shown in Fig. 6.

□

In a more recent paper, a two Degree-Of-Freedom (2 d.o.f.) robust controller was designed [46] and simulated. It is well known that the 2 d.o.f. structure is the most general linear controller structure and corresponds to the observer/controller design [47]. The linear controller is shown in Fig. 7 where

$$K(s) = \frac{1}{(s/w_1)^2+a_1(s/w_1)+1}I$$

$$C_2(s) = \frac{b_2s^2+b_1w_2s+w_2^2}{s/w_2}I$$

where $s^3+b_2s^2+b_1s+1$ is a stable polynomial. Note that $C_2(s)$ is a PID controller.

As was discussed in [9], including the more reasonable quadratic bound will not destroy the L_∞ stability result, but will exclude any L_2 results unless the problem is reformulated and more assumptions are made. In effect, the error will still be bounded but it may or may not have a finite energy. In particular, noisy measurements are no longer tolerated for L_2 stability to hold.

Craig [34] discussed the L_∞ problem in a similar setting, and under certain conditions, was able to show the boundedness of the error signals. In Kuo and Wang [48], the internal model principle developed by Francis and Wonham [49] is used to design a linear controller which minimizes the effects of the disturbance term η. However, since η is a nonlinear function

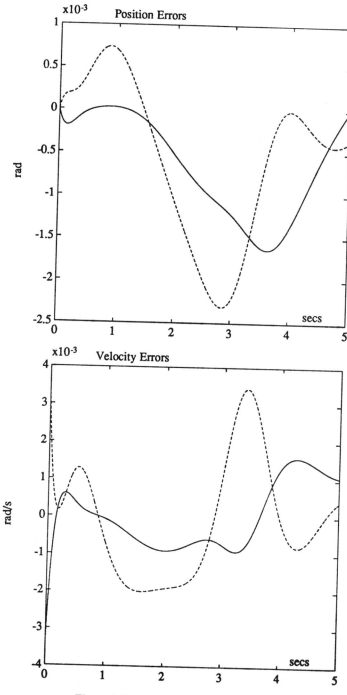

Figure 6: Position and Velocity Errors For Ex. 3

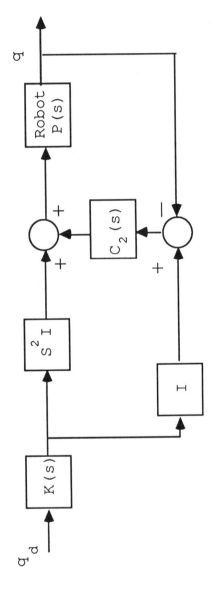

Figure 7: A 2 D.O.F. Control Structure

of e and v, minimizing its effects does not necessarily guarantee closed-loop stability. In Gilbert and Ha [14], Proportional-Integral-Derivative control is applied in order to obtain some sensitivity improvements. Cai and Goldenberg [50] use Proportional-Integral control to improve the robustness properties of the controller. Arimoto and Miyazaki [51] use Proportional-Integral-Derivative feedback control to robustly stabilize robot manipulators.

The feedback-linearization approach has been popular (under different names) in the robotics field. Its main advantage is obviously the wealth of linear techniques which may be used in the linear outer loop. In the presence of contact forces however, this approach becomes much more involved as was discussed in [17]. In addition, many controllers designed using this approach are not practical because they require a large control effort.

In some cases, the previously mentioned local linearization approach was combined with other techniques in order to guarantee robust stability. In particular, Desa and Roth [42] used the internal model principle to minimize the effects of disturbances for a robot model linearized over segments of the total operating time. Here also, closed-loop stability is not guaranteed.

III. PASSIVITY - BASED APPROACH

In this section, we review approaches which rely on the passive structure of rigid robots as described in equations (2) and (6) where $\dot{D}(q)-2C(q,\dot{q})$ is skew-symmetric by an appropriate choice of $C(q,\dot{q})$ [6].

3.1 Passive Controller 1:

Based on the passivity property, if one can close the loop from $\dot{q}$ to τ with a passive system (along with L_2 bounded inputs) as in Fig. 8, the closed-loop system will be asymptotically stable using the passivity theorem [38]. This however, will only show the asymptotic stability of $\dot{e}_1$ and not of e_1. On the other hand, if one can show the passivity of the system which maps τ to a new vector r which is a filtered version of e_1, then a controller which closes the loop between $-r$ and τ will guarantee the asymptotic stability of both e_1 and $\dot{e}_1$. This indirect use of the passivity property was illustrated in [2] and will be discussed next. Let the controller be given by (28-31) where $F(s)$ is a strictly proper, stable, rational function and K_r is a positive definite matrix,

$$\tau = D(q)a + C(q,\dot{q})v + g(q) - K_r(\dot{q}-v) \tag{28}$$

$$v = \dot{q}-r \tag{29}$$

$$r = -[sI+\frac{K(s)}{s}]e_1 = -F(s)^{-1}e_1 \tag{30}$$

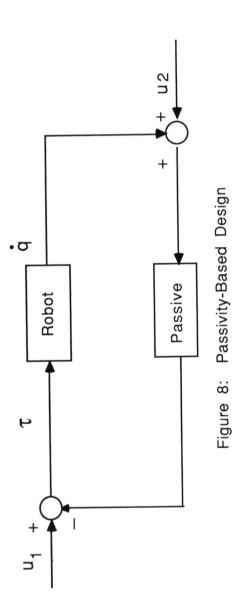

Figure 8: Passivity-Based Design

$$a = \dot{v} \tag{31}$$

Then it may be shown that both e_1 and $\dot{e}_1$ are asymptotically stable. This approach was used in the adaptive control literature to design passive controllers [2] but its modification in the design of robust controllers when D, C and g are not exactly known is not obvious. On the other hand, consider the control law (32) where $\Lambda(s)$ is an SPR transfer function,

$$\tau = -\Lambda(s)\dot{e}_1 + u_2 \tag{32}$$

The external input u_2 has to be bounded in the L_2 norm. Unfortunately, the inclusion of an integrator which reconstructs the error e_1 will destroy the SPR condition. Using the above control law, one gets from Fig. 8

$$r = -\Lambda(s)\dot{e}_1 \tag{33}$$

By an appropriate choice of $\Lambda(s)$ and u_2, one can apply the passivity theorem and deduce that $\dot{e}_1$ and r are bounded in the L_2 norm, and since $\Lambda(s)^{-1}$ is SPR (being the inverse of an SPR function), one deduces that $\dot{e}_1$ is asymptotically stable because

$$\dot{e}_1 = -\Lambda(s)^{-1}r \tag{34}$$

Unfortunately, as discussed above, this will only imply that the position error e_1 is bounded but not its asymptotic stability in the case of time-varying trajectories $[q_d^T \ \dot{q}_d^T]^T$. In the set-point tracking case however, and with gravity precompensation, the asymptotic stability of e_1 may be deduced using LaSalle's theorem [37]. The robustness of the system with the controller (32) is guaranteed as long as $\Lambda(s)$ is SPR and that u_2 is L_2 bounded, regardless of the exact values of the robot's parameters. Note that the controller (32) may be deduced from (15) by choosing the nonlinear controller

$$\tau = D[\ddot{q}_d+v] + C\dot{q} + g$$
$$v = -D^{-1}[\Lambda(s)\dot{e}_1+C\dot{q}+g]$$
$$u_2 = D\ddot{q}_d \tag{35}$$

3.2 Passive Controller 2:

The passivity approach in (32) is then a modified version of the feedback-linearization approaches. In [17,18] however, Anderson demonstrated using network-theoretic concepts, that even in the absence of contact forces, a feedback-linearization-based controller is not passive and may therefore cause instabilities in the presence of uncertainties. His solution to the problem consisted of using Proportional-Derivative (PD) controllers with variable gains $K_1(q)$ and $K_2(q)$ which depend on the inertia matrix $D(q)$, i.e.

$$\tau = -K_1(q)e_1 - K_2(q)e_2 + g, \tag{36}$$

Even though, $D(q)$ is not exactly known, the stability of the closed-loop error is guaranteed by the passivity of the robot and the feedback law. The advantage of this approach is that contact forces and larger uncertainties may now be accommodated. Its main disadvantage is that although robust stability is guaranteed, the closed-loop performance depends on the knowledge of $D(q)$ whose singular values are needed in order to find K_1 and K_2.

IV. VARIABLE - STRUCTURE CONTROLLERS

In this section, we group designs that use variable-structure controllers [52]. The VSS theory has been applied to the control of many nonlinear processes [53]. One of the main features of this approach is that one only needs to drive the error to a "switching surface", after which the system is in "sliding mode" and will not be affected by any modeling uncertainties and/or disturbances [52,53].

4.1 VSS Controller 1:

The first application of this theory to robot control seems to be in the work of Young [19] where the set point regulation problem ($\dot{q}_d = 0$) was solved using the following controller

$$\tau_i = \begin{cases} \tau_i^+ & \text{if } s_i(e_{1i}, \dot{q}_i) > 0 \\ \tau_i^- & \text{if } s_i(e_{1i}, \dot{q}_i) < 0 \end{cases} \tag{37}$$

where $i = 1, \cdots, n$ for an n-link robot, and s_i are the switching planes,

$$s_i(e_{1i}, \dot{q}_i) = c_i e_{1i} + \dot{q}_i, \quad c_i > 0. \tag{38}$$

It is then shown using the hierarchy of the sliding surfaces $s_1, s_2, \cdots, s_n$ and given bounds on the uncertainties in the manipulators model, that one can find τ^+ and τ^- in order to drive the error signal to the intersection of the sliding surfaces after which the error will "slide" to zero. This controller eliminates the nonlinear coupling of the joints by forcing the system into the sliding mode. In [22], a modification of the Young controller was presented. Other VSS robot controllers may be found in [23-25]. Unfortunately, for most of these schemes, the control effort as seen from (4.1) is discontinuous along $s_i = 0$ and will therefore create "chattering" which may excite unmodeled high-frequency dynamics.

4.2 VSS Controller 2:

To address this problem, Slotine modified the original VSS controllers using the so-called "suction control" [20,21]. In this approach, the sliding surface s is allowed to be time-varying and the control procedure consists of two steps. In the first, the control law forces the trajectory towards the sliding surface while in the second step, the controller is smoothed inside a possibly time-varying boundary layer. This will achieve optimal trade-off between control bandwidth and tracking precision, therefore eliminating chattering and the sensitivity of the controller to high-frequency unmodeled dynamics. The controller structure in this case is given by (39) where Λ is a diagonal matrix of positive elements λ_i (which may be time-varying) and $\Phi(.)$ is a nonlinear term determined by the extent of the parametric uncertainties and the suction control modifications [20],

$$\tau = \hat{D}\,[\ddot{q}_d - K_2 \dot{e}_1 - K_1^2 e_1 - \Phi(q,\dot{q},t)] + \hat{h}$$
$$K_1 = \Lambda^2 \,,\, K_2 = 2\Lambda \tag{39}$$

More recently, in [26,27], VSS controllers which avoided the inversion of the inertia matrix were introduced. The VSS approach although theoretically appealing, does not fully exploit the physics of the robots. In addition, in practice and to avoid chattering, the asymptotic stability of the error is sacrificed.

V. ROBUST - SATURATION APPROACH

In this section, we review the research that utilizes an auxiliary saturating controller to compensate for the uncertainty present in the robot dynamics as given by (2) where $C(q,\dot{q})$ is defined in (6).

$$D(q)\ddot{q} + C(q,\dot{q})\dot{q} + F(\dot{q}) + g(q) - \tau_d = \tau$$

or

$$D(q)\ddot{q} + C(q,\dot{q})\dot{q} + Z(q,\dot{q}) = \tau \tag{40}$$

Therefore $Z(q,\dot{q})$ is an n-vector representing friction, gravity and bounded torque disturbances. The controllers introduced in this section are robust due to the fact they are designed based on uncertainty bounds rather than on the actual values of the parameters. The following bounds are needed and may be physically justified. The d_i's and ζ_i's in (41,42) are positive scalar constants and the trajectory error e is defined before.

$$d_1 I_n \leq D(q) \leq d_2 I_n \tag{41}$$

$$||C(q,\dot{q})\dot{q}+Z(q,\dot{q})|| \leq \zeta_0 + \zeta_1||e|| + \zeta_2||e||^2 \tag{42}$$

Note the similarity between (24) and (42).

5.1 Saturation Controller 1:

Based on (41,42), Spong [13] used Lyapunov stability theory to guarantee the ultimate boundedness of e, a concept defined in [56] for example. The control strategy is actually based on the works of Cvetkovic [54] and the linear high-gain theory of Barmish [44]. Spong's controller is representative of this class and is given as follows

$$\tau = \frac{2d_1 d_2}{d_1+d_2}[\ddot{q}_d - K_2 e_2 - K_1 e_1 - v_r] + \hat{C}(q,\dot{q})\dot{q} + \hat{Z}(q,\dot{q}) \tag{43}$$

where

$$v_r = \begin{cases} (B^T Pe)\rho(||B^T Pe||)^{-1} & \text{if } ||B^T Pe|| > \varepsilon \\ (B^T Pe)\rho/\varepsilon & \text{if } ||B^T Pe|| \leq \varepsilon \end{cases} \tag{44}$$

and

$$\rho = \frac{1}{1-\alpha}[\alpha||\ddot{q}_d|| + ||K_1||.||e_1|| + ||K_2||.||e_2|| + \frac{1}{d_1}\phi], \tag{45}$$

$$\phi = \beta_0 + \beta_1||e|| + \beta_2||e||^2 \tag{46}$$

$$\alpha = (d_2-d_1)(d_2+d_1)^{-1} \tag{47}$$

Note that in the equations above, the matrix B is defined as in (15), the β_i's are defined as in (24), and the matrix P is the symmetric, positive-definite solution of the Lyapunov equation (48), where Q is symmetric and positive-definite matrix and A_c is given in (26).

$$A_c^T P + PA_c = -Q \tag{48}$$

In particular, the choice of Q

$$Q = \begin{bmatrix} K_1 & 0 \\ 0 & 2K_2-I \end{bmatrix} ; K_2 > I \tag{49}$$

leads to the following P

$$P = \begin{bmatrix} K_1+0.5K_2 & 0.5I \\ 0.5I & I \end{bmatrix} \tag{50}$$

The expression of P in (50) may therefore be used in the expression of v_r in (44).

Example 4: The following design parameters were chosen in simulating this controller:

$$K_1 = 15I \; ; K_2 = 30I \; ; d_1 = 0.15 \; ; d_2 = 6.5,$$

$$\beta_0 = 50 \; ; \beta_1 = 10 \; ; \beta_2 = 10 \; ; \varepsilon = 0.1,$$

$$\hat{C} = 0 \; ; \hat{Z} = 0,$$

and

$$e_1(0) = e_2(0) = 0.$$

The same trajectory is followed by the two-link robot as shown in Fig. 9

□

5.2 Saturation Controller 2:

Upon closer examination of Spong's controller (43-47), it becomes clear that v_r depends on the servo gains K_1 and K_2 through ρ. This might obscure the effect of adjusting the servo gains and may be avoided as described in [29]. In fact, let the controller be given by

$$\tau = -K_2 e_2 - K_1 e_1 - v_r(\rho, e_1, e_2, \varepsilon) \tag{51}$$

where

$$v_r = \begin{cases} (e/2+\dot{e})\rho(\|e/2+\dot{e}\|)^{-1} & \text{if } \|e/2+\dot{e}\| > \varepsilon \\ (e/2+\dot{e})\rho^2/\varepsilon & \text{if } \|e/2+\dot{e}\| \leq \varepsilon \end{cases}$$

and

$$\rho = \delta_0 + \delta_1 \|e\| + \delta_2 \|e\|^2 \tag{52}$$

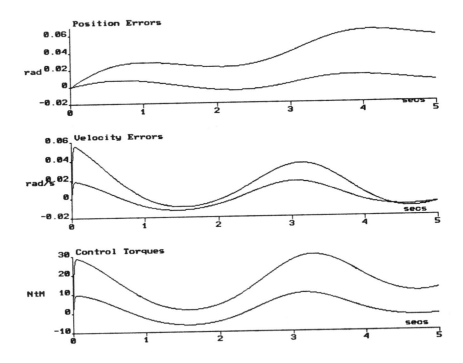

Figure 9: Errors and Torques for Ex. 4

where δ_i 's are positive scalars. Note that ρ no longer contains the servo gains and as such, one may adjust K_1 and K_2 without tampering with the auxiliary control v_r. As was also shown in [29], if the initial error $e(0)=0$ and by choosing $K_2 = 2K_1 = k_v I_n$, the tracking error may be bounded by the following which shows the direct effect of the control parameters on the tracking error,

$$\|e\| \leq \left[\frac{4(2k_v + 3d_2/2)\varepsilon}{k_v d_1} \right]^{1/2} \tag{53}$$

Example 5: In this example, let

$$K_1 = 15I \; ; \; K_2 = 30I,$$

$$\delta_0 = 50 \; ; \; \delta_1 = 10 \; ; \; \delta_2 = 10 \; ; \; \varepsilon = 0.1,$$

The results of the simulation are presented in Fig. 10

$\square$

As an extension to saturation controller 2, Spong introduced a new controller [55] where $\rho = \|Y\| \theta_{max}$. An example showing the behavior of this controller is given next.

Example 6: Let the control parameters be similar to those used in Example 6, and let

$$\theta_{max} = \begin{bmatrix} m_{1max} \\ m_{2max} \end{bmatrix} = \begin{bmatrix} 2 \\ 2 \end{bmatrix}$$

The resulting behavior is illustrated in Fig. 11

$\square$

In [28], Corless presented a simulation of a similar controller using a Manutec R3 robot. Another control scheme was given by Chen in [56]. Chen's controller however, requires acceleration measurements. In [7], Gilbert and Ha used a saturating-type feedback derived from Lyapunov-stability theory in order to guarantee the ultimate boundedness of the tracking error. In [57], Samson derived a "high-gain" controller which also guarantees the ultimate boundedness of the error.

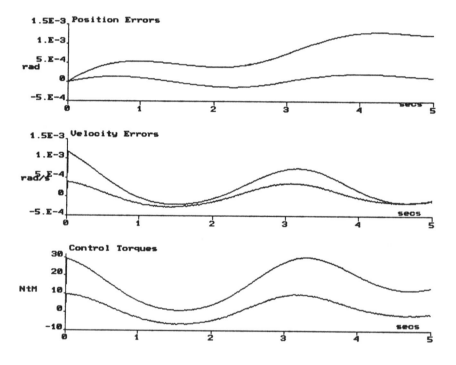

Figure 10: Errors and Torques for Ex. 5

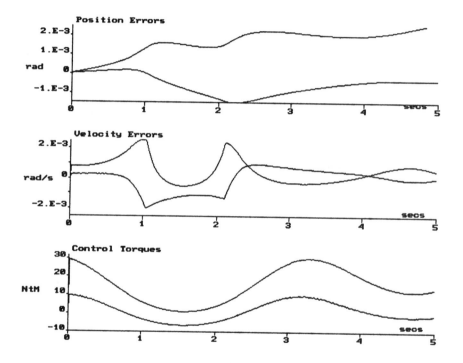

Figure 11: Errors and Torques for Ex. 6

VI. ROBUST-ADAPTIVE APPROACH

In this section, we briefly review some approaches that combine adaptive and robust control concepts. Since so much work has been done in the field of adaptive control of robotic manipulators [2], we only concentrate on schemes that are robust in addition to being adaptive.

6.1 Robust-Adaptive 1:

Let us first review one of the most commonly used robot adaptive controllers. This scheme was derived by Slotine [30] and a simplified version is given by the following where $\hat{\theta}$ is an r vector of the estimated parameters, and $Y(.)$ is an $n \times r$ regression matrix of known time functions.

$$\tau = \tau_a = Y(.)\hat{\theta} - K_2 e_2 - K_1 e_1 \tag{62}$$

$$\dot{\hat{\theta}} = -Y^T(.)[e_1/2 + e_2] \tag{63}$$

If there are no disturbances in the model (2), the tracking error is shown to be asymptotically stable with the above controller. However, the parameter estimate $\hat{\theta}$ in (63) may become unbounded in the presence of a bounded disturbance T_d, or unmodeled dynamics [59]. Robust-Adaptive controllers have attempted to robustify adaptive schemes against such uncertainties.

In [30], Slotine showed that the parameter estimates remain bounded if one uses

$$\tau = \tau_a + k_d sgn(e_1/2 + e_2) \tag{64}$$

where τ_a is given in (62) and k_d is a positive scalar constant satisfying

$$k_d > \|T_d\| \tag{65}$$

Example 7: Let us now add a disturbance torque to each joint given by

$$T_d = 2\sin(10t)$$

and choose

$$k_d = 3$$

The other control parameters remain the same as before and the results of the simulation are given in Fig. 12.

□

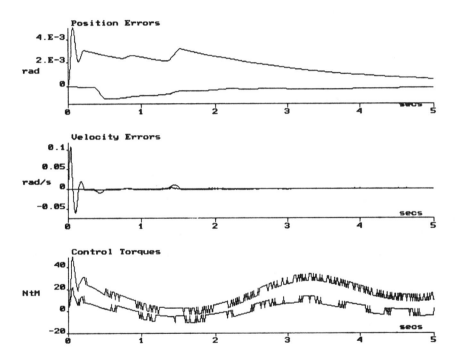

Figure 12: Errors and Torques for Ex. 7

6.2 Robust-Adaptive 2:

More recently [31], Reed introduced the σ–modification method originated by Ioannou [58] in order to compensate for both unmodeled dynamics and bounded disturbances. The control law is now given by

$$\tau = \tau_a = Y(.)\hat{\theta} - K_2 e_2 - K_1 e_1 \tag{66}$$

$$\dot{\hat{\theta}} = -Y^T(.)[e_1/2 + e_2] - \sigma\hat{\theta} \tag{67}$$

where

$$\sigma = \begin{cases} 0 & \text{if } \|\hat{\theta}\| < \bar{\theta} \\ \|\hat{\theta}\|/\bar{\theta} - 1 & \text{if } \bar{\theta} < \|\hat{\theta}\| < 2\bar{\theta} \\ 1 & \text{if } \|\hat{\theta}\| > 2\bar{\theta} \end{cases} \tag{68}$$

and

$$\bar{\theta} > \|\theta\| \tag{69}$$

Using this controller, Reed was able to show that the tracking error and all closed-loop signals are bounded.

Another approach in this section is that of Singh [32] which combines Spong's controller in (44) with adaptive techniques to estimate the uncertainty terms β_0, β_1 and β_2 in (46). Therefore, no prior knowledge about the exact size of the uncertainties is needed.

In [59], Spong and Ghorbel addressed certain instability mechanisms in the adaptive control of robots due to link flexibilities. A composite control law was used to damp out the fast dynamics, then a slow adaptive control law based on the algorithm of Slotine and Li [30] was robustified using the σ– modification [58]. Unfortunately, asymptotic stability is then lost if tracking a time-varying trajectory is desired. The algorithm is modified again using the switching σ– modification to ensure the asymptotic stability to a class of time-varying trajectories.

VII. CONCLUSIONS

The robust motion control of rigid robot was reviewed. Five main areas were identified and explained. All controllers were robust with respect to a range of uncertain parameters although some of them could only guarantee the boundedness of the position-tracking error rather than its asymptotic convergence. In the presence of disturbance torques, a bounded error is the best achievable outcome. In the last section, we also included adaptive controllers that are also robust. The question of which robust control method to choose is difficult to answer analytically but the following guidelines are suggested. The linear-multivariable approach is useful when linear performance specifications (Percent overshoot, Damping ratio, etc.) are available. This approach may however result in high-gain control laws in the attempt to achieve robustness. The passive controllers are easy to implement but do not provide easily quantifiable performance measures. The robust version of these controllers does not exploit the physics of the robot as their adaptive versions do. The variable-structure controllers should not be used when the flexibilities of the links are considerable for fear of exciting their high frequency dynamics. The saturation controllers, are most useful when a short transient error can be tolerated but ultimately, the error will have to be bounded. The robust adaptive controllers require more computing power and an adaptation time. On the other hand, they are most useful when repetitive or long duration tasks are performed. Their performance actually improves with time and they should be used when a high degree of performance is required. It is useful to note, that although the robot's dynamics are highly nonlinear, most successful controllers have exploited their physics and their very special structure [5,62]. This observation should be useful as we try to include force control, and flexibility effects in the current and future robotics research.

VIII. REFERENCES

1.   C. Abdallah, et. al., "Survey of Robust Control for Rigid Robots," *IEEE Control Systems Magazine*, Vol. 11, No. 2, pp. 24-30, Feb. 1991.

2.   R. Ortega, and M.W. Spong, "Adaptive Motion Control of Rigid Robots: A Tutorial," *Proc. IEEE Conf. Dec. & Contr.*, pp. 1575-1584, Austin, TX, Dec. 1988.

3.   P. Dorato, Ed., *Robust Control*, IEEE Press, New York, 1987.

4.   P. Dorato, and R.K. Yedavali, Ed., *Recent Advances in Robust Control*, IEEE Press, New York, 1990.

5.   K. Kreutz, "On Manipulator Control by Exact Linearization," *IEEE Trans. Auto. Cont.*, Vol. 34, No. 7, pp. 763-767, July 1989.

6.   M.W. Spong, and M. Vidyasagar, "Robust Linear Compensator Design for Nonlinear Robotic Control," *IEEE J. Rob. and Autom.*, Vol. RA-3, No. 4, pp.345-351, Aug. 1987.

7.   E.G. Gilbert, and I.J. Ha, "An Approach to Nonlinear Feedback Control with Applications to Robotics," *IEEE Trans. Sys., Man, and Cyber.*, Vol. SMC-14, No. 6, pp. 879-884, Nov./Dec. 1984.

8.   T.J. Tarn, A.K. Bejczy, A. Isidori, and Y. Chen, "Nonlinear Feedback in Robot Arm Control," *Proc. IEEE Conf. Dec. and Contr.*, Las Vegas, NV, Dec. 1984.

9.   N. Becker, and W.M. Grimm, "On L_2 and L_∞ Stability Approaches for the Robust Control of Robot Manipulators," *IEEE Trans. Automa. Contr.*, Vol. 33, No. 1, pp. 118-122, Jan. 1988.

10.   T. Sugie, et. al., "Robust Controller Design for Robot Manipulators," *ASME J. Dynamic Syst., Meas. & Contr.*, Vol. 110, No. 1, pp. 94-96, March 1988.

11.   E. Freund, "Fast Nonlinear Control with Arbitrary Pole-Placement for Industrial Robots and Manipulators," *Int. J. Rob. Res.*, Vol. 1, No. 1, pp. 65-78, Spring 1982.

12.   C. Abdallah, and R. Jordan, "A Positive-Real Design for Robotic Manipulators," *Proc. IEEE Amer. Contr. Conf.*, San Diego, CA, May 1990.

13.   M.W. Spong, J.S. Thorp, and J.M. Kleinwaks, "Robust

Microprocessor Control of Robot Manipulators, *Automatica,* Vol. 23, No. 3, pp. 373-379, 1987.

14.   E.G. Gilbert and I.J. Ha, "Robust Tracking in Nonlinear Systems," *IEEE Trans. Autom. Contr.,* Vol. AC-32, No. 9, pp. 763-771, 1987.

15.   M.W. Spong, "Control of Flexible Joint Robots: A Survey," *Report UILU-ENG-90-2203, DC-116,* Coordinated Science Laboratory, University of Illinois at Urbana-Champaign, Feb. 1990.

16.   M.W. Spong, and M. Vidyasagar, *Robot Dynamics and Control,* John Wiley & Sons, Inc., New York, 1989.

17.   R.J. Anderson, "Passive Computed Torque Algorithms For Robots," *Proc. IEEE Conf. Dec. & Contr.,* pp.1638-1644, Tampa, Fl, Dec. 1989.

18.   R.J. Anderson, *A Network Approach to Force Control in Robotics and Teleoperation,* Ph.D. Thesis, Department of Electrical and Computer Engineering, University of Illinois at Urbana-Champaign, 1989.

19.   K-K.D. Young, "Controller Design for A Manipulator Using Theory of Variable Structure Systems," *IEEE Trans. Sys., Man, and Cyber.,* Vol. SMC-8, No. 2, pp. 210-218, Feb. 1978.

20.   J-J.E. Slotine, "The Robust Control of Robot Manipulators," *Int. J. Rob. Res.,* Vol. 4, No. 2, pp.49-64, Summer 1985.

21.   J.J. Slotine, and S.S. Sastry, "Tracking Control of Nonlinear Systems Using Sliding Surfaces with Applications to Robot Manipulators," *Int. J. Contr.,* Vol. 38, pp. 465-492, 1983.

22.   R.G. Morgan, and U. Ozgunner, "A Decentralized Variable Structure Control Algorithm for Robotic Manipulators," *IEEE J. Rob. & Auto.,* Vol. RA-1, No. 1, pp. 57-65, Mar. 1985.

23.   E. Bailey, and A. Arapostathis, "Simple Sliding Mode Control Scheme Applied to Robot Manipulators," *Int. J. Contr.,* Vol. 45, No. 4, pp. 1197-1209, 1987.

24.   G. Bartolini, and T. Zolezzi, "Variable Structure Nonlinear in the Control Law," *IEEE Trans. Auto. Contr.,* Vol. AC-30, No. 7, pp. 681-684, Jul. 1985.

25.   K.S. Yeung, and Y.P. Chen, "A New Controller Design for Manipulators Using the Theory of Variable Structure Systems," *IEEE Trans. Auto.*

Contr., Vol. 33, No. 2, pp. 200-206, Feb. 1988.

26.   Y-F Chen, T. Mita, and S. Wakui, "A New and Simple Algorithm for Sliding Mode Trajectory Control of the Robot Arm," *IEEE Trans. Auto. Contr.,* Vol. 35, No. 7, pp. 828-829, Jul. 1990.

27.   B.E. Paden, and S.S. Sastry, "A Calculus for computing Filipov's Differential Inclusion with application to the Variable structure Control of Robot Manipulators," *IEEE Trans. Circ. Syst.,* Vol. CAS-34, No. 1, pp. 73-82, Jan. 1987.

28.   M. Corless, "Tracking Controllers for Uncertain Systems: Application to a Manutec R3 Robot," *J. Dyn. Sys., Meas., & Contr.,* Vol. 111, pp. 609-618, Dec. 1989.

29.   D. Dawson, and F. Lewis, "Robust and Adaptive Control of Robot Manipulators Without Acceleration Measurement," *Proc. IEEE Conf. Dec. & Contr.,* Tampa, Fl., 1989.

30.   J.J. Slotine, and W. Li, "On the Adaptive Control of Robot Manipulators," *Int. J. Rob. Res.,* Vol. 3, 1987.

31.   J. Reed, and P. Ioannou, "Instability Analysis and Robust Adaptive Control of Robot Manipulators," *Proc. IEEE Conf. Dec. & Contr.,* pp. 1607-1612, Austin, TX, 1988.

32.   S.N. Singh, "Adaptive Model Following Control of Nonlinear Robotic Systems," *IEEE Trans. Auto. Contr.,* Vol. AC-30, No. 11, pp. 1099-1100, Nov. 1985.

33.   K.Y. Lim, and M. Eslami, "Robust Adaptive Controller Designs for Robot Manipulator Systems," *IEEE Conf. Rob. & Auto.,* San Francisco, April 1986.

34.   J.J. Craig, *Adaptive Control of Mechanical Manipulators,* Addison-Wesley, Massachusetts, 1988.

35.   Meirovitch, *Methods of Analytical Dynamics,* McGraw-Hill, New York, 1970.

36.   R.J. Schilling, *Fundamentals of Robotics: Analysis and Control,* Prentice-Hall, Englewood Cliffs, NJ, 1990.

37.   M. Vidyasagar, *Nonlinear Systems Analysis,* Prentice-Hall, Englewood Cliffs, NJ, 1978.

38.   C. Desoer, and M. Vidyasagar, *Feedback Systems: Input-Output Properties*, Academic press, New York, 1975.

39.   M. Whitehead, and E.W. Kamen, "Control of Serial Maipulators with Unknown Variable Loading," *Proc. IEEE Conf. Dec. & Contr.*, Ft. Lauderdale, Fl, Dec. 1985.

40.   J.Y.S. Luh, "Conventional Controller Design for Industrial Robots: A tutorial," *IEEE Trans. Sys., Man, & Cyber.*, Vol. SMC-13, No. 3, pp. 298-316, May/June 1983.

41.   E.J. Davison, and A. Goldenberg, " The Robust Control of a General Servomechanism Problem: The Servo Compensator," *Automatica*, Vol. 11, pp. 461-471, 1975.

42.   S. Desa, and B. Roth, "Synthesis of Control Systems for Manipulators Using Multivariable Robust Servo Mechanism Theory," *Int. J. Robotic Res.*, Vol. 4, pp. 18-34, Fall 1985.

43.   H. Asada, and K. Youcef-Toumi, *Direct-Drive Robots: Theory and Practice*, The MIT Press, Cambridge, MA, 1987.

44.   B.D.O. Anderson et.al., *Stability of Adaptive Systems: Passivity and Averaging Techniques*, Prentice-Hall Inc., Englewood Cliffs, NJ, 1989.

45.   M. Vidyasagar, *Control Systems Synthesis: A Factorization Approach*, MIT Press, Cambridge, MA, 1985.

46.   T. Sugie, et. al., "Robust Controller Design for Robot Manipulators," *Trans. ASME Dyn. Syst., Meas. & Cont.*, Vol. 110, No. 1, pp. 94-96, March 1988.

47.   T. Kailath, *Linear Systems*, Prentice-Hall, Englewood Cliffs, NJ, 1980.

48.   C.Y. Kuo, and S.P.T. Wang, "Nonlinear Robust Industrial Robot Control," *Trans. ASME J. Dyn. Syst. Meas. Control*, Vol. 111, No. 1, pp. 24-30, March 1989.

49.   B. A. Francis, and W.M. Wonham, "The Internal Model Principle of Control Theory," *Automatica*, Vol. 12, pp. 457-465, 1976.

50.   L. Cai, and A.A. Goldenberg, "Robust Control of Position and Force for a Robot Manipulator in Non-Contact and Contact Tasks," *Proc. Amer. Contr. Conf.*, Pittsburgh, PA, June 1989, pp. 1905-1911.

51.   S. Arimoto, and F. Miyazaki, "Stability and Robustness of PID Feedback Control for Robot Manipulators of Sensory Capability," *Proc. 1st Int. Symp. Robotics Res.*, pp.783-799, 1983.

52.   V.I. Utkin, "Variable Structure Systems with Sliding Modes," *IEEE Trans. Auto. Contr.*, Vol. AC-22, pp. 212-222, Apr. 1977.

53.   R.A. DeCarlo, S.H. Żak, and G.P. Matthews, "Variable Structure Control of Nonlinear Multivariable Systems," *IEEE Proc.*, Vol. 76, No. 3, pp. 212-232, March 1988.

54.   V. Cvetkovic, and M. Vukobratovic, "One Robust, Dynamic Control Algorithm for Manipulation Systems," *Int. J. Robotics Res.*, Vol. 1, pp. 15-28, 1982.

55.   M. Spong, "On the Robust Control of Robot Manipulators," *Preprint*, April 1991.

56.   Y.H. Chen, "On the Deterministic Performance of Uncertain Dynamical Systems," *Int. J. Contr.*, Vol. 43, No. 5, pp. 1557-1579, 1986.

57.   C. Samson, "Robust Nonlinear Control of Robotic Manipulators," *Proc. 22nd IEEE Conf. Dec. Contr.*, San Antonio, TX, Dec. 1983, pp. 1211-1216.

58.   P.A. Ioannou, and P.V. Kokotovic, "Instability Analysis and Improvement of Robustness of Adaptive Control," *Automatica*, Vol. 20, No. 5, pp. 583-594, 1984.

59.   M.W. Spong, and F. Ghorbel, "Robustness of Adaptive Control of Robots," Proceedings of the *Symposium on the Control of Robots and Manufacturing Systems*, Arlington, Texas, November 1990.

60.   S. Sastry and M. Bodson, *Adaptive Control: Stability, Convergence and Robustness*, Prentice-Hall, Englewood Cliffs, NJ, 1989.

61.   P. Ioannou, and K. Tsakalis, "A Robust Direct Adaptive Controller," *IEEE Trans. Auto. Contr.*, Vol. AC-31, No. 11, pp. 1033-1043, Nov. 1986.

62.   J-J.E. Slotine, "Putting Physics in Control - The Example of Robotics," *IEEE Contr. Syst. Mag.*, Vol. 8, No. 7, pp. 12-17, Dec. 1988.

Introduction to Non-linear Control Using Artificial Neural Networks

H. M. Wabgaonkar
A. R. Stubberud

ECE Dept., University of California, Irvine
Irvine CA USA 92717

Introduction

With the increasing demands for performance and versatility being placed on engineering systems, far greater requirements are being imposed on the design of control systems. The most crucial among these is the requirement for such systems to perform autonomously under uncertain environments. The causes for this uncertainty or imprecision may be several. First, the control process or the plant may be such that a precise mathematical model cannot be developed or is too costly to develop. Secondly, even if the model is known to a reasonable degree, there may be significant delays between an application of an input and indication of its effect. Also, there could be significant non-linearities in the system dynamics, large parameter variations, faulty sensors and actuators, etc. A control system can cope with these problems by reconfiguring its control law so as to meet the performance requirements. To do this, the controller must first learn about the plant, the environment, and its own capabilities. A control system with this learning and decision making capability can be thought as being intelligent (Stubberud[1], Bavarian[2]). Thus, the intelligence of such systems is directed towards delivering a specific performance level even in the presence of high uncertainty.

Indeed, the goal of traditional control systems design is to synthesize a system with acceptable performance over a certain prespecified range of parameter variations, environmental conditions, and input signals. The degree to which a system performs over a range of uncertainty is referred to as its robustness or insensitivity (Bavarian[2]). Classical designs that incorporate fixed-gain controllers provide robustness over a relatively narrow range of uncertainty. For systems which must perform over a wider range of uncertainty, adaptive control/signal processing techniques have been developed (e.g., Goodwin and Sin[3], Widrow and Stearns[4]). Such systems typically employ a controller in an inner loop, whose parameters are adjusted by an adaptation mechanism operating in an outer loop. The adaptation mechanism tracks changes in plant parameters, environmental conditions, etc. It then reconfigures the control law so that the system delivers the desired performance. The primary objective of an intelligent control system is the same as that of the traditional and adaptive systems. However, an intelligent system is expected to tolerate a level of uncertainty or imprecision much higher than that of the traditional and adaptive systems. Thus, an intelligent system differs from adaptive systems only in degree but not in kind. In order to achieve its goal, an intelligent system needs to be endowed with several features such as:

1) ability to deal with qualitative, unquantified and possibly inconsistent data emanating from diverse sources;

2) ability to deal rapidly with large amounts of data;

3) ability to satisfy simultaneously a large number of constraints which may not be specified or quantified completely or which may be conflicting;

4) ability to generalize i.e., the ability to deal with a novel situation based on the ones experienced earlier.

Control systems that currently have these features typically depend on human intervention to function properly. However, human intervention may not be acceptable under many circumstances. For example, the speed of human response can be a limiting factor in real-time applications. Secondly, continual process monitoring and supervision may be stressful to the operator; or an error on the part of the operator can lead to severe damage or degradation in service. Hence, it becomes necessary to develop techniques that will replace human intervention as much as possible, while still retaining its desirable features. In this context, the role of Artificial Intelligence (in which we include connectionist or neural computing) becomes quite clear. Capturing human expertise to form and utilize knowledge-bases, qualitative reasoning, emulating human cognitive processes in general, etc. belong to the domain of Artificial Intelligence (AI). With significant advances in this field beyond simple game playing tasks, it is only natural for control engineers to exploit these advances in AI to synthesize intelligent control systems.

1 AI and Connectionist Computing

All the approaches in AI are basically mechanistic-they assume that intelligence can be understood, at some level, as the operation of a physical mechanism upon signals or entities that represent information. The assumption here is that there is nothing that is fundamentally beyond human understanding or physical laws. However, this premise does not imply at all that the processes of intelligence are simple or that they are well understood. Within this broad assumption of the mechanistic nature about intelligence, two divergent views have emerged. Underlying the first category is a powerful abstraction of human thought process without any specific regard for the conformity with neurophysiological detail of human information processing. We will call this category the 'Symbolic Processing Paradigm'. The second one is based on a parallel distributed processing model of human cognition; it borrows heavily from cognitive psychophysiology. We will refer to this category as the 'Connectionist or Neural Processing Paradigm'. While there has been some philosophical debate about the exact relationship between these two paradigms (e.g., Fodor and Pylyshyn[5], and Smolensky[6]), our interest here is not of that type, but to bring out what we consider is relevant to enhance the performance of control systems by improving their robustness and insensitivity properties.

The symbolic paradigm relies on obtaining an abstract stylized representation of the world through a proper selection of symbols (Charniak and McDermott[7]). The key idea is to obtain useful semantic inferences about the world through syntactic manipulations of the symbols. The symbols are manipulated by employing a set of rules which contain the symbols. This leads to a somewhat natural division between knowledge representation, acquisition, and utilization. Designs of such systems have evolved to incorporate many complex data structures and schemes for their manipulation, such as in parsing, recursion, complex scene analysis, planning, etc. However, over years, discovering general-purpose, universally applicable rules has been found to be an illusory goal. As a result, symbolic systems have been developed so as to solve problems in very narrow specialized domains. The domain has to be narrow enough so that, within the domain, the discovered rules are general enough. However, as the domain is expanded, the initial advantages of having such rules is lost, since more and more exceptions to the rules may be encountered. Secondly, in real life, a need to fuse these various specialists so as to form a meaningfully functional aggregate often arises, but this aspect cannot be taken care of a priori while synthesizing the individual expert systems. A very desirable feature of such systems is their ability to provide an external user with an explanation for the reasoning undertaken by the system to reach a certain goal or inference. However, it is this very insistence on the rigid, unambiguous, logically sound 'theorem-proving' style knowledge processing that has made these systems preclude the possibility of incorporating practical, common sense reasoning.

The connectionist or the PDP approach can be characterized as a bottom-up approach (Rumelhart et al.[8]). In this approach, one begins with very simple elements (neurons) and connects them to form a complex whole- called the neural network. The collective actions (and interactions) of the simple elements leads to an interesting emergent knowledge-processing behavior. The term connectionism refers to the storage of information in the paths or the links between the constituent elements or neurons of the network. The words Parallel and Distributed have their obvious interpretations. Information processing takes place at several sites simultaneously. Thus, the abstract architecture of neural networks bears a close resemblance to human neuro-biological systems. In addition to the innate parallelism, properties such as graceful degradation under noisy inputs and faults, learning generalization and adaptability, associative and content-addressable memory are among the desirable properties usually attributed to human information processing that connectionist models aspire to capture (Pfeifer et al.[9]). There may be information processing tasks that involve sequential processing, abstract reasoning, complex data structures, explainability of behavior, etc. For such applications neural networks at the current state of affairs may not be well suited. Additionally, a very large-scale network may suffer from undesirable crosstalk between its units; there may be communication/connection cost overheads between the units, and finally, one must consider the cost of computations local to each neural site. However, many interesting possibilities exist, and many useful applications need to be explored within the domain of control systems, before the above issues begin to pose serious obstacles. The central issue in connectionist models is learning or training of a neural network to perform a given information processing task. Currently, two training paradigms exist. The first one is called unsupervised training in which the network has no external teacher to indicate what is the right action. The network picks up useful features (usually statistical) from the inputs presented to it and then uses these extracted features to perform useful information processing. While this paradigm may be useful for data compression and clustering, it is not appropriate for its inclusion directly in typical control systems. The other paradigm is the supervised training paradigm in which the network is presented an input and also the corresponding desired output or action. The network is trained to remember the associations of this type for a collection of input-output pairs (Rumelhart et al.[10]). This computational approach is very much relevant to control systems synthesis as we indicate below. The following discussion is therefore confined to this approach.

2 Parameter Estimation and Associative Memories

In their typical applications, neural networks accept feature vectors as their inputs and perform further operations on the input so received. The feature vector inputs typically represent a list of measured quantities or observations about the state of an observed system or an object under study. The information processing operation performed by the network results in another pattern or list of numerical quantities being produced at the output of the network. Thus, the information processing action of a neural network can be viewed as an application of a mapping between two spaces- the space of the input feature patterns and the other space of the output feature patterns. Further, this mapping can be static or dynamic depending on whether the mapping itself changes with time. *Whether this mapping is static or dynamic, the fundamental issue in neural network synthesis is the realization of an appropriate mapping between two finite-dimensional spaces.* The exact mathematical form of the mapping is specified by the network architecture, i.e., by specifying how the various neurons or the units in the network are connected, and also, by specifying the mathematical operations that take place at each neuronal site. In general, the overall network mapping can be represented as linear combinations and compositions of linear and non-linear functions. With this view, it is possible to hypothesize some commonalities between pattern classification, control theory and signal processing on one side, and the connectionist associative memories on the other side.

The technique of system identification plays an important role in control theory as well as in signal processing (Goodwin and Sin[3], Widrow and Stearns[4]). This technique involves the construction of a mathematical model whose input-output behavior matches with that of a given system. In other words, the model produces nearly the same output signal as the given system being modelled, when both of them receive the same input signal. One specific way of building such a model is by parameter estimation. The structure of the mathematical model is fixed and only a finite set of parameters associated with that structure are estimated so that the model and the system match. Different models from the same structural class may be generated by using different parameter values. Thus, a common technique in system identification is to first specify a model structurally, and then, to adjust its parameters so that the model and the system being modelled have similar input-output behavior. This adjustment of the model parameters is usually done on-line by employing an iterative estimation procedure. To summarize, the model parameters are tuned so that the model learns the appropriate association between various input patterns and the corresponding output patterns. The association is governed by the observed system that the model is to imitate.

The above interpretation of parameter estimation is closely related to associative memories in connectionist computing. The relationship between the two was explored in the first author's earlier work (Wabgaonkar[11]). A detailed exposition on this topic appears in Barto[12]. An associative memory is a device that stores the association between a pair of quantities (Hinton and Anderson[13]). Depending on the size or the capacity of the memory many such pairs may be stored in the memory. After an association is learned, presentation of one of the items from a pair leads to the recall of the other item of the pair at the output of the associative memory. The traditional (non-connectionist) characterization of memory is in terms of a place for storing information. Retrieval of stored information amounts to going to the right place and recovering the item from that location. The power of information processing involving such memories ensues from the complexity of the items stored. Some of the standard non-connectionist data organizations used for information storage have been hierarchical trees, semantic networks, frames, scripts, plans, etc. (Schank and Abelson[14]). The interpretation of the associative memory paradigm in connectionism is somewhat different. Information is thought of as being 'evoked' rather than 'found' (Rumelhart and Norman[15]). In an abstract way, it is possible to say that the input key to the network itself acts as an address. Generally, many neurons participate in the encoding as well as the recall of a stored pair. The same units tend to react to similar input patterns which leads to interactions among stored patterns which are similar. The interaction between various stored patterns is responsible for some desirable and as well as some undesirable features. Ability to "fill in" or complete an incompletely specified input as well as the ability to deal with novel inputs are among the desirable features. However, interference or crosstalk between the units can lead to slow training. And as new associations are stored in the memory, the old ones may be forgotten. Thus, if in a certain application, 100 pattern associations were stored, and if the need arises to store a new 101st association, it may become necessary to teach the first 100 associations all over again.

The use of pattern classification and associative memory based techniques in control systems design is not new. Many attempts along these lines were made in the sixties and early seventies (e.g., Fu[16], Mendel and Zapalac[17]). However, most of these involved large amounts of memory and were heuristically oriented. As the use of serial digital computers became more and more common, new theoretically rigorous techniques (with provable convergence properties) were developed for problems that involved usually a few control variables. The implementation of such techniques on the digital computer opened up a broad range of applications which were not otherwise tractable to analog computers or other relay-based control schemes. This led to a shift away from the heuristic techniques and their virtual abandonment until recently. However, as we described earlier, the study of synthesis of intelligent control systems has rekindled the interest in such systems. The quintessential features of connectionist models or their precursors in control systems are simple heuristic collective computations and incorporation of experimental procedures. This somewhat goes against the grain of orthodox control engineering practice. Thus, there is a trade-off. If we demand rigid mathematical formulations throughout, it stifles the very spirit of new development and severely constrains the progress. On the other hand, in traditional control systems engineering methodology, experimentation and heuristic approaches will not gain easy acceptance. An approach to solve this dilemma is to examine the existing control techniques critically and correlate them with the new emerging techniques. A careful reconciliation of the two can lead to enhanced performance. It is with this view that we examine the mathematical basis for synthesizing associative memories. The next section deals with some of the representative results related to function approximation in neural networks.

3 Function Approximation in Neural networks

In the last section, we introduced the concept that neural networks in a broad range of their applications can be viewed as devices for implementing appropriate mappings between two finite dimensional spaces. One of these spaces is the space of the input patterns on which the neural network acts. The other space is that of the output patterns, those which result from the action of the network on the presented input patterns. Neural networks are synthesized so that the mappings realized by them closely approximate useful functions between the input and the output pattern spaces. The mapping realized by a neural network is governed by the structural attributes of the network. Hence, it becomes necessary to examine the structure of the network to understand the properties of the realized approximation.

The processing units (neurons) in a neural network are arranged in layers. A neural network has a layer (called the input layer) of units which receive the external input pattern to be processed, a layer (called the output layer) of units whose outputs constitute the network's output pattern, and possibly other layers (called hidden layers) of neurons sandwiched between the input and the output layers. Each processing unit in a given layer receives outputs from other units via paths or links that connect the outputs of those units to the input of the processing unit. As mentioned earlier, information in the network is stored in the form of weights or connection strengths on these paths. Thus, each of the inputs received by the processing unit is appropriately weighted by the path weight associated with the particular path. The actual computation performed at each neuronal site is very simple. To minimize computational overheads, the same type of computation is performed at each of the neuronal sites. Typically, the inputs are summed and the sum is operated upon by a non-linear mapping called the neuronal activation function (e.g., $f_i(s) = [1/(1+e^{-s})]$, sometimes, linear mappings are also employed, particularly for neurons that are in the output layer). The result of this computation at a given neuronal site is treated as the output of the neuron. This output then travels on the paths that connect the output of the neuron to the inputs of other neurons. In general, the mappings represented or realized by a neural network can be thought of as superpositions (i.e., compositions and sum of the compositions) of non-linear functions of fewer variables. To be concrete, the mapping realized by a network that consists of an input layer, a single hidden layer of P non-linear units, and a single output unit can be written as

$$y = f(x) \tag{1}$$

where

$$f(x) = \sum_{i=1}^{P} w_{1i}^2 \, f_i \left(\sum_{j=1}^{n} w_{ij}^1 \, x_j \right). \tag{2}$$

In above equations, y is the scalar network output for an n-dimensional vector input $x = (x_1, x_2, ..., x_n)$. The quantity w_{ij}^1 is the path weight from the jth unit of layer one (the input layer) to the ith unit of layer two (the hidden layer).

The function f_i represents the non-linear activation function of the ith hidden layer neuron. Similarly, w_{1i}^2 represents the path weight from unit i of the second layer to the only output unit, unit one of the the output layer. The outer sum over the index variable i represents the fact that the output unit of the network is linear.

With this background, it is clear that the neural network synthesis problem for a network with n inputs and m outputs can be posed as a problem of generating a suitable approximation (in terms of superpositions of functions of fewer variables) to a multivariate function $g: D \subset E^n \rightarrow E^m$. Here, D is the domain of the input patterns, it is assumed to be a compact subset of the n-dimensional Euclidean space E^n. We now describe some of the important results related to the approximation problem under consideration.

A classical theorem related to the approximation of a multivariate function is the Kolmogorov Superposition Theorem (KST). In the year 1900, at the International Conference of Mathematicians in Paris, Hilbert presented 23 problems which he thought were important for the advancement of mathematics. The thirteenth of these problems was related to the conjecture that not all continuous functions of three variables are representable as superpositions of continuous functions of two variables. This particular conjecture was refuted in 1957 by Kolmogorov[18] and his pupil Arnold[19]. The resulting representation theorem is thus called the KST. Since the publication of the original results, the subject has attracted attention of a number of mathematicians. There are many versions of the theorem and many related results. Some of the useful references on this topic are: Lorentz([20],[21]), Sprecher[22], Vitushkin[23], and Hedberg[24]. The complete proof of the theorem is quite lengthy and intricate; therefore we present only the theorem statement. For a formal treatment, the above references are recommended.

Let I be the closed interval [0, 1] and let S be the n-cube I^n, i.e., $0 \le x_p \le 1$; $p = 1,...,n$. Let C(I) be the space of continuous functions $\phi: I \to I$; C(S) the space of continuous functions $f: S \to R$; and C(R) the space of continuous functions from R to R.

Theorem   1:(KST) If $n \geq 2$, there exist n real numbers $\lambda_1, ..., \lambda_n$ and elements $\phi_1, ..., \phi_{2n+1}$ of C(I) with the following property: for each f in C(S) there exists g in C(R), such that:

$$f(x_1, ..., x_n) = \sum_{q=1}^{2n+1} g(\lambda_1 \phi_q(x_1) + ... + \lambda_n \phi_q(x_n)). \tag{3}$$

It is to be noted that the functions ϕ_q are independent of f, while the function g depends on f. Hecht-Nielsen was perhaps the first to relate this theorem to neurocomputing[25]. Unfortunately, an implementation based on the KST is beset with many practical difficulties. In order to implement the function approximation scheme based on the KST, it necessary to be able to calculate/store the functions ϕ_q and the function g either digitally or in an analog fashion. As it turns out, due to the conditions imposed on the functions ϕ_q, these functions cannot be very smooth; they can at best be of the class Lip(1). In fact, the construction of the functions ϕ_q can imitate that of another set of functions ω_q. The construction is quite complicated, it will not be described here. It appears in

Lorentz[20]. The functions ω_q are continuously increasing but with zero derivative almost everywhere; sometimes, they are referred to as the Cantor functions (Natanson[26], Lorentz[20]). A digital or analog storage of such functions is not currently possible and even if it were, it would be fragile and very sensitive. An important question that may be raised at this point is this: Can we search for other candidate functions ϕ_q with smooth properties?

Unfortunately, the answer is negative. For example, a result due to Kolmogorov[27] is as follows. Let W_q^n be the space of functions f defined on the n-cube S, whose derivatives up to and including order q are continuous and are bounded by some constant. Let $\gamma = n/q$ be defined as the characteristic of the members f of W_q^n . Then,

Theorem 2:(Kolmogorov): Not every function of a given characteristic $\gamma_0 = n_0/q_0$ can be represented by superpositions of functions of characteristic $\gamma = n/q < \gamma_0$, $q \geq 1$.

Hilbert's conjecture was based on the thought that a 'hard' function cannot be represented by 'simpler' ones.

However, being a function of n variables, n>1, is not by itself an adequate measure of hardness. The characteristic γ defined above is a more appropriate measure. Underlying the above theorem is the concept of metric entropy due to Kolmogorov[27]. It is a measure of massivity of a set and is related to the characteristic γ. Thus the set with the characteristic γ_0 is more massive than the set with characteristic γ (in the metric entropy sense) and hence some of the members cannot be represented by 'simpler' functions. In his thesis, one of the authors (Wabgaonkar[28], Chapter 3) attempted to generate the functions ϕ_q using the so-called 'Devil's Staircase' function which has been studied extensively in the theory of chaotic systems. However, at that time, the method appeared to be computationally very demanding. It must be noted that the KST seeks an exact equality rather than an approximation. If this demand of exact equality is relaxed, it is possible to obtain some interesting results. We describe some of these results very briefly.

First, we review the work of Cybenko[29]. It was one of the early results to be published on function approximation. It makes use of the well-known Hahn-Banach Theorem of real analysis (e.g., Royden[30], pp. 223). To describe the theorem, let I^n be the n-dimensional unit cube $[0, 1]^n$. The space of continuous functions on I^n is denoted by $C(I^n)$; and we use $\|f\|$ to denote the supremum (or uniform) norm of an element f in $C(I^n)$. The space of finite, signed, regular Borel measures on I^n is denoted by $M(I^n)$.

The main goal is to show that, under suitable conditions, sums G of the form:

$$G(x) = \sum_{j=1}^{N} \alpha_j \ \sigma(y^{j^T}x + \theta_j) \qquad (4)$$

are dense in $C(I^n)$ with respect to the supremum norm. Here, $\sigma : R \rightarrow R$ is a (univariate) function to be specified in the following; y^j is an n-vector of the path weights; and α_j and θ_j are real scalars. First, we need two definitions:

Definition C.1: The function σ is *discriminatory* if for a measure μ in $M(I^n)$,

$$\int_{I^n} \sigma(y^T x + \theta) \ d\mu(x) = 0 \qquad (5)$$

for all y in R^n and θ in R, implies that $\mu = 0$.

<u>Definition C.2</u>: The function σ is *sigmoidal* if lim $\sigma(t) = 1$ as $t \to \infty$, and lim $\sigma(t) = 0$ as $t \to -\infty$.

The required result is derived by means of the following Theorem C.1 and Lemma C.1:

<u>Theorem C.1(Cybenko):</u> Let σ be any continuous

discriminatory function. Then finite sums of the form (4) are dense in $C(I^n)$.

<u>Lemma C.1:</u> Any bounded measurable sigmoidal function is discriminatory.

Thus, a neural network whose neurons possess measurable sigmoidal activation functions can be used for function approximation. Another interesting observation is that the structural form of the right hand side of the Equation (4) is very similar to that of (2). Thus, Cybenko's work essentially establishes the adequacy of a net with a single hidden layer of continuous sigmoidal units, and with an output layer of linear units.

White and co-workers (referred to as HSW[31]) established similar results using another approach. They use the Stone-Weierstrass Theorem (Royden[30], pp. 212) to prove the adequacy of the network structure same as the one considered by Cybenko. The concept of a set being dense in another one plays a central role in any work based on the Stone-Weierstrass Theorem, it does so in the present context as well. We have to state the following definitions before describing the results: For any r, the set of all affine transformations A: $R^r \rightarrow R$ is denoted by A^r. Specifically, an *affine transformation* A is given by: $A(x) = w^T x + b$, w and x are in R^r, and b is in R. Next, a function ψ defined on R and taking values in [0, 1] is called a *squashing function* if it is non-decreasing and has Lim ψ (x) = 1 as $x \rightarrow +\infty$, and Lim ψ (x) = 0 as $x \rightarrow -\infty$. Then,

<u>Definition W.1</u>: A subset S of a metric space (X,d) is *d-dense* in a subset T of X if for every $\varepsilon > 0$, and for every t in T, there is an s in S, such that $d(s,t) < \varepsilon$.

<u>Definition W.2</u>: Let $C^r = C(R^r)$ be the space of continuous real-valued functions f : $R^r \rightarrow R$. A subset S of C^r is said to be

uniformly dense in C^r on compacta if for every compact subset K of R^r, S is d_K-dense in C^r, where for f, g in C^r, $d_K(f,g) = \sup_x |f(x)-g(x)|$, x in K.

With these definitions we state the central result:

Theorem   W.1(HSW): Let g be any continuous, non-constant function from R to R. Then, the set G is uniformly dense in C^r on compacta, where the set G consists of all functions h of the form:

$$h(x) = \sum_{j=1}^{q} \beta_j \prod_{k=1}^{m_j} g(A_{jk}(x)) \tag{6}$$

where r, m_j, q are in N (the set of natural numbers), β_j is in R, A_{jk} is in A^r, and x is an r-dimensional vector.

HSW also consider function approximation within the setting of measure spaces. For this, let B^r be the Borel sigma-field of R^r, and M^r the space of all Borel-measurable functions from R^r to R. Next, given a probability measure μ on (R^r, B^r), a metric is defined on $M^r \times M^r$ to R^+ by:

$$\rho_\mu(f,g) = \inf\left\{\varepsilon > 0 : \mu(x \in R^r : |f(x)-g(x)| > \varepsilon) < \varepsilon\right\} \tag{7}$$

for f and g in M^r. With this definition of the metric, the following result is established:

Theorem   W.2(HSW): For every non-constant continuous function g, every r, and every probability measure μ on (R^r, B^r), the set G is P_μ-dense in M^r.

Finally, it follows from Theorem W.1 that the set G of trigonometric polynomials   with $g(\cdot) = \cos(\cdot)$, is uniformly dense in C^r on compacta. By the repeated application of the identity $\cos(a+b) + \cos(a-b) = \cos(a)\cos(b)/2$, each of the polynomials in the above set can be written as a sum:

$$s(.) = \sum_{t=1}^{T} \alpha_t \cos(A_t(.)); \ \alpha_t \in R, \ A_t \in A^r. \tag{8}$$

But, for each such sum s(.), there exists a function f(.) = $\sum_k \gamma_k \psi(A_k(.))$, ψ an arbitrary squashing function, such that: $\sup_x |s(x)-f(x)| < \varepsilon$, as x ranges over K, an arbitrary compact subset of R^r. Now, let F be the set of all functions of the form that is the same as that of the above mentioned function f. Then, we have the following important result:

<u>Theorem   W3   (HSW)</u>: For every squashing function ψ, every r, and every probability measure μ on (R^r, B^r), the set F is uniformly dense on compacta and ρ_μ -dense in M^r.

We see that the elements of the set F have the same form as the one represented by the neural net with a single hidden layer of squashing units, and with linear units in the output layer. HSW extend their result further to obtain convergence in L^p spaces. However, all these results are existential and non-constructive. Finally, we consider the work of Carroll and Dickinson.

Carroll and Dickinson[32] demonstrated a semi-constructive technique based on the Radon Transform theory for implementing an arbitrarily close approximation to any L^2 function defined on a compact subset of R^n, such as, on $[-1, 1]^n$. The key concept in their work is to identify the functional form of a linear output, single hidden-layer net with a finitely parametrized approximate form of the back-projection operator. The back-projection operator is a component of the inverse Radon Transform(Deans[33]).

Let $D = C^{\infty}(S)$ denote the space of functions
continuously differentiable to all orders, and taking finite
values on a compact subset S of R^n. The Radon Transform
operator $\Re$ takes a member f (defined on R^n) of D to a
function $\hat{f}$ defined over $R \times S^{n-1}$, where S^{n-1} denotes the
n-dimensional unit sphere in R^n. The transform is given
by:

$$\Re(f) = \hat{f}(\alpha,u) = \int_{u^T x = \alpha} f(x) \, d\mu(x) \tag{9}$$

where μ is a measure on a hyperplane in R^n. The
hyperplane is indexed by the unit vector u normal to the
hyperplane, and by the least distance α of the hyperplane
from the origin.

The inverse Radon Transform is obtained by the
composition of two operators: $\Re^{-1} = B \circ F$. The first operator
F takes (n-1) derivatives (and for n even, the Hilbert
Transform) with respect to α, thus producing the so-called
filtered back-projection data denoted by $F(\hat{f}(\alpha,u)) = h(\alpha,u)$.
The second operator B, the back-projection operator, takes
the filtered data back to the original space:

$$f(x) \; = \; Bh(\alpha,u) = \int_{\|u\|_2 = 1} h(u^T x, u) \; d\mu(u) \tag{10}$$

where $d\mu(u)$ is the unit surface area measure on the

sphere S^{n-1}. The above integral can be evaluated by

discretization:

$$f(x) = Bh(u^T x, u) \equiv \sum_{i=1}^{K} \mu_i \; h(u^{i^T} x, u^i) \tag{11}$$

If each of the above terms $h(u^{i^T} x, u^i)$ is represented by a

linear combination of sigmoidal functions, then:

$$h_i = h(u^{i^T} x, u^i) = \sum_{j=1}^{m_i} \alpha_{ij} \; \sigma(u^{i^T} x + \beta_{ij}). \tag{12}$$

By combining the above equations, the overall output y of

the net in response to an n-dimensional input vector x can

be written as:

$$y(x) = \sum_{i=1}^{N} a_i \; \sigma(w^{i^T} x + b_i) \tag{13}$$

where w^i is the n-dimensional weight vector from the

input to the ith hidden unit; a_i is the weight (called the

output weight) connecting that unit to the output of the

net. This is exactly the functional form of a linear output,

single hidden layer neural net.

The above procedure involves two types of error. The first type corresponds to the finite discretization of the back-projection operator leading to the functions h_i. The second one involves the approximation of each h_i in terms of the linear combinations of the sigmoids. Carroll and Dickinson demonstrate that the errors produced by the discretization and by the approximation add in the RMS, with a constant dependent only on the dimension of the input space. Finally, we must mention an important limitation of this method. To approximate an L^2 function one must find a smooth function, find the corresponding Radon Transform, and generate the filtered back-projection data functions h_i. Each of these tasks is far from trivial. Secondly, as mentioned earlier, the process of generating the functions h_i involves taking partial derivatives of order (n-1), n being the dimension of the input space. Thus, the method will not be satisfactory in the case of noise-corrupted data since differentiation will accentuate high frequencies

Another relevant view on neural approximation of functions is proposed by Poggio and coworkers (Poggio and Girosi[34]). They view the approximation problem as a multidimensional surface fitting problem. Their theory makes use of such classical approximation tools as generalized splines and regularization techniques. Their theory subsumes the popular radial-basis functions, and some of the pattern recognition techniques such as Parzen windows and potential functions. However, due to lack of space, we will not review their work further. To conclude this section, we see that a net with three layers of units-the input layer, a single hidden layer of non-linear units, and an output layer of linear units is sufficient to approximate a multivariate function on a compact domain. However, the number of units required in the hidden layer and the path weights connecting the various units cannot be determined a priori by the techniques we have considered so far. One possibility for synthesizing a network is to start with a network that has a large number of hidden layer units and randomly chosen weights. The various weights can then be determined by 'training'. The process of training involves:

1) presenting to the net a set (the training set) of exemplary inputs x^i and the corresponding desired outputs y^i, i =1,2,.., N, and

2) adjusting the weights such that the network stores the association between the presented input and output patterns. After training, the trained network should produce an output $y = y^i$ if the input x to the network is set to x^i.

The adjustment of the path weights is done to minimize some suitable performance index. Usually, this performance index is of the least squares type, for example, it is chosen to be the sum of squared differences between the desired outputs from the training set and the actual network output. A popular training algorithm for the adjustment of the weights is the Backpropagation (BP) algorithm or the generalized delta rule. It adjusts the weights by performing a gradient descent in the space of the weights so as to minimize a least squares error criterion (Rumelhart et al.[8], Chapter 8). The algorithm employs the chain rule for the calculation of derivatives to efficiently compute the gradient of the performance index with respect to the weights. Following Ruck et al.[35], one of the authors (Wabgaonkar[28], Chapter 1) examined the connection between the BP algorithm and the Extended Kalman Filter (EKF) algorithm. In fact, the EKF algorithm can be shown to subsume the BP algorithm. In the next section, we focus on neural network training using the Kalman Filter algorithm.

4 Neural Network Training:

In this section, we consider the problem of neural network training. We consider two cases of this problem. In the first case, we assume that the data association (the training set) is specified in terms of a finite discrete set of input-output pairs. For this particular case, the (linear) Kalman Filter algorithm is applicable. The second case we consider is the one in which the training set is non-discrete (a continuum). This can happen, for example, when the input domain over which the function f (to be stored) is defined, is itself a continuum. In this case, one possibility is to select a set of 'representative' input-output pairs, thus reducing the problem to the first case. The second alternative is to employ the (non-linear) EKF algorithm for neural network training. We now consider each of these cases in the following.

Case (A) : Application of the Kalman Filter Algorithm -In this section, we consider the problem of the estimation of the output weights in the proposed 3-layer neural network structure. Recall that this structure has an input layer of units, a single hidden-layer of non-linear processing units, and an output layer of linear processing units. For the present case, we assume that the net structure has a single output unit, corresponding to a scalar network output. In order to store a set of N input-output data pairs $\{P_i = (x^i, y_i), i = 1, 2,..., N, x^i \in D \subset E^n, y_i \in R\}$, the network must have N hidden-layer processing units and the network output y, in response to an input pattern x, is governed by the following equation:

$$y = f(x) = \sum_{i=1}^{N} a_i \, g_i(x), \qquad x \in D \subset E^n, \qquad (14)$$

in which the function g_i represents the activation function of the ith hidden-layer unit. We assume that the functions g_i are chosen such that the interpolation matrix $G = [(g_i(x^j)]$ is non-singular. The authors in their earlier work had indicated how this particular choice can be made by employing a Reproducing Kernel Hilbert Space (RKHS) framework (Wabgaonkar and Stubberud[36]). For the sake of brevity, these results are not discussed here. Further, we assume that the functions g_i have been completely specified with the implication that the quantities $v_i = g_i(x)$, $i = 1, 2,..., N$ can be evaluated for any pattern x in the input domain D. Thus, the only quantity that needs to be computed in (14) is the vector a of the coefficients or the 'output' weights a_i. It may not be possible to invert the interpolation matrix G directly, for example, when the number N of the pattern-pairs is large. Hence, it becomes necessary to estimate the vector of weights using alternative procedures. In this section, we propose to use the Kalman Filter algorithm (Wabgaonkar and Stubberud [37]) for estimating the output weight vector a.

Due to our above mentioned assumption that the functions g_i have been completely specified, the estimation problem under consideration is a *linear state estimation* problem. We treat the output weight vector as a state vector to be estimated, leading to the following equations that describe the 'state and observation dynamics':

$$a(k+1) \; = \; a(k) \tag{15}$$
$$d(k) \; = \; y(k)+v(k) \; = \; f_k(a(k),x(k))+v(k) \tag{16}$$

in which a(k) denotes the output weight vector and d(k) is the desired output, both at time (instant) k. The network output y is represented by a possibly time-varying function $f_k(\cdot,\cdot)$ whose arguments are the output weight vector a and the input pattern x to the network. For the present case, the function f_k is linear in the output weight vector $a(k) = [a_1(k), a_2(k),..., a_N(k)]^T$:

$$f_k(a(k),x(k)) \; = \; \sum_{i=1}^{N} a_i(k)\, g_i(x(k)) \tag{17}$$

It is assumed that the network output is corrupted by Gaussian white noise v with covariance matrix R. At the beginning of each training cycle, the input x is set to the input component x^i of one of the input-output pairs $P_i = (x^i, y_i)$ in the training set, and the desired output d is set to the corresponding output component y_i. During each training cycle, the filter is made to iterate one or more number of times while the values of x and d are held fixed. These values are changed again at the beginning of the next training cycle. The Kalman Filter algorithm for the 'system' represented by the last three equations is as follows:

$$\hat{a}(k+1) = \hat{a}(k) + K(k)\{d(k) - f_k(\hat{a}(k), x(k))\} \qquad (18)$$

$$K(k) = P(k)H^T(k)\{R(k) + H(k)P(k)H^T(k)\}^{-1} \qquad (19)$$

$$P(k+1) = P(k) - K(k)H^T(k)P(k) \qquad (20)$$

In the above equations, $\hat{a}(k)$ denotes the estimate of a(k), K is the Kalman gain, P is the error covariance matrix, and H is the Jacobian of f_k (Equation 17) with respect to the weight vector a, evaluated at the current estimate $\hat{a}(k)$. In the present linear case, the ith element of H is simply $g_i(x)$. It can be seen that the most intensive computational requirement is that of inverting the matrix $\{R + HPH^T\}$ in the calculation of the gain. However, for the case of a single output, it simply means taking the reciprocal of a scalar. In the general case of m outputs, it represents an inversion of the m×m matrix whose size remains independent of the number of the weights involved; however, if the matrix R is diagonal, each output element can be processed independent of the others and again only a scalar needs to be inverted.

It should be noted that the form of the estimate update equation (18) is similar to many other estimation algorithms such as the delta rule. However, the Kalman algorithm is essentially a second order algorithm in the sense that that it makes use of the covariance information which is a second order statistic. Secondly, the algorithm is global implying that a change in the current estimate of a certain path weight potentially affects the changes in all the other weight estimates. The global coupling is provided through the off-diagonal entries in the covariance matrix (which appears in the computation of the Kalman gain) and this usually makes the algorithm converge more rapidly than the first order algorithms. This should be contrasted with the popular Backpropagation algorithm where such an explicit global coupling is missing. In fact, as discussed in Ruck et al.[35], and (Wabgaonkar[28], Chapter 1), the derivation of the BP algorithm from the Extended Kalman Filter algorithm requires that the covariance matrix be held fixed ($P = pI$, p a positive scalar) throughout the training session. We now present an example.

Example 1

We consider the 3-input EXOR problem. This is the problem of associative recall corresponding to the 3-

variable parity function f: $E^3 \rightarrow$ R; specified in terms of the following N = 8 data-pairs: the output of the network is to be -1 for an odd number of -1's in the input patterns, i.e., for patterns $x^1 = [1, 1, -1]^T$, $x^3 = [1, -1, 1]^T$, $x^5 = [-1, 1, 1]^T$, and $x^7 = [-1, -1, -1]^T$. The output of the network is to be +1 for an even number of -1's in the input patterns, i.e., for $x^2 = [1, 1, 1]^T$, $x^4 = [1, -1, -1]^T$, $x^6 = [-1, 1, -1]^T$, and $x^8 = [-1, -1, 1]^T$. Based on the theory described in Wabgaonkar[36] and other references cited therein, we choose the functions g_i as:

$$g_i(x) = \exp(x^T x^i), \quad i = 1, 2, ..., 8. \tag{21}$$

Note that, with this choice for the functions g_i, the weights from the input units to the hidden units have been chosen to be the input patterns from the set of exemplars. Thus, we do not have to estimate these weights which are otherwise non-linearly related to the output of the net. This is precisely the reason for the linear nature of our estimation problem. Secondly, the interpolation matrix G = $[g_i(x^j)]$ is invertible, its determinant being 2.1219 E +10. Upon imposing the interpolation conditions, the output weight vector a is obtained by the direct inversion of the G matrix as: a = 0.077015 $[-1, 1, -1, 1, -1, 1, -1, 1]^T$. For the

Kalman Filter algorithm, R was set to 0.1, the matrix P(0) was set to 100 I_8, and each of the eight components of the initial estimate of the weight vector was randomly chosen between -1 and +1 (uniform distribution). The actual training was performed by presenting one input-output exemplary pair at a time to the network and by adjusting the weights after each such presentation. Although it is possible to iterate many times while holding the input to the network and the corresponding desired output fixed to their respective exemplary values, it was not required in the present case. The network weights were changed only once after each such presentation. In Fig. 1(a), we show the entire training session. The final value of k is 40, which implies that each pattern-pair was shown 5 times. The Fig. 1(b) focuses on the later part of the training session in which the transients have been mostly deleted. From these plots, it can be seen that the training error is made very small by the filter right after the very first showing of the 8 pattern-pairs. After three such showings of each of the training pairs, the error was seen to have been reduced to the order of 10^{-4} - a very small number in

comparison with the magnitude of the desired output,
which is 1. The behavior of the weight vector estimate is
shown in Fig. 2(a), (b). In the part (a), four of the
estimates can be seen to have converged to -0.077015. In
the other part, the estimates converged to +0.077015 . The
Kalman filter estimates are identical (up to six decimal
digits) to the ones obtained by direct inversion of the
interpolation matrix. This extraordinary success of the
training procedure is largely due to the strong nonsingularity of the interpolation matrix G, which in turn
arises from the choice of the underlying RKHS function
members. In fact, during the simulation studies, the filter
algorithm was found to be quite insensitive to the choice of
the initial guess for the estimates, as well as to choice of
the covariance matrices R and P.

As an another interesting case, we choose the
constituent functions g_i to be:

$$g_i(x) = \tanh(x^T x^i), \quad i = 1, 2, ..., 8. \tag{22}$$

Note that in the above equation the hyperbolic tangent
functions are not biased. Normally, bias b_i is added to the
argument $(x^T x^i)$ of the the $\tanh(\cdot)$ function and the bias is

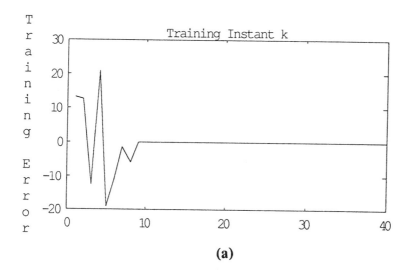

(a)

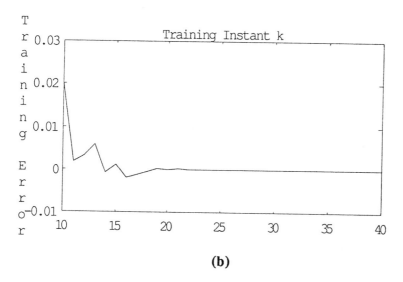

(b)

Fig. 1: Plot of training error v/s training instant for the Kalman Filter algorithm for the 3-input EXOR problem (exponential functions). Part(a) depicts complete training session, while part (b) depicts near steady state.

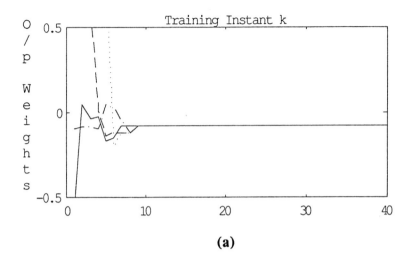

(a)

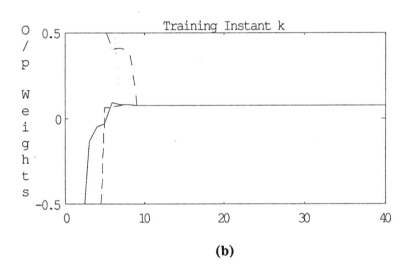

(b)

Fig. 2: Plot of estimates of weights v/s training instant for Kalman Filter algorithm for the 3-input EXOR problem (exponential functions). Part (a) shows four of the estimates, the remaining four appear in part (b)..

also estimated. However, this leads to a non-linear estimation problem since each such bias b_i is non-linearly related to the network output. However, the focus in this section is on linear estimation. Therefore, we have not included it in the above formulation. In this particular case, the interpolation matrix G is singular, its eigenvalues being 2.6346 E -14, -1.0747 E -15, 5.4189 E -14, -4.8957 E -14, 3.5133 E 0, 3.5133 E 0, 3.5133 E 0, -2.5795 E 0. A pseudo-inverse of the matrix G was calculated using singular value decomposition; the corresponding weight vector estimate $a^\#$ was found to be: $a^\# = 3.8768$ E -1 [1, -1, 1, -1, 1, -1, 1, -1]. The above mentioned Kalman filter algorithm was run with the same initial conditions and the same values of R and P(0) as in the case of the exponential functions. The behavior of the training error and the weight estimates is shown in Fig. 3 and Fig. 4. The estimates from the filter converge rapidly; however, the estimate of the weight vector is $\hat{a} = 4.0649$ E -1, -8.5287 E -1, 9.9499 E -1, -5.6930 E -1, 2.0605 E -1, 2.1964 E -1, -7.7516 E -2, -3.6886 E -1 . Although this is not identical to the pseudo-inverse estimate, the

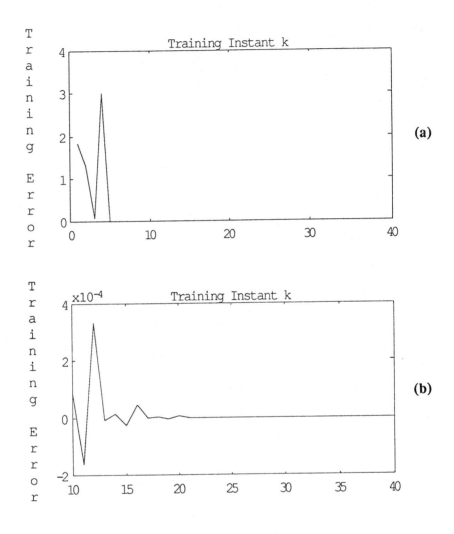

Fig. 3: Plot of training error v/s training instant for Kalman Filter algorithm for the 3-input EXOR problem (unbiased tanh functions). Part (a) depicts the whole training session, while part (b) depicts near steady state.

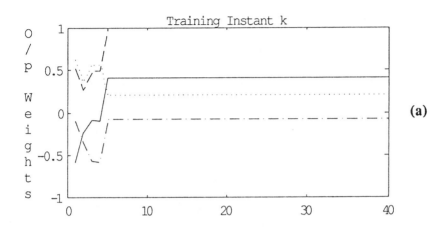

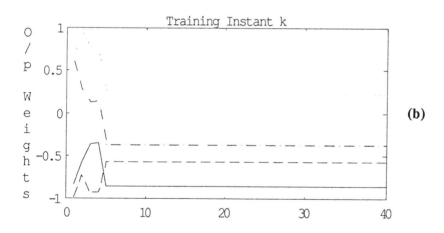

Fig. 4: Plot of weight estimates v/s training instant for Kalman Filter algorithm for the 3-input EXOR problem (unbiased tanh functions). Part (a) depicts four of the estimates, the remaining four appear in part (b).

filter performs satisfactorily as far as the drastic reduction in the training error is concerned.

Two computational issues related to this algorithm must be mentioned. The first one, is the inversion of an m-dimensional matrix for the case of m neural network outputs. This was discussed earlier in the last section. Another issue is that of storage of the covariance matrices P and Q. If the number of parameters to be estimated is p, the storage required is of the order of p^2. This can be a large number. However, if this number is not too large (which is many times the case in the control systems context), the Kalman Filter algorithm offers an important alternative for estimating the output weights.

Case (B): Application of the Extended Kalman Filter:Algorithm

-In many situations, a neural network may be required to approximate a given multivariate function such that the function is defined over a continuum rather than on a discrete set. In other words, the function is not specified in terms of a finite discrete set of input-output pairs, but may be given in some other form such as a closed form expression. In this particular case, it is possible to pick a representative (finite and discrete) set of exemplary input-output pairs from the potentially uncountably infinite set. The authors indicated how this can be done by applying orthogonalization to the neuronal activation functions within an RKHS framework (Wabgaonkar and Stubberud[38]). Aside from being more efficient, redundancy elimination improves the numerical conditioning of the underlying estimation problem. Also, recalling the discussion about the Equation (21), we see that the path weights from the input units to the hidden units can be chosen to be the input exemplary patterns themselves, if the set of the input-output exemplars is a finite discrete set of reasonable cardinality. Thus, redundancy detection leads to linear estimation problems. However, the method is computationally intensive. We do not pursue this approach in the following. Therefore, we assume that it is not be possible to choose *a priori* (i.e., without training) the path weights from the input layer to the hidden layer. Therefore, it becomes necessary to estimate these path weights also. However, these weights are non-linearly related to the output of the neural network and hence, we have a non-linear estimation problem to be solved. We propose to use the Extended Kalman Filter algorithm for this purpose.

The Extended Kalman Filter training algorithm (see, for example, Gelb[39]) is a modification of the (linear) Kalman Filter algorithm. It involves linearization of the non-linearities in the model (whose parameters are to estimated) at the current estimates of the parameters. The algorithm was first discussed by Singhal and Wu[40] in connection with neural network training. The algorithm is essentially the same as represented by the Equations (18)-(20). The only modification is to the observation equation- the function $f_k(a(k), x(k))$, in the present case, depends non-linearly on the parameters or the weights $a(k)$. Of course, the function should be differentiable with respect to the weight vector a since the algorithm involves the computation of the Jacobian of the function with respect to the vector a. We now present two examples to illustrate this algorithm. The first example deals with the short-time prediction of a chaotic time-series. The second example deals with a simple non-linear control problem. The latter will set the stage for a general formulation for the adaptive control of non-linear dynamical systems which is presented in the next section.

Example 2:

Our first example concerns the prediction of a chaotic time-series represented by the autonomous quadratic logistic map: (Devaney[41], Moody and Darken[42])

$$y_p(k+1) = 4\, y_p(k)\, (1-y_p(k)) \tag{23}$$

given the initial condition $y_p(0)$. For the purpose of simulation, the initial condition was set to 0.3. A neural network with a single neuron at the input and a single linear neuron at the output was employed. The hidden layer had seven neurons with the so-called Hardy's multiquadratics as the neuronal activation functions (Hardy[43]). The neural network is represented by:

$$y_n(k) = \sum_{i=1}^{7} w_i\, ((v_i\, x(k)-q_i)^2 + c^2)^{1/2} \tag{24}$$

in which $y_n(k)$ represents the network output for an input $x(k)$ to the neural network. The quantity c was arbitrarily chosen to be 1. Also, the vector of the 'q_i', q, was set to [-5, -2, -1, 0, 1, 2, 5] arbitrarily. Thus, the weight vector a consists of the output weights w_i, and the weights from the input to the hidden layer, v_i. Thus, a total of 14 weights needed to be estimated. The network output depends linearly on the output weights w_i, and non-linearly on the (hidden) weights v_i. The goal of training is to acquire the weights in such a way that the network predicts the time-series to a satisfactory degree of accuracy. By this type of prediction we mean that the network output $y_n(k)$ is sufficiently close to the 'next' value of the time-series $y_p(k+1)$, given that the network input $x(k)$ is the same as the current value $y_p(k)$ of the time-series. The objective of the time-series prediction is achieved by capturing the quadratic non-linearity on the right-hand side of (23) by the linear combination of the multiquadratics on the right-hand side of (24).

The various quantities related to the EKF algorithm (see (18)-(20)) were chosen as follows. The initial measurement noise covariance R(0) was set to 10. It was decreased exponentially at each training instant i according to:

$$R(i) = R(0) \exp(-(i-1)/(0.125\,NT)) \tag{25}$$

where NT represents the total number of training instants for which the training is to be performed. This number was set to 250. The initial (approximate) error covariance matrix P(0) was set to 200 I_{14}. The initial guess for each element of the weight vector was chosen randomly between 0.05 and -0.05. The algorithm was found to perform better with a small 'state noise' covariance matrix Q(i) which was chosen to be:

$$Q(i) = Q(0) \exp(-(i-1)/(0.125\,NT)) \tag{26}$$

in which, Q(0) was chosen to be 0.001 I_{14}. At each training instant i, up to a maximum of 4 filter iterations to adjust the weights were performed for a given data pair $\{x(i),y_p(i+1)\}$ provided that the training error $y_p(i+1)-y_{n,i}(j)$ exceeded a tolerance value of 0.01. Here, $y_{n,i}(j)$ represents the output of the network at the training instant i, during the iteration j. The iterations were terminated and the next data pair was generated, if either the number of iterations exceeded 4, or the error fell below 0.01. For each training instant i (i.e., for each data pair), the covariance matrix R(j) was further adjusted during the iterations j as follows:

$$R(j+1) = R(i) \exp(-(j-1)/1.25) \tag{27}$$

provided that it was greater than or equal to 0.05. If it fell below 0.05, it was reset to 0.05. In Fig. 5(a), the training error $e(i) = [y_p(i+1)-y_n(i)]$ is plotted for the entire training session. The steady state part of the training session appears in Fig. 5(b). The waveforms for $y_p(i+1)$ (indicated by a solid line) and $y_n(i)$ (indicated by '+') are shown in Fig. 6. The chaotic nature of the time-series can be seen clearly in this plot. The evolution of the estimates of the output weights w_i, as the estimation algorithm proceeds, appears in Fig. 7(a). The plot for the hidden weights v_i is depicted in Fig. 7(b). The training algorithm makes rapid changes in the weights in the beginning. As training proceeds, the estimation error decreases and the changes tend to be less drastic. However, due to the chaotic nature of the times-series that is being predicted, parameter adjustment continues to occur although infrequently.

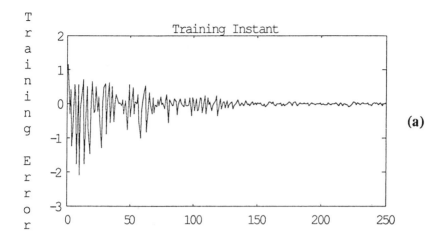

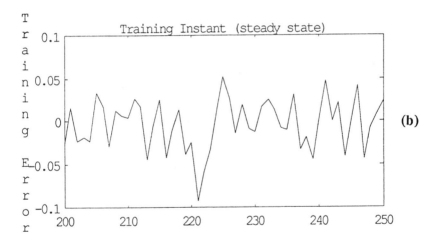

Fig. 5: Plot of training error v/s training instant for EKF algorithm for Chaotic Time-series Prediction (multiquadratic functions). Part (a) refers to whole training session, while part (b) refers to near steady state.

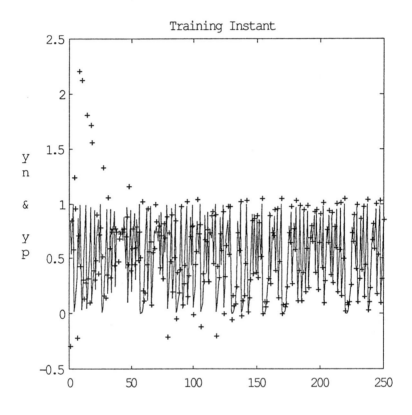

Fig. 6: Plot of chaotic signal and its neural prediction v/s training instant (EKF algorithm with multiquadratic functions). Solid curve refers to chaotic signal, '+' symbols indicate its neural prediction at training instants.

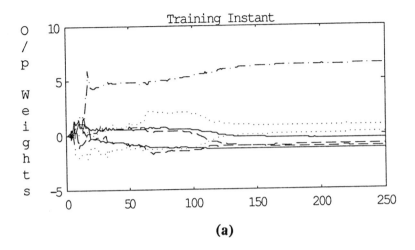

(a)

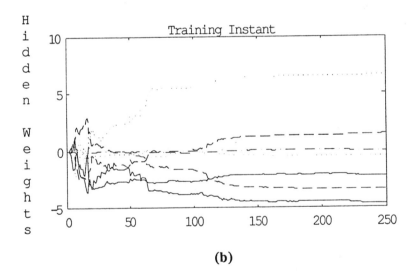

(b)

Fig. 7: Plot of weight estimates v/s training instant for EKF algorithm for Chaotic Time-series Prediction (multiquadratic functions). Part (a) depicts the output weights w_i, while part (b) depicts the 'hidden' weights v_i.

Example 3:

In this example, we illustrate an application of the EKF training technique to solve the Model Reference Adaptive Control (MRAC) problem (see, for example, Narendra et al.[44]) for a simple non-linear plant given by:

$$y_p(k+1) = f(y_p(k)) + u(k) \tag{28}$$

in which, $u(k)$ represents the input to the plant, $y_p(k)$ represents the output of the plant, both at time k. The non-linearity in the plant is represented by the function $f(\cdot)$, it is given by:

$$f(y_p(k)) = y_p(k)/(1 + y_p^2(k)). \tag{29}$$

The exact functional form of the non-linearity is assumed to be unknown to the control system designer. For the purpose of designing the control, it is required that the plant stably follow the model given by:

$$y_m(k+1) = 0.6 y_m(k) + r(k), \quad -10 \le r(k) \le 10, \forall k \tag{30}$$

in which $r(k)$ represents the external input, and $y_m(k)$ represents the model output. For the purpose of control, if the plant input $u(k)$ were set to be

$$u(k) = -f(y_p(k)) + 0.6 y_p(k) + r(k) \qquad (31)$$

then, the error $e(k+1) = y_p(k+1) - y_m(k+1)$ would be dynamically stable, i.e., it will decay to zero, with

$$e(k+1) = 0.6 e(k). \qquad (32)$$

However, the non-linearity is unknown, and hence must be captured by a neural network. Thus, if we have a neural network approximation $g(\cdot)$ to the non-linearity $f(\cdot)$ over the given input range, we can substitute it for the actual plant non-linearity. Therefore, the control law becomes:

$$u(k) = -g(y_p(k)) + 0.6 y_p(k) + r(k). \qquad (33)$$

This is the control law that was implemented in the simulation. For the purpose of approximating the non-linearity, a neural network with a single input unit and a single non-linear output unit was employed. The hidden layer consists of 20 neurons with non-linear activation functions. The neuronal activation function for all the neurons is chosen to be the biased hyperbolic tangent function. The overall network output $y_n(k) = g(x(k))$ for an input $x(k)$ to the network is given by:

$$y_n(k) = \tanh[\, d + \sum_{p=1}^{20} w_p \{\tanh(v_p x(k) + b_p)\}] \qquad (34)$$

in which the 61 parameters d, w_p, v_p, and b_p are to be estimated using the EKF algorithm.

At each training instant i, the input $r(i)$ is generated by randomly choosing a value between -10 and +10 (uniform distribution). This input is applied to the plant. Together with the current plant output $y_p(i)$, it is used to generate the next plant output $y_p(i+1)$, according to the Equation (28). The current plant output $y_p(i)$ is also applied to the input of the neural network so as to generate the network output $y_n(i)$. The parameters of the network are adjusted iteratively at each training instant so that the sum $(y_n(i)+r(i))$ approximates the next plant output $y_p(i+1)$ as closely as possible. The neural network training procedure based on the EKF algorithm is essentially the same as the one described in the earlier Example 2 except for the values of some of the statistical quantities. In the present case, the initial covariance matrix R(0) was chosen to be 10. The value of Q(0) was chosen to be 0.01 I_{61}. The total number NT of training instants was 300. At each training instant, up to a maximum of 4 iterations were performed subject to the criterion mentioned in the earlier example. The initial guess for each of the weights was drawn from a Gaussian distribution with zero mean, and a variance of 0.25. The initial approximate error covariance matrix P(0) was chosen to be 100 I_{61}. In Fig. 8(a), the the training error e(i+1) = [y_p(i+1)-(y_n(i)+r(i))] is plotted at each training instant, for the entire training session. The training error under steady state conditions is plotted in Fig. 8(b). It fluctuates between +0.05 and -0.05. The actual non-linear function f($\cdot$) (solid line) of Equation (29) is compared in Fig. 9 with its neural representation g($\cdot$) (dotted line) as represented by Equation (34). It can be seen that a reasonable approximation has been achieved over the entire input range between -10 and +10. The response of the autonomous MRAC system is shown in Fig. 10. For this case, the external input r(i) was set to zero. The initial condition for the plant y_p(0) was set to -5, and that for the model, y_m(0), was set to +10. It is this difference in the initial conditions that drives the system. It is seen that the error e(i) decays to zero. To test the forced response of the system, two different external inputs were applied. In

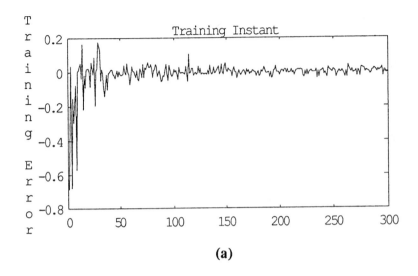

(a)

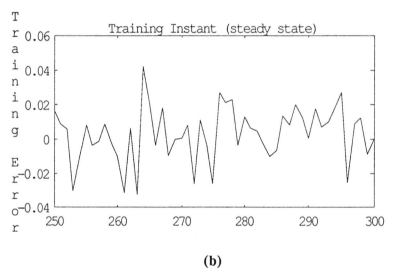

(b)

Fig. 8: Plot of training error v/s training instant for EKF algorithm for MRAC (tanh functions). Part (a) shows entire training session, part (b) focuses on near steady state.

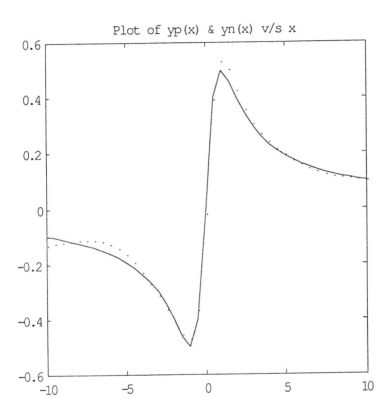

Fig. 9: Plot of actual non-linear function and its neural approximation v/s input, for MRAC (EKF algorithm, tanh functions). Solid curve refers to actual non-linearity, the dotted one to its neural approximation.

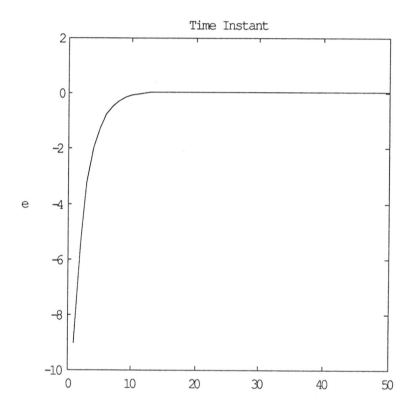

Fig. 10: Autonomous response of MRAC system. The model-following error e = (y_m - y_p) is plotted against time. In this case, the reference input is set to zero. Initial conditions on plant and model differ, this difference drives the closed-loop system.

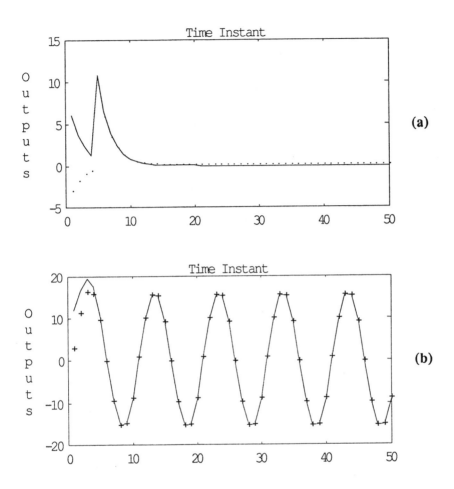

Fig. 11: Forced response of MRAC. In part (a), an impulse of strength 10 is applied at k = 5. The solid curve shows model's response, the dotted one shows that of the controlled system. In part (b), the reference signal is a sinusoid of amplitude 10 and period 10. Model's response is shown by solid curve, while that of the controlled system by '+'.

Fig. 11(a), the system is subjected to an impulse input of size 10 at i = 5. In Fig. 11(b), the applied input is a sinusoidal function given by: r(i) = 10 sin(2*Pi*i/10). In both cases, the controlled plant follows the model in a satisfactory way. Thus, the EKF algorithm can be used for training neural networks so as to capture non-linearities in plants. The neural networks trained in this fashion can in turn be used for stable model following control. This topic is examined further in the next section for more general non-linear models.

5 Elements of Non-linear Control:

In the last section, we considered the problem of approximating static (i.e., non-time-varying) non-linear functions by using neural networks. The EKF algorithm was employed for training a network so as to capture the unknown non-linear function. In this section, we extend these ideas for the control of non-linear dynamical systems. The key idea is that the training technique for the static case can be effectively applied to this case of time-invariant dynamical systems as well, provided the systems to be controlled are represented by non-linear state space models.

Specifically, let the non-linear time-invariant system (the plant) to be controlled be represented by:

$$x(k+1) = f(x(k), u(k)) \tag{35}$$

where x(k) is the n-dimensional state vector of the plant at the instant k, and u(k) is the p-dimensional input vector to the plant at the instant k. We assume that the function f is unknown, but for its dimensionality. As a first step towards control synthesis, the function f is approximated to the desired degree of accuracy using a neural network (see Fig. 12(a)). Of course, we implicitly assume that the function f can be represented as a suitable superposition (see Section 3) of the functions that act as the neuronal activation functions. This implies that there exists a parameter vector w of path weights such that the neural network N_f can represent f. We denote this as:

$$N_f(x, u, w) = f(x, u); \quad \forall x \in K_x \subset R^n, u \in K_u \subset R^p \tag{36}$$

where K_x and K_u are suitably chosen closed and bounded subsets of the spaces R^n and R^p, respectively. In practice, N_f will only approximate f, thus we can write:

$$f(x,u) = N_f(x,u,w_f) + \varepsilon_f(x,u); \quad \forall x \in K_x, u \in K_u \tag{37}$$

where w_f is a finite-dimensional, fixed (but unknown, as yet) parameter vector, the value of which defines the neural network N_f, and ε_f is the (unknown) approximation error. For the purpose of training the network, at each training instant k, a training data pair (x(k),u(k)) is presented to the network and the weight vector w_f is adjusted iteratively until a certain preset number of iterations for the given training pair is reached or the magnitude of the error ε_f for the given training pair is reduced below a predetermined threshold. The actual adjustment is carried out using the EKF algorithm, it is identical to the one considered in the last section.

Having captured the unknown plant dynamics in the neural network N_f, we come to the second step of the control synthesis procedure (see Fig. 12(b)). We require that the controlled plant follow a desired model given by:

$$x_m(k+1) = A x_m(k) + B r(k) \tag{38}$$

where x_m is the m-dimensional state-vector of the desired

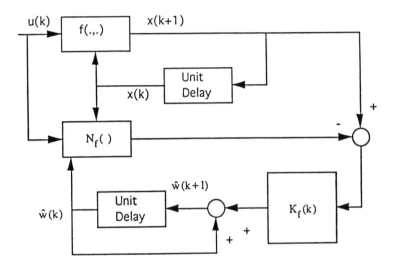

Fig. 12(a): EKF-based training of the neural network N_f to capture plant dynamics.

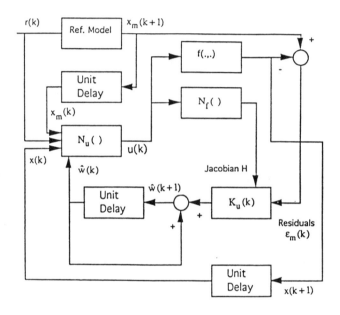

Fig. 12(b): EKF-based training of controller neural network N_u. Note that the network N_f is used for calculation of the Jacobian H. The plant itself is used to calculate the residuals.

model, and the matrices A and B are completely specified. The q-dimensional vector input r(k) to the model represents the known, external reference signal. Note that it is not necessary that the desired model be linear, although frequently it is the case. It is assumed that the model state vector x_m and the reference input are confined to some known closed and bounded subsets K_m and K_r of the spaces R^m and R^q, respectively. In the following, we assume that m = n, i.e., the number of the states of the reference model is the same as the number of plant states. The control problem which we address is that of generating a plant control input sequence u(k), such that

$$x(k+1) = x_m(k+1), \quad \forall k. \tag{39}$$

At this point, we make a key assumption that there exists a function g: $R^m \times R^n \times R^q \rightarrow R^p$ with

$$u(k) = g(x_m(k), x(k), r(k)) \tag{40}$$

such that the equality of (39) is satisfied. We approximate this function using a second neural network N_u, i.e.,

$$g(x_m, x, r) = N_u(x_m, x, r, w_u) + \varepsilon_u(x_m, x, r); \quad \forall x \in K_x,$$
$$\forall x_m \in K_m, \forall r \in K_r. \tag{41}$$

The parameter vector w_u represents the fixed (as yet unknown) weights that specify the controller neural network N_u. The p-vector ε_u is the approximation error. Thus, the controller neural network has (x_m, x, r) as its input and the plant control input u as its output. Additionally, the control input u is also supplied to the already trained neural network N_f. The 'plant' equations for the training of the controller network are given by

$$w_u(k+1) = w_u(k) \tag{42}$$

$$x_m(k+1) = f(x(k), N_u(x_m(k), x(k), r(k), \hat{w}_u(k))) + \varepsilon_m(k) \tag{43}$$

where the quantity ε_m represents the fictitious observation noise which is assumed to be Gaussian and white, with covariance matrix $R(k)$. The EKF algorithm in this case is as follows

$$\hat{w}_u(k+1) = \hat{w}_u(k) + K_u(k)\varepsilon_m(k) \tag{44}$$

$$K_u(k) = P(k)H^T(k)\{R(k) + H(k)P(k)H^T(k)\}^{-1} \tag{45}$$

$$P(k+1) = P(k) - K_u(k)H^T(k)P(k) \tag{46}$$

where $\hat{w}_u(k)$ is the estimate of w_u at the training instant k. The quantity K_u is the Kalman gain and P is the approximate error covariance matrix. The symbol H represents the approximate Jacobian of f with respect to the weight vector w_u evaluated at the current estimate $\hat{w}_u(k)$. Note that the exact mathematical form of f is unknown. Therefore, the Jacobian of f with respect to w_u is approximated by that of N_f with respect to w_u. Also, instead of using ε_m in Equation (44), its approximate value can be used, it is given by

$$\hat{\varepsilon}_m(k) = x_m(k+1) - N_f(x, N_u(x_m(k), x(k), r(k), \hat{w}_u(k)), w_f)$$

(45)

Finally, the model in Equation (42) can be modified to achieve better control over the estimation process by incorporating a noise term:

$$w(k+1) = w(k) + \varepsilon_w(k)$$

(46)

in which ε_w is assumed to be white and Gaussian. The magnitudes of the elements of its covariance matrix are made to decay exponentially with time as training proceeds. This leads to a better controlled evolution of the covariance matrix P. We now illustrate the algorithm with a numerical example.

Example 4

This example deals with the control of the following second order non-linear system

$$x_1(k+1) = x_2(k) \tag{47}$$

$$x_2(k+1) = 2(0.5 - e^{-[x_1(k)+x_2(k)+u(k)]^2}). \tag{48}$$

We assume that the order of the system and its structure are known, while the actual mathematical form of the non-linearity is unknown. The corresponding model for training the neural network N_f is identical in structure to the above model, the only change is that the right hand side of (48) is replaced by $N_f(x_1(k),x_2(k),u(k))$. Thus, the neural network N_f has three input units and a single output unit. In addition to the input and the output layers, the neural network also has two intermediate layers of hidden units that are sandwiched between the input and the output layers. The output unit has linear characteristics, while all of the intermediate (hidden) layers have units with biased hyperbolic tangent characteristics. The network has a feedforward structure, i.e., the input units are connected to the units in the first hidden layer; the units in the first hidden layer are connected to the units in the second hidden layer; and the units in the second hidden layer, in turn, are connected to the output unit. The number of units in the first hidden layer was arbitrarily chosen to be 8, and that of the units in the second hidden layer was arbitrarily chosen to be 6. For this particular configuration, the total number of adjustable parameters including the path weights and the bias-weights turns out to be 92. Thus, the unknown parameter vector w_f is 92-dimensional. For the purpose of simulation, the initial guess of each element of the corresponding estimate vector was chosen to be zero-mean Gaussian with variance of 0.01. The initial value of the covariance matrix P was chosen to be 100 I_{92}. The initial measurement noise covariance R(0) was set to 10. It was decreased exponentially at each training instant i according to:

$$R(i) = R(0) \exp(-(i-1)/(0.125\,NT)) \tag{49}$$

where NT represents the total number of training instants for which the training is to be performed. This number was set to 250. As mentioned above, the algorithm was found to perform better with a small 'state noise' covariance matrix Q(i) which was chosen to be:

$$Q(i) = Q(0) \exp(-(i-1)/(0.125 \, NT)) \qquad (50)$$

in which, Q(0) was chosen to be 0.01 I_{92}. The training algorithm was started with zero initial state. The plant input was chosen to be uniformly distributed between +2 and -2. At each training instant i, the randomly chosen plant input from the uniform distribution was generated and applied to the plant so as to obtain the 'next' plant states according to Equations (47) and (48). The couplet $\{u(i), x_2(i)\}$ constitutes the training pair at the training instant i. At each training instant i, up to a maximum of 4 filter iterations to adjust the weights were performed for a given data pair $\{u(i), x_2(i+1)\}$ provided that the training error $x_2(i+1)-y_{n,i}(j)$ exceeded a tolerance value of 0.08. Here, $y_{n,i}(j)$ represents the output of the network at the training instant i, during the iteration j. The iterations were terminated and the next data pair was generated, if either the number of iterations exceeded 4, or the error fell below 0.08. For each training instant i (i.e., for each data pair), the covariance matrix R(j) was further adjusted during the iterations j as follows:

$$R(j+1) = R(j) \exp(-(j-1)/1.25) \tag{51}$$

provided that it was greater than or equal to 0.05. If it fell below 0.05, it was reset to 0.05. In Fig. 13, the training error $e(i) = x_2(i+1) - y_n(i)$ is plotted for the entire training session. It can be seen that the training error is substantially reduced after about 100 training instants.

Having captured the plant dynamics in the neural network N_f, we consider the next phase- controller training. Under the closed-loop conditions, the plant is required to follow the reference model given by

$$x_{m1}(k+1) = x_{m2}(k) \tag{52}$$
$$x_{m2}(k+1) = (3/32)x_{m1}(k) + (1/4)x_{m2}(k) + (1/4)r(k). \tag{53}$$

The natural modes of the reference model are at -0.2057, 0.4557. The controller neural network N_u was chosen to have a feedforward structure identical to that of the neural network N_f. However, the number of units in the first hidden layer of N_u was arbitrarily chosen to be 10, and that for the second hidden layer was chosen arbitrarily to be 8. This led to a 156-dimensional parameter vector w_u that had

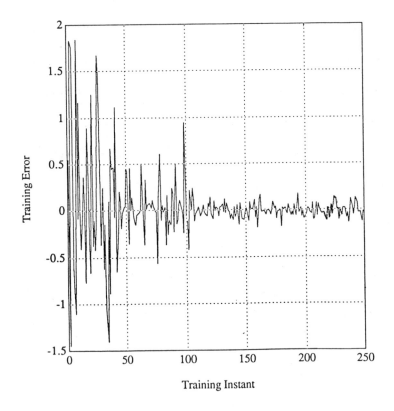

Fig. 13: Plot of training error v/s training instant during plant identification phase.

to be estimated. The various quantities of interest related to the EKF algorithm were chosen as follows. For the purpose of simulation, the initial guess of each element of the estimate of the vector w_u was chosen to be zero-mean Gaussian with variance of 0.001. The initial value of the covariance matrix P was chosen to be 50 I_{156}. The initial measurement noise covariance R(0) was set to 2. The matrix Q(0) was chosen to be 0.01 I_{156}. The exponential decays in the R and Q matrices were identical to the one described in the plant identification phase. For the purpose of controller training, the plant as well as the reference-model initial states were set at the origin. For training, the reference input signal r was chosen to be uniformly distributed between +1 and -1. Similarly, the output of the controller neural network N_u was constrained to lie between +3.5 and -3.5. At each training instant i, the reference input was generated randomly from the distribution and applied to the reference model to produce the reference model's next state. Thus, the couplet $\{r(i), x_{m2}(i+1)\}$ constituted the training data pair at the training instant i. To generate the corresponding plant response, the plant control input u(i) needed to be calculated. This was done by applying the input $\{x_{m1}(i), x_{m2}(i), x_1(i), x_2(i), r(i)\}$ to the controller network N_u. The network operates on its input to generate the plant input u(i). This allows the computation of the next plant state and also of the tracking error (residuals) $\varepsilon_m(i)$; the latter being required for the Kalman gain adjustment (see Equation (43)). If the current value of the tracking error was found unacceptable, the network parameters had to be adjusted. As explained earlier, a key step in the adjustment procedure was the approximation of the Jacobian of the plant non-linearity with respect to the vector w_u, by the Jacobian of N_f with respect to w_u. In Fig. 14(a), we show the tracking error during the training phase. The signals $x_{m2}(i+1)$ and $x_2(i+1)$ are plotted in the Fig. 14(b). The control input u(i) is plotted in Fig. 15. After training, the closed performance of the system was tested for the sinusoidal input $r(i) = \sin(2*\pi*i/25)$. The tracking error i.e., the difference between $x_{m2}(i+1)$ and $x_2(i+1)$ is plotted in Fig. 16(a), the two signals are shown in Fig. 16(b). It can be   seen   that   the   system   behaves

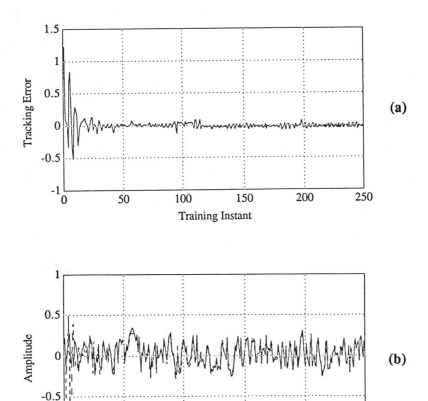

Fig.14: Part (a) shows the plot of tracking error v/s training instant during controller training phase. Part (b) depicts $x_{m2}(i+1)$ (solid) and $x_2(i+1)$ v/s training instant during the same phase.

satisfactorily. The control input u(i) during the controller training phase is shown in Fig. 15. After training, the closed performance of the system was tested for the sinusoidal input r(i) = $\sin(2*\pi*i/25)$. The tracking error i.e., the difference between $x_{m2}(i+1)$ and $x_2(i+1)$ is plotted in Fig. 16(a), the two signals are shown in Fig. 16(b). It can be seen that the system behaves satisfactorily. The control input u(i) and the reference signal r(i) are shown in Fig. 17. During this post-training phase also, the shape of the plant control input is the same as that of the reference signal but it has a different amplitude and phase. In conclusion, it can be seen that the control system has satisfactory closed-loop behavior.

6 Discussion

In this paper, we have argued that the incorporation of artificial neural networks into engineering systems can lead to enhanced performance of control systems. This approach can form the basis for the synthesis of intelligent systems which are highly adaptive and reconfigurable. The analysis and synthesis of artificial neural networks can be examined from within the domain of multivariate function interpolation and approximation. For a broad class of problems, the celebrated Kalman filter and its non-linear extension, the Extended Kalman Filter, can be employed for neural network training. We have tried to illustrate training schemes that involve the Kalman algorithm within the context of controller synthesis for non-linear systems.

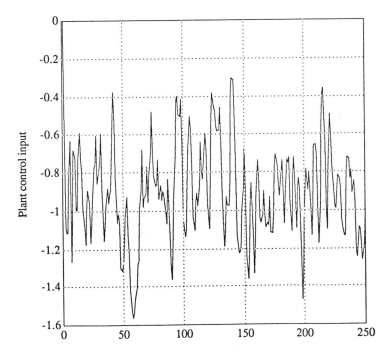

Fig. 15: Plot of output u(i) of controller neural network N_u v/s training instant during controller training.

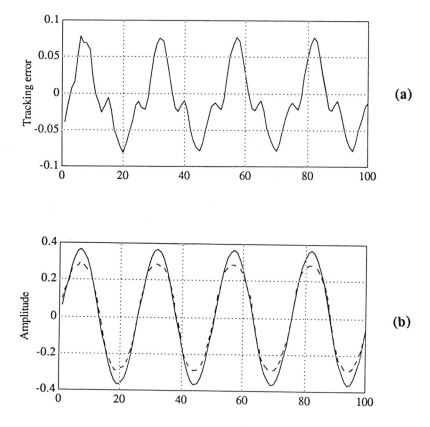

Fig. 16: Part (a) shows tracking error v/s time during post-training test. Part(b) shows $x_{m2}(i+1)$ and $x_2(i+1)$ v/s time during post-training test.

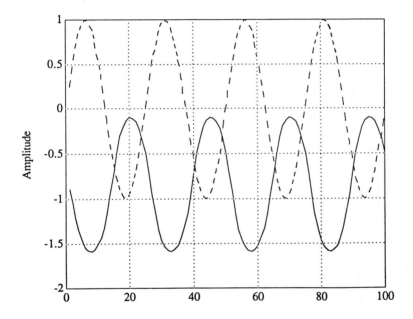

Fig. 17: Plot of control input u(i) (solid) and reference signal r(i) (dotted) during post-training test.

The results presented are preliminary in nature and further work is in progress. A natural extension of the plant model (Equation 35) would be to incorporate non-linear measurements, with the additional measurement equation

$$z(k+1) \; = \; h(x(k+1)) + v(k+1) \tag{54}$$

where z(k+1) represents a vector of measurements at the instant k+1. The function h(·) represents the known non-linear transformation associated with the (calibrated) transducer. Random measurement noise with known mean and covariance is represented by v(k+1). This extension requires the inclusion of an EKF to estimate the plant state, in addition to the EKF needed for neural network training.

An important design consideration in systems that involve neural networks is the choice of regions over which the neural approximations are valid. Operation outside these regions can lead to unreliable behavior. Under these conditions, adaptation of the nets to the appropriate operating regions is recommended. Secondly, our current interest is in the use of neural networks that can complement the existing linear controllers (that are based on linearized models) so as to enhance the overall system performance. This will allow us to exploit the theory of linear control systems which is very well developed. It is hoped that the suggested approach of employing neural networks in auxiliary control loops around the main linear controller in the system should lead to more robust behavior. Further extensions of our approach are under development so as to be able to deal with non-linear systems which have non-linear transducers. The transducers are assumed to be well calibrated so that their non-linear behavior is completely specified. Results on these topics will be published elsewhere.

References:

[1] Stubberud A. R., "Intelligent control", presented at the Symposium on the Frontiers of Intelligent Systems, Univ. of California-Irvine, Irvine, CA, USA, 1986.

[2] Bavarian B., "Introduction to neural networks for intelligent control", IEEE Control Sys. Mag., pp. 3-7, April 1988.

[3] Goodwin G. C. and Sin K. S., Adaptive filtering prediction and control, Prentice-Hall, Englewood Cliffs, NJ, 1984.

[4] Widrow B. and Stearns S. D., Adaptive signal processing, Prentice-Hall, Englewood Cliffs, NJ, 1985.

[5] Fodor J. A. and Pylyshyn Z. W., "Connectionism and cognitive architecture: a critical analysis", Cognition, no. 28, pp. 3-71, 1988.

[6] Smolensky P., "On the proper treatment of connectionism", The Behav. Brain Sci., no. 11, pp. 1-23, 1988.

[7] Charniak E. and McDermott D., Introduction to artificial intelligence, Addison-Wesley, Reading, MA, 1986.

[8] Rumelhart D. E., McClelland J. L., & the PDP Research Group, Parallel distributed processing: explorations in the microstructure of cognition, vol. 1: foundations, MIT Press, Cambridge, MA, 1986.

[9] Pfeifer R., Schreter Z., Fogelman-Soulie F., and Steels L., "Putting connectionism into perspective", in Connectionism in Perspective, Pfeifer R. et.al.(eds.), Elsevier Science Publ., pp. xi-xxi, 1989.

[10] Ref. 1.8 above, pp. 54-57.

[11] Wabgaonkar H., "Introduction to intelligent control", Ph. D. Exam. Report, ECE Dept., Univ. of California-Irvine, Irvine, CA, USA, 1988.

[12] Barto A. G., "Connectionist learning for control", in Neural Networks for Control, Miller III, W. T., et.al.(eds.), MIT Press, Cambridge, MA, pp. 5-58, 1990.

[13] Hinton G. E., and Anderson J. R.,(eds.), Parallel models of associative memory, Lawrence Earlbaum Assoc., Hillsdale, NJ, 1989.

[14] Schank R. C. and Abelson R. P., "Scripts, plans, goals, and understanding, Lawrence Earlbaum Asso., Hillsdale, NJ, 1977.

[15] Rumelhart D. E., and Norman D. A., "A comparison of models", in Ref. 1.13, pp. 15-21, 1989.

[16] Fu K. S., "Learning control systems- review and outlook", IEEE Trans. Auto. Con., pp. 210-221, 1970.

[17] Mendel J. M. and Zapalac J. J., "The application of techniques of artificial intelligence to control system design", in Advances in Control Systems, vol. 6, Leondes C. T. (ed.), pp. 1-94, 1968.

[18] Kolmogorov A. N., "On the representation of continuous functions of several variables by superposition of continuous functions of one variable and addition", Dokl. Akad. Nauk. SSSR, vol. 114, pp. 369-373, 1957.

[19] Arnold V. I., "On functions of three variables", Dokl. Akad. Nauk. SSSR, vol. 114, pp. 953-956, 1957.

[20] Lorentz G., Approximation of functions, Chelsea Publ., New York, NY, Second Ed., Chap. 11, 1986.

[21] Lorentz G., "Metric entropy, widths and superposition of functions", Amer. Math. Monthly, vol. 69, pp. 469-485, 1962.

[22] Sprecher D. A., "An improvement in the superposition theorem of Kolmogorov", J. Math. Anal. Appl., vol. 38, pp. 208-213, 1972.

[23] Vitushkin A. G., "Some properties of linear superpositions of smooth functions", Dokl. Akad. SSSR Nauk., vol. 156, pp. 1003-1006, 1964.

[24] Hedberg T., "The Kolmogorov superposition theorem", in Topics in Approximation Theory, H. S. Shapiro, Lecture Notes in Math., no. 187, Springer Verlag, Berlin, pp. 267-275, 1971.

[25] Hecht-Nielsen R., "Kolmogorov's mapping neural network existence theorem", Proc. IEEE First International Conf. on Neural Networks, San Diego, CA, 1987, vol. III, pp. 11-14, 1987.

[26] Natanson I. P., Theory of functions of a real variable, vol. 1, F. Ungar Publ., New York, pp. (49-50, 213-214), 1961.

[27] Kolmogorov A. N. and Tihomirov V. M., "ε - entropy and ε - capacity of sets in function spaces", Uspehi, vol. 14, no. 2, pp. 3-86, 1959.

[28] Wabgaonkar H. M., "Synthesis of discrete associative memories by multivariate interpolation", Ph. D. Thesis, University of California- Irvine, ECE Dept., 1991.

[29] Cybenko G., "Approximation by superposition of a sigmoidal function", CSRD Report 856, Univ. of Ilinois,

Urbana, IL, also published in Math. Cont. Signals Sys., 1989.

[30] Royden H. L., Real analysis, Third ed., Macmillan, New York, NY, 1988.

[31] Hornik K., Stinchcombe M. and White H., "Multilayer feedforward networks are universal approximators", UCSD Dept. of Economics Discussion Paper, Univ. of California-San Diego, CA, June 1988.

[32] Carroll S. and Dickinson B., "Construction of neural nets using the Radon transform", Proc. IJCNN, Washington, D. C., pp. 607-611, 1989.

[33] Deans S. R., The Radon transform and some of its applications, Wiley, New York, NY, 1983.

[34] Poggio T. and Girosi F., "Networks for approximation and learning", Proc. IEEE, vol. 78, no. 9, pp. 1481-1497, 1990.

[35] Ruck D. W., Rogers S. K., Maybeck P. S., Kabrisky M., "Back propagation: a degenerate Kalman filter?", Preprint, 1990.

[36] Wabgaonkar H. M. and Stubberud A. R., "How to store incorporate new pattern pairs without having to

teach the previously acquired pairs", (accepted for publication), Neural Computation, 1991.

[37] Wabgaonkar H. M. and Stubberud A. R., "Approximation and estimation techniques for neural networks", IEEE Conf. Dec. and Con. (CDC), Honolulu, Hawaii, pp. 2736-2740, 1990.

[38] Wabgaonkar H. M. and Stubberud A. R., "Input data redundancy in interpolation-based neural nets", presented at the IJCNN, Seattle, WA, 1991.

[39] Gelb A.(ed.), "Applied optimal estimation", M.I.T. Press, Chapter 6, 1974.

[40] Singhal S. and Wu L., "Training multilayer perceptrons with the extended Kalman algorithm", in Advances in Neural Information Processing Systems I, Touretzky D. S. (ed.), Morgan Kaufmann, pp. 133-140, 1989.

[41] Devaney R. L., Introduction to chaotic dynamical systems, Addison-Wesley, pp. 31-39, 1989.

[42] Moody J. and Darken C., "Fast learning in networks of locally-tuned processing units", Neural Computation, vol. 1, pp. 281-294, 1989.

[43] Hardy R. L., "Multiquadratic equation of topography and other irregular surfaces", J. Geophy. Res., 76, pp. 1905-1915, 1971.

[44] Narendra K. S. and Parthasarathy K., "Identification and control of dynamical systems using neural networks", IEEE Trans. Neural Networks, vol. 1, pp. 4-27, 1989.

[45] A. R. Stubberud, H. M. Wabgaonkar, and S. A. Stubberud, "A neural network based system identification technique", Proc. IEEE 30th Conf. on Decision and Control, Brighton, U.K., 1991.

INDEX

A

Adaptive control system
  decentralized, 125–171
    appendix to, 169–171
    general considerations for, 125–129,
      167–168
    with known parameters, 132–139
    mathematical preliminaries in, 129–131
    problem statement in, 131–132
    reference control scheme for, 139–145
    stability analysis in, 145–167
      aggregation of subsystem properties,
        163–167
      properties resulting from adaptive law,
        145–153
      properties resulting from specific
        control structure, 153–163
    with unknown parameters, 139–145
  multivariable, 98–123
    error model and adaptive law in, 104–108
    general considerations for, 99–101,
      118–121
    plant and controller structure in, 101–104
    stability and robustness analysis in,
      108–118
Algorithm
  adaptive, 125–171. *See also* Adaptive control
    system, decentralized
  branch and bound, 3–50
    convergence of, 9–11
    general considerations for, 3–9
    parameter problems and, 13–50
      complexity in, 21
      computation of bounds in, 21–32
        loop transformation and, 21–22
      examples of, 32–47
        for analysis, 33–40

        for design, 40–44
        for maximization of multiple
          objectives, 44–47
      general considerations for, 47, 50
      H^2 norm in, 19–20
      H^∞ norm in, 18–19
      open-loop systems in, 15
      stability degree in, 17–18
      well-posedness in, 16, 20–21
    simultaneous maximization of multiple
      objectives in, 11–13
  interior point, 3
  recursive, robust, 172–218. *See also* Bilinear
    systems, robust recursive methods
    for state and parameter estimation of
  sliding control, 299–320. *See also* Nonlinear
    control systems, robust, sliding control
    design in
Array processors, broadband, robust, 321–386.
  See also Broadband beamforming, robust

B

Beamforming, broadband, 321–386. *See also*
  Broadband beamforming, robust
Bilinear systems, robust recursive methods for
  state and parameter estimation of, 172–218
  appendix to, 207–216
  extended least squares method in, 199–202
    convergence analysis in, 200–201
    results of stimulation in, 201–202
  general considerations for, 173–174, 207
  instrumental variable method in, 186–190
    convergence analysis in, 187–189
    results of stimulation in, 189–190
  least squares method in, 181–186
    convergence analysis in, 183–184

Bilinear systems, least squares method in,
    (*continued*)
    results of stimulation in, 184–186
output error method in, 190
    convergence analysis in, 192–193
    results of stimulation in, 193–194
prediction error method in, 202–203
problem formulation in, 177–181
pseudolinear regression method in, 199–202
    convergence analysis in, 196–197
    results of stimulation in, 197–199
robust bootstrap method in, 174–199
    convergence analysis in, 196–197
    results of stimulation in, 197–199
robust identification and, 174–177
Branch and bound algorithm, 3–50. *See also*
    Algorithm, branch and bound
Broadband beamforming, robust, 321–386
general considerations for, 321–325
linearly constrained power minimization in,
    331–333
mathematical models in, 325–331
    broadband processor and signals and,
        325–329
    frequency response and presteering and,
        329–331
maximally flat presteered broadband
    processors in, 333–344
    maximally flat spatial response and,
        334–335
    necessary and sufficient conditions,
        335–337
    zero order plus, 337–344
notation and symbols in, 378–379
partitioned processor formulation in,
    367–377
    general considerations for, 367–370
    linearly constrained robust processor and,
        370–374
    numerical results and, 374–377
quadratic approach to, 345–367
    general considerations for, 345–347
    generalized response deviation constraint
        and, 351–358
        channel phase errors, 357–358
        deviations in array geometry, 353–357
        directional mismatch, 352–353
        exact presteering, 350–351
    numerical results and, 361–367
    optimization of new processor and,
        358–361
    response constraint and, 347–351
        coarse presteering, 349–350

        exact presteering, 350–351
        no presteering, 349

C

Control system. *See also* Adaptive control
    system; Robust control system
analysis and design of, global optimization
    in, 1–55. *See also* Algorithm, branch
    and bound

D

Discrete-time model, in combined filtering and
    parameter estimation, 57–98. *See also*
    Robust analysis, for combined filtering and
    parameter estimation

F

Failure detection and identification, analytical
    redundancy in, 289–292

I

Instrumental variable method, in state and
    parameter estimation of bilinear systems,
    186–190
Intelligence, artificial, 430. *See also* Neural
    networks, artificial

K

Kalman-Bucy filter, 59, 67, 86–87
    in robust solution methods, 90
Kalman filter, artificial neural networks and,
    462–494. *See also* Neural networks,
    artificial
Kharitonov's theorem, 15, 21

L

Least squares method, in state and parameter
    estimation of bilinear systems, 181–186
Linear systems
    parameter-dependent, global optimization in
        analysis and design of, 13–21

with unknown disturbances, 261–298
  applications of, 278–292
  decentralized estimation in, 286–288
  general considerations for, 261–262, 292, 294
  model uncertainties and, 273–278
  optimal modal robust servomechanism problem in, 279–286
  proportional-integral observer in, 268–273
  sensor failure detection and identification problem in, 289–292
  unknown input observer in, 262–268
Loop transformation, 21–32
  parameter problems and, 21–22
Lyapunov design methods, 387–388. *See also* Rigid robot, robust control of
Lyapunov method, 300–302, 312, 317. *See also* Nonlinear control systems, robust, sliding control design in

M

Multi-input, multi-output plants, continuous and discrete time, multivariable model reference adaptive control schemes and, 98–123. *See also* Adaptive control system, multivariable

N

Neural networks, artificial, nonlinear control and, 427–522
  connectionist computing and, 431–436
  function approximation and, 443–462
  general considerations for, 427–431, 570–575
  neural network training and, 462–494
    Kalman filter algorithm, 462–476
    extended, 476–494
  nonlinear dynamical systems and, 494–510
  parameter estimation and, 436–443
Noise distribution, with heavy tails, 73–90. *See also* Robust analysis, for combined filtering and parameter estimation
Nonlinear control systems, robust, sliding control design in, 299–320
  general considerations for, 299–304, 317–318
  multiple-input-multiple-output (MIMO) systems in, 304, 310–317

control law and system stability, 313–315
  sliding surface and sliding condition, 311–312
  tuning mechanisms, 316–317
  single-input-single-output (SISO) systems in, 304–310
    control algorithm and system stability, 307–308
    relative stability and tuning mechanisms, 308–310
    second-order sliding condition, 305
    sliding surface and error dynamics, 306–307

O

Open-loop system, parameter problems and, 15
Optimization, global, 1–55
  branch and bound algorithms in, 3–4. *See also* Algorithm, branch and bound
  convex, 2
  interior point algorithms in, 3
  local versus, 2
  non-convex, 2
  parameter problems and, 13–50. *See also* Algorithm, branch and bound
  simulated annealing in, 3
Outliers, 173
Output error method, in state and parameter estimation of bilinear systems, 190

P

Parameter problem
  bilinear systems and, robust recursive methods for, 172–218. *See also* Bilinear systems, robust recursive methods; for state and parameter estimation of
  decentralized adaptive control for, 125–171. *See also* Adaptive control system, decentralized
  global optimization and, 1–55. *See also* Algorithm, branch and bound
  loop transformation and, 21–22
  multivariable adaptive control system for, 98–123. *See also* Adaptive control system, multivariable
  open-loop system and, 15
  robust analysis in discrete time and, 57–98. *See also* Robust analysis, for combined filtering and parameter estimation
  well-posedness and, 16, 20–21

R

Rigid robot, robust control of, 387–426
  coriolis and centripetal terms in, 389–390
  disturbance torques in, 390–391
  feedback-linearization approach in, 407
  friction terms in, 390
  general considerations for, 387–392,
    420–421
  gravity terms in, 390
  inertia matrix in, 389
  linear-multivariable approach in, 392–407
    dynamic compensators and, 401–407
    static compensators and, 396–401
  passivity-based approach in, 407–410
  robust-adaptive approach in, 418–420
  robust-saturation approach in, 411–417
  variable-structure controllers in, 410–411
Robot, rigid, robust control of, 387–426. *See
  also* Rigid robot, robust control of
Robust analysis
  for combined filtering and parameter
    estimation, 57–98
    in discrete time, 67–90
      with heavy-tailed observation noise
        distribution, 73–90
        case with limiter, 78–83
          solution methods, 80–83
          theoretical considerations, 78–80
        case without limiter, 74–80
          solution methods, 77–83
          theoretical considerations, 74–77
      random coefficients and, 83–90
        case without limiter, 83–87, 87–90
          solution methods, 86–87, 89–90
          theoretical considerations,
            83–86, 87–88
      previous results of, 67–93
    general considerations for, 57–59
    previous results of, 61–73
      continuous-time filtering and, 61
      in discrete time, 69–73
      with unknown parameters, 67–69
      without unknown parameters and,
        61–67
    simulation of, 90–95
  in multivariable adaptive control system,
    108–118
  parametric. *See also* Robust analysis, for
    combined filtering and parameter
    estimation
    bilinear systems and, 172–218. *See also*
      Bilinear systems, robust recursive

methods for state and parameter
  estimation of
  global optimization in, 3–50. *See also*
    Algorithm, branch and bound
Robust broadband beamforming. *See also*
  Broadband beamforming, robust
Robust control system
  nonlinear, sliding control design in,
    299–320. *See also* Nonlinear control
    systems, robust, sliding control
    design in
  in presence of uncertainty, 219–259
    conditions for stability robustness and,
      240–243
      structured small gain theorem, 240–243
    duality in linear programming and,
      250–254
      general case, 253–254
      ℓ^1 problem as linear program, 251–253
    general class of structured uncertainty
      and, 236–237
    general considerations for, 219–221,
      254–255
    ℓ^1 norm in, 223–224
    ℓ^1 theory and, 220–221
    linear fractional transformation and,
      234–236
    nominal performance objectives in,
      226–234
      command following with saturation and,
        228–229
      disturbance rejection and, 227
      duality between stability and
        performance and, 233–234
      robust stability and, 229–230
      stable coprime factor perturbations and,
        231–233
      unstructured multiplicative
        perturbations and, 230–231
    performance robustness versus stability
      robustness and, 238–240
    stability and parameterization in, 224–226
    standard concepts in, 221–223
    synthesis of ℓ^1–optional controllers and,
      243–250
      bad rank case, 247–250
      characterization of subspace S, 244
      good rank case, 245–247
Robust identification. *See also* Bilinear systems,
  robust recursive methods for state and
  parameter estimation of in state and
  parameter estimation of bilinear systems,
  174–177

Robust recursive methods, for state and parameter estimation of bilinear systems, 172–218. *See also* Bilinear systems, robust recursive methods for state and parameter estimation of

Robust state estimation theory, techniques in, 261–298. *See also* Linear systems, with unknown disturbances

S

Second-order sliding condition, in sliding control algorithm, 305, 317–318. *See also* Nonlinear control systems, robust, sliding control design in

Sensor failure detection and identification, analytical redundancy in, 289–292

Servomechanism problem, robust, 279–286

Simulated annealing, 3

Sliding control design, in robust nonlinear control systems , 299–320. *See also* Nonlinear control systems, robust, sliding control design in

Spacial spectrum estimation, 321–386. *See also* Broadband beamforming, robust

State estimation, robust, 261–298. *See also* Linear systems, with unknown disturbances

V

Variable-Structure Control, sliding control techniques in, 299–300. *See also* Nonlinear control systems, robust, sliding control design in

W

Well-posedness, parameter problems and, 16, 20–21

ISBN 0-12-012753-9